Manuals in Archaeological Method, Theory and Technique

Series Editors:

Charles E. Orser, Jr., *Illinois State University*, Normal, Illinois
Michael B. Schiffer, *University of Arizona*, Tucson, Arizona

For futher volumes:
http://www.springer.com/series/6256

Debra L. Martin • Ryan P. Harrod
Ventura R. Pérez

Bioarchaeology

An Integrated Approach
to Working with Human Remains

 Springer

Debra L. Martin
Department of Anthropology
University of Nevada
Las Vegas, NV, USA

Ryan P. Harrod
Department of Anthropology
University of Nevada
Las Vegas, NV, USA

Ventura R. Pérez
Department of Anthropology
University of Massachusetts
Amherst, MA, USA

ISSN 1571-5752
ISBN 978-1-4614-6377-1 (Hardcover)   ISBN 978-1-4614-6378-8 (eBook)
ISBN 978-1-4939-2119-5 (Softcover)
DOI 10.1007/978-1-4614-6378-8
Springer New York Heidelberg Dordrecht London

Library of Congress Control Number: 2012956309

Printed on acid-free paper

Springer is part of Springer Science+Business Media (www.springer.com)

This book is dedicated to those that went before us, and those still with us
George A., Juj, Deb Jr., Mike, Dodie, Lira, Lola, Juniper, and Little Bear (DLM)
Stephanie, Kael, Amara, Anne, Nicole, and Gizmo (RPH)
Kathleen and Millie (VRP)

Contents

List of Figures

List of Tables

Chapter 1
The Practice of Bioarchaeology

Bioarchaeology is the study of ancient and historic human remains in a richly configured context that includes all possible reconstructions of the cultural and environmental variables relevant to the interpretations drawn from those remains. As a field of study within anthropology, it is informed by a wide range of scientific methods and theories drawn from all of its subdisciplines (archaeology, biological anthropology, cultural anthropology, applied/practicing anthropology, and linguistics). However, bioarchaeologists borrow as needed from fields such as medicine, forensics, anatomy, epidemiology, nutrition, geosciences, and demography. At the heart of a bioarchaeological project is the scientific study of human remains using the archaeological record to enhance what can be known about the past. Bioarchaeologists frame hypotheses about human behavior that can be verified (or not verified) by the empirical data sets that are generated from the human remains in their contexts. At its very best, bioarchaeology illuminates human behavior and why certain patterns emerge in some cultures at particular times. Bioarchaeology uniquely provides time depth and a cross-cultural perspective on humans as both biological and cultural beings. Bioarchaeology is rooted in the human experience spanning vast swaths of time across the entire planet. By reconstructing biological identity and cultural context, bioarchaeology can illuminate the complexities that lie at the heart of human behavior. Because bioarchaeology *is* anthropology (Armelagos 2003), it has the potential as few other disciplines do to reveal important dimensions of the human life history that are currently unfathomable.

1.1 Bioarchaeology is Anthropology

Because bioarchaeology is situated within the subdiscipline of biological anthropology, which itself is part of the overarching and inclusive discipline of anthropology, it is focused on the scientific study of humankind. Within that broad mandate, bioarchaeology seeks to *explain* human behavior within an evolutionary and biocultural

D.L. Martin et al., *Bioarchaeology: An Integrated Approach to Working with Human Remains*, Manuals in Archaeological Method, Theory and Technique, DOI 10.1007/978-1-4614-6378-8_1, © Springer Science+Business Media New York 2013

framework. Bioarchaeological studies are pieces of the puzzle that help explain why there is disease and early death in some populations but not others. And, when and under what circumstances are violence and warfare used to promote societal goals? Bioarchaeologists also grapple with similarities and differences within the sexes, exploring the underlying reasons for behaviors such as patriarchy, the sexual division of labor, differential patterns in mortality and morbidity, and age-related health problems. Bioarchaeology has most recently shined a spotlight on children. What is the role of children in society? What are the risks for early death? What do high rates of infant mortality reveal about the subsistence pattern of a group?

Addressing these kinds of complex questions requires using and synthesizing information from many different areas within anthropology along with an impressive array of method, theory, and data from other fields. As anthropologists, bioarchaeologists are highly specialized within their area, yet must remain generalists in the ways that observations of the human condition are interpreted. Under "What is Anthropology" from the web pages of the American Anthropological Association is this framing of the social context within which anthropological research is done: *Anthropologists collaborate closely with people whose cultural patterns and processes we seek to understand or whose living conditions require amelioration. Collaboration helps bridge social distances and gives greater voice to the people whose cultures and behaviors anthropologists study, enabling them to represent themselves in their own words. An engaged anthropology is committed to supporting social change efforts that arise from the interaction between community goals and anthropological research. Because the study of people, past and present, requires respect for the diversity of individuals, cultures, societies, and knowledge systems, anthropologists are expected to adhere to a strong code of professional ethics* (American Anthropological Association 2012).

Bioarchaeology at its very best fits well into these tenets and this approach to bioarchaeology is emphasized throughout this text. At the same time, it must be recognized that there are many different kinds of bioarchaeology being practiced within the United States as well as throughout the rest of the world. Some studies are more descriptive and particularistic, and others may utilize only a limited amount of information regarding cultural context. Yet, in every published bioarchaeological study, there is likely something that is useful or that can be combined with other studies to further the goal of understanding human behavior.

However, if bioarchaeology is to have a part in making the world a better place for present and future generations, it must strive to be integrative (combining data sets and applying theory), engaged (pursuing some larger goal of value to a broader group including non-academics), and ethical (maintaining a self-reflective and genuine concern regarding the scientific process). Admittedly, not every bioarchaeology study can address or attain all three aspects, especially with collections in repositories where much of the contextual information is lost. The idea is to attempt to use as many levels of analysis as possible. There is much to be gained by incorporating integration, engagement, and ethics into bioarchaeological studies whenever possible. It will set in motion ways of asking questions and framing problems and that is what this text is advocating.

1.1.1   Bioarchaeology as Integrative, Engaged, and Ethical

Bioarchaeology is a way of conducting scientific studies that is aligned with the overall mission of the field of anthropology and therefore must be at all times informed theoretically, methodologically, and ethically in the same ways that anthropological research is. Armelagos (2003) presents a compelling overview of the history of bioarchaeology and the ways that it bridges different data sets. He shows how early research into the origins of plant domestication and its effects on human biology and demography set the stage for understanding a complex web of factors that underlie human well-being. His case study details the integrative nature of bioarchaeological investigations and the ways that data from the human remains can be enhanced with data from the archaeological reconstruction of subsistence activities and diet. Further, isotopic analyses conducted on the human remains are productively integrated with analyses on the floral and faunal component of the diet as revealed through ethnobotanical and zooarchaeological reconstructions. Thus, bioarchaeology has the potential to integrate diverse data sets in innovative ways.

The questions framed by bioarchaeologists are those being asked by scientists the world over. Why are there wars? Why is there starvation? Why are women raped? How do we stop the spread of diseases? If even tangentially, it is useful to conduct bioarchaeological studies with some larger series of questions to which the data may be relevant. Bioarchaeological research can be very useful for understanding and solving problems in the modern world. But this only becomes possible if the research design is broadly integrative from the start.

Thus, an integrative and engaged bioarchaeology seeks to broaden the ways questions are framed. A broad, cross-cultural, historically situated study of human behavior is an important scholarly activity that contributes to explaining the complex human behaviors that underlie the pressing and persistent problems today. Bioarchaeology may be the only disciplinary effort that can provide information on the origin and evolution of violence, disease, inequality, and diet, to name just a few. As Armelagos concluded, ". . . We are enriched when essential insights drawn from the past provide a prologue to the future. . ." (2003:34).

Locating the origins of contemporary problems is productive to do because it isolates the very specific, historically contingent factors that help to situate and explain human behavior. Often, to be able to understand a complex behavior in its specific manifestation (e.g., culturally determined age at weaning or the age at which males go off to war), it is useful to look deep into the past to see when those behaviors first show up and what the circumstances were that favored them. Bioarchaeological studies have the potential to situate modern-day problems within a larger temporal and spatial framework. Using these cross-cultural and deeply temporal analyses, bioarchaeology contributes to understanding human variation within and across different cultures as well as non-Western ways of dealing with and adapting to challenges.

Along with promoting integration and engagement in bioarchaeological research, there must also be self-reflection and concern about the ethical implications of scientific research conducted using human remains. What are the ethical concerns that bioarchaeologists regularly confront? In the USA and in many countries of the world where archaeology is conducted, indigenous groups have pressed for account-ability regarding the excavation and curation of human remains. Research based on human remains must consider how this type of scientific analysis affects the people who view the remains as ancestors. The Native American Grave Protection and Repatriation Act (NAGPRA) in the USA and similar kinds of injunctions, legisla-tion, and mandates in other countries have forever changed the way that burials and human remains are approached. From the moment of discovery through to analysis and interpretation, NAGPRA and NAGPRA-like mandates have brought bioarchae-ologists and indigenous or descendant populations together in often surprising and productive ways that could not have been predicted (Martin and Harrod 2012:31).

Today, virtually no analysis is done on human remains without proposals that must pass through and be approved by many different groups. Prior to any analyses, con-sensus and some form of cooperative effort between bioarchaeologists and other stakeholders must be obtained. From museum and governmental entities to tribal rep-resentatives and indigenous committees, research proposals, excavation permits, and access to repositories are strictly controlled. The product of this more collaborative and consensual effort is not only a much deeper engagement with descendant com-munities but also a more detailed understanding of the human remains themselves.

This book is not for novices who seek to learn how to analyze bones and mum-mies. It is for practicing bioarchaeologists and students of bioarchaeology who are already familiar with the study of human osteology as well as the discipline and the methodologies used in the field and laboratory. It is written with scholars, professors, and instructors of bioarchaeology in mind as well as students (both undergraduate and graduate) who desire to see the larger picture of what bioar-chaeology encompasses. It may be particularly useful for those seeking help or wanting fresh ideas in designing and carrying out research projects (both in field and in laboratory settings).

For individuals new to bioarchaeology and to the comprehensive study of the human skeleton, we recommend that they begin with White et al. (2012) *Human Osteology, 3rd Edition*. This is the most up-to-date, relevant, and comprehensive manual for training in human osteology and bioarchaeology. Another important book to consult would be Clark Larsen's (1997) *Bioarchaeology, Interpreting Behavior from the Human Skeleton*, which lays down the foundation for integrating method and data. Buikstra and Beck's (2006) edited volume, *Bioarchaeology, The Contextual Analysis of Human Remains*, provides one perspective on the intellec-tual trajectory of historic and contemporary practitioners of skeletal biology and physical anthropology. Other more critically engaged intellectual histories can be found in Blakey (1987), Armelagos (2003), Rankin-Hill and Blakey (1994), and Martin (1998).

This text does not replicate the above efforts nor does it provide an exhaustive overview of the literature in bioarchaeology, rather it is meant to be a handbook or manual of "best practices" culled from the more recent scholarship in bioarchae-ology. It provides a concise overview of the major and important areas that make up bioarchaeology. Seminal works from selective sources are discussed in each chapter that will be useful in framing a bioarchaeological study. The references that are cited are those that will be most helpful in teaching bioarchaeology or in formulating projects. The cited literature is only meant to give readers a start in the right direction.

What follows is an overview of well-established trends along with new directions in bioarchaeology, and a guide for incorporating as much information as possible into bioarchaeological studies. It is a "how to" for individuals seeking the broadest overview of what modern bioarchaeology strives to be. Admittedly, there is a bias toward bioarchaeology as it is practiced in the USA, but modern bioarchaeology is a globalized enterprise and when possible perspectives that are more international are included. There is also a focus on skeletonized human remains (vs. mummified remains).

Practicing an integrative, engaged, and ethical bioarchaeology is important because the stakes are high. The world is in trouble and people are dying. Sectarian warfare, gangs, poverty, increasing numbers of refugees, resource depletion, environmental degradation, and all of the "isms"—racism, sexism, ageism, homophobia, and classism—are worries that dog us all. Bioarchaeological research can provide a perspective on all of these problems that is both unique and critically needed. Data on human biology that traverses time and space places all of these problems in perspective. Bioarchaeological data also provide a much-needed dimension to understanding human diversity and illustrating the plasticity of human behaviors. Armed with these kinds of data, the problems mentioned above can be challenged with non-Western approaches and more broadly conceived perspectives. Bioarchaeological data on human adaptation also highlights just how historically situated and culturally constructed many of the problems are.

1.1.1.1 Goals of Bioarchaeological Research

Bioarchaeology integrates information from the human remains (such as age at death, sex, stature, pathology, physique, and trauma) with other aspects of the environment and culture that the person lived in (population density, environmental factors, weather patterns, local food sources, living quarters, and family structure). As such, it is distinguishable from the traditional study of human remains (both skeletonized and mummified) which focuses on the medical and forensic aspects of osteology and paleopathology. Bioarchaeology is also distinguishable from mortuary archaeology and funerary studies, because these tend to focus only on the ancient and historical grave architecture and accouterments. Yet, bioarchaeologists try to include information about the mortuary context in with the biological and cultural observations as well. Thus, bioarchaeology is not simply archaeology with an emphasis on the human burials. Rather the goal of bioarchaeological scholarship is to incorporate all of the possibilities for reconstructing meaning from patterns of death in a creative and innovative way.

The operational way that bioarchaeology meets the goal of making meaning out of the study of the dead is to use a mode of inquiry composed of five crucial aspects. (1) A bioarchaeology study must start with a research question that can be answered with the available empirical data. That is, a bioarchaeological study is not purely descriptive, but rather it seeks to use data from a variety of sources (e.g., bones, artifacts, archaeological reconstruction) to hone in on a particular set of hypotheses or questions. (2) A bioarchaeological study must at all times be concerned with the ethical dimension of the project. This translates into bioarchaeology having its own ethos for inclusion and collaboration of invested parties at every juncture if appropriate. At this time, bioarchaeology is in an early stage of formulating an ethos and a guiding set of principles that go beyond legislated mandates. Since the passage of legislation under the Native American Graves and Repatriation Act (NAGPRA), there are basic guidelines for dealing with human remains in the USA, but bioarchaeology needs to go beyond NAGPRA. (3) A bioarchaeological study must include systematic, rigorous, replicable, and scientifically sound data (both quantitative as well as qualitative data) from the human remains, particularly with respect to providing identity and life history of the individuals. (4) A bioarchaeological study must include detailed (when possible) data on the mortuary context and funerary items. (5) Finally, a bioarchaeological study must link interpretations to broader theoretical issues regarding human behavior. Bioarchaeologists must be forthcoming in their study with what particular kinds of theory inform their decisions about how to best carry out the project and how to best interpret the findings. The findings then need to be linked to broad and encompassing questions in anthropology. Bioarchaeology integrates research questions, ethics, and skeletal, mortuary, and archaeological data *with* theory.

1.2   Methodological and Theoretical Approaches

There is a long history in biological anthropology of decoupling the study of past populations from their living descendants. At the turn of the century in the USA, there was interest in sending biological anthropologists (as well as ethnologists and archaeologists) to the western states where they were trained in anthropological method (see Martin 1998 for a full exploration of this). These early studies set in motion a particular way of imagining indigenous peoples of the Americas (vis-à-vis the precolonial past). It created a methodological science for the study of ancient people through their ancestor's bones and artifacts, but it also disconnected those studies from the concerns and struggles of contemporary native people.

For example, early scholars, in the absence of scientifically rigorous empirical data, tended to visualize the ancient world as analogous to the historic and modern condition of living native people. This in turn led the native community to be wary of archaeological and biological studies of their ancient ancestors and to perceive the results of those excavations as both irrelevant and disrespectful. The enactment of legislation prohibiting the excavation and analysis of human remains without the consent and collaboration of direct descendants resulted. In 1990, NAGPRA was passed, and it changed the way excavations and analyses were done.

Bioarchaeology as a scientific discipline took shape in the early 1980s just as processual archaeology began to provide a set of scientific principles and a focus on ecological explanations (Binford and Binford 1968). Concurrently, human adaptability developed within biological anthropology as a means of combining interests in evolutionary change with concern for the various adaptive problems faced by humans today, especially those living in limited and ecologically marginal environments (Buikstra and Cook 1980; Larsen 1987; Goodman et al. 1988). With questions focusing on how humans manage to survive and adapt (behaviorally, physiologically, developmentally, or genetically) to environmental constraints and stressors, human adaptability clearly shared an ecological perspective with processual archaeology.

1.2.1 Integrating Human Remains with Archaeological Context

Newer and innovative approaches to skeletal analysis have come out of the merging of biological anthropology and the study of human remains with archaeology. Buikstra (1977) made an early call for a program of skeletal analyses that situated itself within a multi-methodological framework that investigates the skeleton as well as a number of other important cultural and environmental variables. Twenty years later, Larsen (1997) compiled a textbook dedicated to delineating the history, method, and data that are all parts of what became a new paradigm for skeletal analysis. Armelagos (2003:28) noted that the development of bioarchaeology in 1980s brought skeletal biology out of a stagnant state, as skeletal biology in 1970s and 1980s was a theoretically and methodologically impoverished field. Bioarchaeology provided a fresh new way to frame questions about human evolution and adaptation, and it linked analysis of remains to archaeology.

As developed by Buikstra, Larsen, Armelagos, and others, bioarchaeologists have a deep commitment to scientific inquiry. Formulating testable hypotheses and using multiple methodologies and interdisciplinary approaches is a major departure from the more descriptive techniques used by the early biological anthropologist who worked primarily in laboratory (not archaeological) settings. Bioarchaeology moves skeletal analysis beyond description and blends the methods and data from skeletal biology with archaeology. Related disciplinary approaches from ethnography, taphonomy, forensics, medicine, history, and geology likewise present potential complementary and revealing data sets that can be used in conjunction with skeletal analyses and interpretation.

Walker (2001) presented a review of bioarchaeological method, theory, and data regarding violence in past populations. He designed a highly useful flow diagram (2001:577) that aids in using strong inference in the interpretation of injury and violence from skeletal remains. With a concern for analysis of skeletal remains in addition to cultural context, biological processes, and taphonomy, Walker demonstrated that bioarchaeology can and should be used to independently test hypotheses.

The difference, then, between human osteology/skeletal analysis and bioarchaeology is that the latter employs interdisciplinary and cross-cultural research tools that can aid in the analysis of a wide range of questions about human behavior. Its goal is to interpret the biological data in relation to social and ecological contexts such as changes in diet, increases in population size and density, shifts in power and stratification, and differential access to resources. By using multiple working hypotheses, scientific methodology, and strong inference, it provides a means for utilizing multiple lines of evidence in interpretations. Most importantly, variability is not factored *out*; rather, it is considered and weighed against all other available lines of evidence so that it is accounted for (instead of being discounted).

1.2.2   Integrating Human Remains with Ethical Considerations

Another dimension of bioarchaeological work is the conscious adherence to both the legal mandates (i.e., NAGPRA) and ethical concerns (see Ferguson et al. 2001 for an excellent overview of what constitutes ethical and moral obligations in this area) regarding the study of human skeletal remains, grave offerings, mortuary contexts, and burial data (discussed at length in Chap. 2). As Ferguson and colleagues working with the Hopi Nation today note, "One thing archaeologists should keep in mind is that the disturbance of human remains is agonizing for Hopi people . . . In presenting results of mortuary studies, archaeologists need to understand that for the Hopis, the heartfelt spiritual concerns about the disruption of graves far outweighs any [of the] scientific studies . . . what archaeologists find to be interesting results and findings are colored by the desecration of the graves that led to those results" (2001:22).

It is incumbent upon bioarchaeologists to use information that has been collected in a way that does not trivialize or diminish the lives of the living descendants. Collaboration and consultation with representatives from Native American groups regarding data derived from excavations in the USA is one way to rectify the past and to begin to build an ethos of inclusiveness. Good science does not have to be exclusionary. Tribal collaboration in scientific research and in review of scholarly work *prior to publication* has not been the norm, but a (very small) trend in doing so among bioarchaeologists has resulted in a much richer interpretation because of the inclusive nature of the enterprise (see, e.g., Ferguson et al. 2001; Kuckelman et al. 2002; Ogilvie and Hilton 2000; Spurr 1993). Without collaboration and the nuanced layering of additional knowledge from those most closely related to the ancient people being studied, the interpretations that scientists formulate may be grounded in expert technique and state-of-the-art methodology but utterly wrong or incomplete in the interpretation. Bioarchaeologists must also accept that Native Americans (and indigenous people the world over) may sometimes refuse to participate or may withhold information deemed esoteric and inappropriate for publication. Bioarchaeology working within these kinds of frameworks is transforming the biological anthropology of the past into a more dialogical and relational process, similar to Wood and Powell's (1993) proposal for archaeology (discussed further in Chap. 2). This will necessarily include decision-making and research as a collaborative and reflexive process among co-participants. It is important not to overlay modern analogs about the world over the past to extract meaning. Without the textured layering of other kinds of empirical data drawn from environmental and cultural reconstructions, oral traditions, ethnography, and the ideas of those most closely related to the people being studied, bioarchaeologists will be destined to create a series of scenarios that, although grounded in theoretical modeling and empirical observation, are wanting in exactness and authenticity.

Because bioarchaeology is a research program that is *de facto* interdisciplinary, it ensures that data collection and, more importantly, data interpretation will be scrutinized and challenged by people from a number of backgrounds with a variety of viewpoints on the appropriate use and meaning of data derived from human skeletal remains. For these reasons alone, bioarchaeology in the twenty-first century must be an inclusive scientific enterprise.

1.2.3 Integrating Human Remains with Environment and Culture

While there were only a few books with the word bioarchaeology in the title before the 1980s, there are now several dozen. *Social Bioarchaeology*, a text edited by Agarwal and Glencross (2011), provides a number of case studies that present the study of human remains using a biocultural framework. The editors pushed the authors to integrate studies of human remains into other dimensions such as cultural and environmental reconstructions. This volume is one of the best for articulating the need to continue to build a biocultural synthesis (see Goodman and Leatherman 1998) and to truly provide models that do not privilege biological data over cultural data. This is a powerful tool for bioarchaeologists, and it is one of the more dominant perspectives used within the field.

Bioarchaeologists must constantly remind themselves that those events that are regarded in scientific terms (such as age, sex, disease state, or population density) at the population level are life-history events on the individual level (Swedlund 1994). Births, deaths, puberty, marriages, rites of passage, and disease are all biological transition points that find meaning and expression through ritual, ceremony, ideology, and other cultural practices. These transitions also provide the timing for generational histories and points of focus for kin and group identities. Taken in their cumulative context, they provide the data for estimation of those larger evolutionary processes of growth, regulation, population composition, and epidemiological profile. These things provide a tangible record and graphic reminder of how well a society is doing. The loss of an infant to a family or an epidemic episode to the larger group presents a concrete experience requiring ideological and adaptive adjustments. To capture all of this in a model is challenging, but it can be done.

1.2.3.1   Using a Biocultural Model

The linking of demographic, biological, and cultural processes within an ecological framework is essential for dealing with the kinds of questions that interest archaeologists and biological anthropologists today. These include, for example, understanding the relationship between political centralization and illness, the impact of population reorganization or collapse on mortality, and the relationship between social stratification, differential access to resources, and health. These kinds of problems demand a multidimensional approach because they cross over numerous disciplinary boundaries.

The interpretation of data derived from human skeletal remains requires an evaluation of the individual's resistance to stressors (by examining the presence, severity, and status of skeletal lesions), the source of the stressor (environmental or cultural), and the effect of the buffering systems. The concept of adaptation to stress is complex. Stress, as used in bioarchaeology, is the physiological disruption that results from any insult (Goodman et al. 1988:177). Most importantly, stress can be measured and evaluated based on empirical evidence garnered from the human skeletal remains.

Methods for the analysis of human remains have advanced tremendously in the last 10 years, and this has increased the capacity for researchers to obtain biological information on diet and health that was previously unavailable. Historically, skeletal analyses were primarily descriptive, with the goal of identifying the geographic distribution and evolution of disease through time and establishment of genetic relationships between groups. Recent emphasis on the interactions between biology and culture in the disease process has proven to be extremely useful and yields direct information concerning the human condition (see Zuckerman and Armelagos 2011 for a complete review and history of the biocultural model).

The study of stress in ancient populations requires an understanding of skeletal responses to change within the context of those variables that affect the skeletal system's ability to respond. Quantifiable changes in the skeleton and dentition reflect disturbances in growth and development, as well as in bone maintenance and repair. The cultural and noncultural stressors that cause observed bone changes can often be inferred. Occurrence of stress markers at different stages in the life cycle can be examined and compared to the mortality rates of the group as a whole.

A deceptively simple model (Fig. 1.1) provides a very useful framework for integrating information regarding human adaptability and health with the larger biocultural and ecological context. In this model, the physical environment is viewed as the source of resources essential for survival. If there are constraints on the resources, then the ability of the population to survive may be limited accordingly (Fig. 1.1, box 1). The adaptation of human populations is enhanced by a cultural system, which buffers the population from environmental stressors (Fig. 1.1, box 2). The technology, social organization, and even the ideology of a group provide a filter through which environmental stressors pass. However, cultural practices can also be the source of stress as well. For example, the development of agriculture in North American allowed for greater production of calories relative to

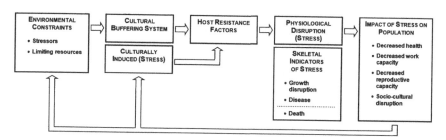

Fig. 1.1 Biocultural model of stress

human expenditure; however, the resulting increased population density and ecological changes associated with intensive farming had a negative influence on overall health. Cohen and Armelagos (1984) and Cohen and Crane-Kramer (2007) provide numerous case studies from the New and Old World regarding the impact of agriculture on human health.

When thinking about all of the possible ways that individuals can be physiologically stressed (discussed in Chaps. 6 and 7), it is important to acknowledge that the impact of stress will be different depending on the age, sex, and overall health of the individual being stressed (Fig. 1.1, box 3). These host resistance factors include not only age and sex but also the overall physical condition of individuals. Infants and the elderly may be harder hit by a seasonal drought that decreases food supplies than a healthy adult. A female who has lost a lot of blood during a difficult childbirth experience will be hit harder by food shortage or cold stress than a female who has not just given birth. Someone suffering from dysentery will have a lower resistance to contagious infections than someone who is healthy. Thus, host resistance is both biological but also cultural in nature because such things as wealth can buffer some people from dying of disease and subordinate status can predispose groups to greater morbidity and mortality. One really good example of this is articulated by Gravlee (2009) in a study entitled, "How Race Becomes Biology: Embodiment of Social Inequality." This study demonstrates how host resistance is always part of a larger political economy in which some bodies/hosts are "worth more" than others and thereby receive access to food, medicine, treatments and protection. Gravlee states that ". . . racial inequality becomes embodied—literally—in the biological well-being of racialized groups and individuals" (2009:47). One can extend this kind of reasoning to other biologically based phenomenon such as age and sex across the life history of individuals that also are affected by inequality and differential access to resources.

Because skeletal tissue typically responds in a nonspecific and generalized way to stress and disease, the diagnosis of specific causes is often not possible. Fortunately, what has the greatest explanatory power is not the specific disease agent but rather the severity, duration, and temporal course of physiological disturbances. These general stressors may be read and deciphered from skeletal changes (Fig. 1.1, box 4). The response to stress is often a stereotypic physiological change that results from the biological effort to adjust and overcome the stress, and this is frequently manifested in relatively permanent osteological indicators. Information derived from human bones and teeth (and mummified soft tissue) provides a large body of evidence for how well individuals in antiquity were doing prior to death.

Although it is crucial to document these physiological changes at the individual level, from an anthropological perspective it is even more important to realize that health and adaptation fit into a larger network of relations that extends beyond the individual to the population and community (Fig. 1.1, box 5). For example, undernutrition of individuals can be established by examining the bones and teeth of individuals. Severe or prolonged undernutrition in large numbers of people within a group has the potential to negatively impact work capacity, fertility, and mortality. It is also associated with disruptions to the social, political, and economic structure of single communities and has the potential to destabilize whole regions as well.

Although ecological stress can be sometimes causally related to biological stress, ecological factors are not the only source of stress. Warfare can become pervasive due to shifts in ideology and power, and this can be a source of biological stress and mortality as well. The model in its most simplistic form may seem to be largely processual in suggesting unicausal variables and a simple feedback loop. However, the model can easily accommodate much more complex (and post-processual) cultural factors as causal mechanisms creating biological stress.

The feedback from box 5 back into boxes 1 and 2 represents the ways that cultural and population-level changes can further cause changes in the environmental (both the physical and the culturally constructed) systems. During these times, the subcomponents of cultures, including the economic, political, and social systems that are inextricable linked with the ability to respond to stressors, could be further impacted as well.

Although this generalized model may strike some as being static and containing simple factors within boxes, as a heuristic device, it is invaluable to bioarchaeologists. And, with the recognition that conditions are historically contingent, relational, and highly dynamic, the model can be adapted to particular moments in time and space. The biocultural model is only as dynamic and complex as the researcher using it makes it. When there is a great deal of archaeological information on the environmental and cultural context, the model will have many extra features to it that can be utilized in visualizing all of the possibilities for dynamic forces and processes at work.

1.2.3.2   Adapting the Biocultural Model to Specific Research Questions

Many bioarchaeologists have produced creative variations on the above biocultural model (Fig. 1.1) that is adapted to very specific questions about various cultures, geographic regions, and time periods. For example, Sheridan (2000) utilized the biocultural model and applied it to a historic study of the Byzantine culture. Because she was analyzing a historic archaeological site, she had many other sources of information to build into the model. The questions about adaptation to that particular period in time structured the way she included information on the environment, culture, and biological remains. For the Byzantine St. Stephen's site, her version of this heuristic tool reshaped the model into a more circular set of boxes, each having multiple feedback loops. In her study, because there are historical documents and a wealth of archaeological data on the site, data from the human remains were contextualized using a great deal of information. Sheridan writes that ". . . pilgrimage and liturgical records, arts and iconography, legal and medical documents, material culture, and a variety of dating methods have all contributed to our understanding of the cultural context, and so, to the biocultural setting" (2000:576).

There are any number of ways that the basic biocultural model can be modified and enhanced to take into account particular data that is available. Because Sheridan had access to written records for this historic archaeological site, those data could be factored in. What is most important in bioarchaeology is for researchers to utilize a biocultural approach that systematically organizes the kinds of information that is accessible and crucial to the questions under study. This is best summarized by Zuckerman and Armelagos (2011:28) when they write that ". . . biocultural approaches in bioarchaeology have revolutionized the field, facilitating its transition from a descriptive enterprise to (a) socially, culturally, and politically informed dynamic force in biological anthropology . . . ultimately the biocultural approach may also enable researchers to understand challenges to human health in the past as well as in the present and future."

1.2.4 Integrating Human Remains with Contextual Data

There is often confusion in thinking about the difference between heuristic models and frameworks on the one hand, and hypotheses and theories on the other. Models, such as the biocultural model discussed here and utilized in many bioarchaeological studies, are constructed to approximate the real world and each bioarchaeologist will need to input which variables are thought to be the most important in capturing the world of the people that they are studying. When the pioneers of bioarchaeological method first started envisioning the use of models, such as Buikstra (1977:82) and Goodman and Armelagos (1989:226), they found it was important to understand both the environmental and the cultural context that people were living in. The human remains were then to be analyzed and interpreted using these models as a way to not lose sight of the most important variables that impact human behavior in a given setting. These models were also an extremely useful way to coordinate and organize a large body of information coming from many different sources (e.g., archaeological data on the site and the mortuary context, biological data from the analysis of the bones and teeth, and ethnohistoric information).

The beauty of these generalized models is that one can fill in the environment-culture-biology boxes with any kind of data that one deems important to consider. However, it needs to be stated that no one researcher can ever collect every single piece of important information regarding a culture. Researchers always make decisions about which data to collect and include and which data they simply will not include. An unfocused approach to collecting data from human remains is perhaps the least useful of all. Collecting a few indicators such as age, sex, and prominent pathologies will not permit very much of an interpretation.

That is why it is always a best practice to start a bioarchaeological study with a question or a series of hypotheses that can be answered or addressed by the available empirical data from a particular archaeological case study (discussed in Chap. 3). If the interest is in children, it makes sense that the study would draw from a case study where there were children represented in the human remains. If the focus of the study were going to be on the role of violence, there would need to be biological and archaeological information on trauma and artifacts such as weaponry. If the interest is on the difference between male and female work patterns, there would have to be a sample size that was large enough to compare adult males to females in terms of entheseal changes (places where muscle use and size alter bone morphology) and use patterns.

Usually bioarchaeologists and archaeologists formulate these questions and hypotheses based on having been exposed to particular theories in the literature about aspects of human behavior for which there is a body of data that suggests particular kinds of associations and relationships. Theories are formed by seeing consistent ways that key factors seem to always underlie a particular behavior or response. Theories are broad observations about the human condition that generate a lot of discussion and interest because they hold explanatory power.

A biocultural model, however, is not a theory. A theory is what helps a researcher make sense of and interpret data that has been carefully collected and arranged using something like a biocultural model. Approaching a study with a set of theories about the evolution of sex differences and the sexual division of labor will necessarily focus the study on particular data sets. There will be more of an emphasis on collecting data from the human remains and the archaeological context that relate to sexual differences that may be codified by type of burial, number and kind of grave goods, and other attributes of social identity. Theories about the evolution of sexual division of labor or theories about the cultural construction of male and female identity are two different but equal ways of thinking about the role of sex in a variety of social and environmental settings.

Theories about human behavior come from many different intellectual traditions within the social and natural sciences, and bioarchaeologists can benefit from being familiar with theories about human behavior from these different disciplines. While archaeologists have taken advantage of the use of theory, biological anthropologists and bioarchaeologists have a much shorter history of doing so. Even with the major influences of evolutionary theory within biological anthropology, it does not form the basis for bioarchaeological studies to the extent that theories about inequality or gender do.

The important thing for bioarchaeological research that aims to be integrated across biocultural domains, engaged in thinking broadly about human behavior, and ethical in the way it incorporates diverse points of view, is that it be informed by a well-articulated theory. Without the guiding light of theory, data collected from human remains will be less useful in explaining why humans do what they do, and it will be less interesting to non-anthropologists and the lay public. Bioarchaeological studies that do not employ theory in the interpretation fall short of being useful to a much wider scientific and humanities audience as well (see Chap. 3 for a full discussion of theory in bioarchaeology).

1.2.4.1   Case Study: Ancient Arabia

Combining theory with a biocultural model provides a way to make bioarchaeology relevant to many other aspects of the human condition. For example, Baustian 2010 adapted the biocultural model (Fig. 1.2) in order to examine the causes of preterm and neonatal infant death in the region of the Arabian Peninsula during the Bronze Age (circa 2200–2000 bc). Research into the early inhabitants of this region (using ethnohistoric and archival documents) revealed a strong preference for polygamy and early marriages. This initial research led her to incorporate theoretical scholarship about the origin and evolution of kinship and family structure in groups indigenous to the Arabian Peninsula. In this way, she saw that in order to interpret the skeletal data from the large number of preterm and term infants found in a tomb she would have to examine maternal-infant health in a much broader context. By including theoretical notions about kinship and marriage with ethnographic information on forms of social organization, she was able to look at the bigger picture for ancient Arabia. The addition of clinical literature on maternal-infant health helped to further shape the underlying biocultural factors that may be at work.

Thus, to develop how the data from the skeletal remains could be contextualized, she used theory about kinship along with ethnohistoric and medical information. The medical literature clarified the range of possible effects of polygamy, consanguineal marriages, and arranged early marriages regarding the causes of premature births and infant mortality. As can be seen in Fig. 1.2, very specific information from various published sources were utilized to build the model, and then the data were interpreted using theories about human behaviors (such as polygamy and consanguinity) that explain the high rates of prematurity and infant death.

In building this model, Baustian reconstructed the major features of the environment, culture, and biology as they related to the archaeological site, Tell Abraq, from which the bones were retrieved. Based on research from the archaeological reconstruction of the ancient geographic and ecological environment, it was well documented that there would have been high fluoride levels in the ground water, that mosquitoes and sand flies would have been ubiquitous, and that proximity to wells and other sources of fresh (nonsalinated) water would have been challenging aspects of the physical environment.

The warm and humid environment would also have played a significant role in types of housing structures and indeed the building of a large high mud brick platform (called a tell) contained a well at its center and fortification around the platform. This archaeological data confirmed the constraints on fresh water placed on the inhabitants. Cultigens such as dates and grains were a large part of the daily diet. These aspects formed part of the cultural activities at the site, which may have buffered inhabitants from dietary problems.

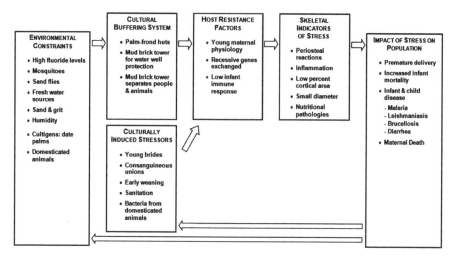

Fig. 1.2 Biocultural model of stress applied to ancient populations in the Arabian Peninsula

The theoretical and anthropological research on polygamous kinship structures suggests some culturally induced stressors may have played a role in overall patterns of morbidity and mortality. Arranged marriages and young brides, consanguineous (first cousin) unions, and early weaning are all behaviors associated with polygamy, and they are clinically associated with higher rates of premature births and maternal and infant distress. Also, close proximity to domesticated animals (in this case sheep/goats and camels) produces higher rates of zoonotic infections, which can be passed on to humans.

Because of her interest in maternal-infant health, host resistance factors that would be important to consider would be the young maternal physiology, the passing of recessive genes in first cousin unions, and the immature immune systems of premature infants. Focusing in on infant human remains, she collected data from the bones on infectious disease, inflammations, aspects of growth and development, and nutritional stress (see Chaps. 6 and 7 for these methods).

All of these factors, taken together, were interpreted in light of how they might be influencing the population in terms of increased infant death and the possibility of increased maternal deaths. In putting all of these data from the archaeological reconstruction of the site with the bioarchaeological data from the human remains, Baustian was able to formulate a viable explanation for the high number of premature and newborn infant deaths at this site. Theories about the origin and nature of polygamous kinship arrangements provided a very broad dimension for her discussion of infant mortality and female morbidity in the ancient Arabian context in particular but also about the role of culture in mother-infant well-being in general.

Without the use of a biocultural model to hone in on the most important features of the environment and the culture with respect to infant health, and without having incorporated theoretical notions about kinship and family structure, this study would have only been able to document that there were a significant number of premature

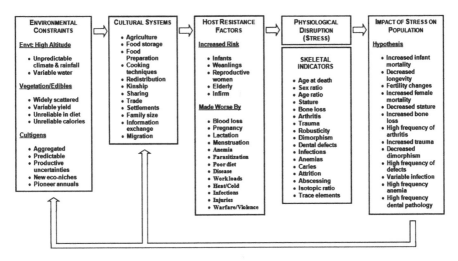

ENVIRONMENTAL CONSTRAINTS	CULTURAL SYSTEMS	HOST RESISTANCE FACTORS	PHYSIOLOGICAL DISRUPTION (STRESS)	IMPACT OF STRESS ON POPULATION
Envt: High Altitude	• Agriculture • Food storage • Food Preparation • Cooking techniques • Redistribution • Kinship • Sharing • Trade • Settlements • Family size • Information exchange • Migration	**Increased Risk**		**Hypothesis**
• Unpredictable climate & rainfall • Variable water		• Infants • Weanlings • Reproductive women • Elderly • Infirm	**SKELETAL INDICATORS**	• Increased infant mortality • Decreased longevity • Fertility changes • Increased female mortality • Decreased stature • Increased bone loss • High frequency of arthritis • Increased trauma • Decreased dimorphism • High frequency of defects • Variable infection • High frequency anemia • High frequency dental pathology
Vegetation/Edibles			• Age at death • Sex ratio • Age ratio • Stature • Bone loss • Arthritis • Trauma • Robusticity • Dimorphism • Dental defects • Infections • Anemias • Caries • Attrition • Abscessing • Isotopic ratio • Trace elements	
• Widely scattered • Variable yield • Unreliable in diet • Unreliable calories		**Made Worse By**		
Cultigens		• Blood loss • Pregnancy • Lactation • Menstruation • Anemia • Parasitization • Poor diet • Disease • Workloads • Heat/Cold • Infections • Injuries • Warfare/Violence		
• Aggregated • Predictable • Productive uncertainties • New eco-niches • Pioneer annuals				

Fig. 1.3 Biocultural model of stress applied to precontact populations in the American Southwest

and full-term infants in the ancient tomb. The biocultural model in addition to social theory permitted a much more broad interpretation of infant deaths and, ultimately, revealed the complex underlying factors that explained those deaths.

1.2.4.2   Case Study: Ancient America

A completely different set of factors were included in a bioarchaeological study of an ancient Pueblo group from the American Southwest (circa ad 900–1100). For the human remains from numerous habitation sites from a region referred to as Black Mesa in northeastern Arizona, Martin et al. (1991) were able to draw on an extensive archaeological and ethnographic literature on the ancient environment and on historic cultural practices. The biocultural model for this study was constructed in such a way to highlight and contextualize questions about survival and adaptation in a very harsh and marginal desert environment where resources were limited (Fig. 1.3). The high altitude contributed to unpredictable resources as well as ephemeral and variable sources of water. The pinion-pine environment yielded patchy and unpredictable edible resources for the inhabitants. Maize agriculture presented a way to have a predictable and aggregated source of food but was hampered by short growing seasons and lack of rainfall.

Theories about human evolution and adaption in desert environments were used to help think about the ways that these specific Southwest peoples could have coped over a 200-year period to droughts, seasonal shortages in food, and ephemeral water supplies. Agriculture is dependent on combined supplies of winter and summer rainfall, both of which are in short supply. The environment of Black Mesa is highly risky, with few good locations for agriculture and with wide fluctuations in climatic and hydrologic variables for successful harvests.

Cultural innovations were reconstructed from the archaeological patterning of artifacts. Food storage, food preparation items, cooking vessels, and trade items all suggested buffers that could enhance survival in such a marginal environment. The archaeological reconstruction of settlement patterns, storage features, and trade suggests many ways that people living at Black Mesa could buffer themselves from droughts and food shortages. With smaller family sizes, frequent seasonal relocation, and the formation of alliance and kin ties across the region, the inhabitants would have been able to share resources and information that would offset the environmental instability.

A complete analysis of all individual human remains from the site revealed a population that suffered from some nutritional stress that may have contributed to infant mortality and adult morbidity. But the biological data did not support severe malnutrition or starvation. Instead, a picture emerged of endemic, mild to moderate nutritional stress that had an impact on mostly infants. The generally mild nature of the pathologies suggests an ability to respond to and recover from bouts of physiological disruption. Taken together, the data sets on childhood growth and development suggest that even though Black Mesa was a harsh and marginal environment for farming, the adult strategies of seasonal mobility, maintaining a diverse diet, and political autonomy may have protected the children in ways not possible in larger communities in the region. With their adaptability and organizational strategizes, the ancestral Pueblo of Black Mesa created a stable and sustainable lifestyle for over 200 years. Placed in a broader context, this case study revealed that humans can use a variety of cultural tactics to deal with challenging environmental stresses.

Given the looming and large problems likely to increase with desertification of large parts of the world and global warming, information on how populations can remain relatively healthy in very hot, dry, and unpredictable desert conditions is important to the larger national conversation going on. Bioarchaeological data are crucial to understand the impact on the group as a whole and, most importantly, to have the data on growth and health from all of the individuals who did not survive to old age. It is that segment of the population, those dying young, that reveals compelling data on what the limits of human adaptability are.

1.3   Summary

It may seem ironic that studies centering on patterns of disease and death of past people provide important information that adds to our ability to explain human behavior and to better understand the human condition. Anthropology is one of the few disciplines that is inherently interdisciplinary and holistic in its approach to answering questions about *why* humans do what they do as well as the impact of their behaviors on their survival. Death is the end result of an accumulated set of biological, behavioral, and cultural responses to challenges in the social and physical environment. Individuals are constantly adjusting to their environments, and the success of those adjustments is reflected in their ability to survive (at the individual level) and reproduce (at the population level).

Skeletal material is a very distinctive part of the archaeological record because it is the only record of humans as biological entities interacting within a cultural and environmental context. Using biocultural models as a heuristic method to clarify the relationship among important factors that affect human health and viability has proven to be the only way of getting beyond purely descriptive studies. These models are useful because they permit the inclusion of factors that the researcher can obtain and factors that have the most importance for the questions and hypotheses being addressed.

How does one know which theory to use to interpret the findings? Theories about human behavior come from a lot of different intellectual traditions (see Chap. 3). From political-economic theory, which focuses on the production of inequality, to evolutionary theory with its focus on natural selection and to gender theory where there is an emphasis on the ways that male–female relationships work in different contexts, there is an abundance of ways to approach setting up a series of questions and hypotheses that bioarchaeological data can contribute to. New theories about identity illustrate the ways that identity itself as socially constituted, embodied, and experienced. The so-called third wave gender theory goes so far as to destabilized and abandon the categories of male–female as natural *a priori* analytical concepts. New theories are emerging that challenge researchers to examine when, where, and how these categories came to exist and what implications they have for human experience and systems of inequality. Practice theory emphasizes the consideration of the lived experience of individuals within broader sets of political and social constraints and opportunities. Marxist theory seeks to understand power relations and control of production and distribution of resources. There is really no end to the kind of theory one might use to filter and highlight the study being conducted, but the important thing here is that there be a body of theory about human behavior that directs the nature of the study.

Analysis of human remains used as anthropological inquiry takes advantage of the interdisciplinary nature of the bioarchaeology to provide time depth and geographic variability to the understanding of short- and long-term consequences and mechanisms of change and human response systems. Bioarchaeology studies add a dimension of history and context to research, and they have the potential to link past processes of human existence with present conditions. Bioarchaeology studies are being used as a means for addressing larger social issues such as the specific relationship between colonization and disease, disability, and death. The lack of understanding about patterns of disability and disease as it relates to inequality and access to resources is not unique to ancient people.

Thus, bioarchaeology studies are more useful and relevant when they start with a research question that can be answered with the available empirical data. In addition to this, bioarchaeological studies must at all times be concerned with the moral dimensions of the project. This may constitute working with tribal representatives on the specifics of the project or it may represent presenting the study to a museum advisory council for permission. Studies must utilize methods that are systematic, rigorous, replicable, and scientifically sound. Lastly, bioarchaeological studies must always seek to link interpretations of the data to broader theoretical issues regarding human behavior. It is crucial to articulate the larger body of information that exists on the topic and the ways that others have begun to understand patterns and processes based on these theories. The findings then need to be linked to broad and encompassing questions in anthropology.

Joanna Sofaer captures the essence of this kind of integrated research when she says "... we cannot take an empiricist view and assume that osteological data speak for themselves ... as the body is simultaneously biological, representational and material" (2006:11). Bioarchaeology is informed by the use of frameworks, models, and especially theories that can help think through different ways of understanding the data, as well as applying the data to larger questions and problems.

References

Agarwal, S. C., & Glencross, B. A. (2011). *Social bioarchaeology*. Malden: Wiley-Blackwell.

American Anthropological Association. (2012). *What is Anthropology?* http://aaanet.org/about/WhatisAnthropology.cfm. Accessed February 15, 2013.

Armelagos, G. J. (2003). Bioarchaeology as anthropology. In S. D. Gillespie & D. L. Nichols (Eds.), *Archaeology is anthropology* (pp. 27–41). Washington, DC: Archaeological Papers of the American Anthropological Association, No. 13.

Baustian, K. M. (2010). *Health Status of Infants and Children from the Bronze Age Tomb at Tell Abraq, United Arab Emirates*. MA thesis, University of Nevada, Las Vegas, Las Vegas.

Binford, S. R., & Binford, L. R. (1968). *New perspectives in archaeology*. Chicago: Aldine Transaction.

Blakey, M. L. (1987). Intrinsic Social and Political Bias in the History of American Physical Anthropology: With special reference to the work of Aleš Hrdlička. *Critique of Anthropology, 7*(2), 7–35.

Buikstra, J. E. (1977). Biocultural dimensions of archaeological study: A regional perspective. In R. L. Blakely (Ed.), *Biocultural adaptation in prehistoric America* (pp. 67–84). Athens: Southern Anthropological Society Proceedings, No. 11, University of Georgia Press.

Buikstra, J. E., & Beck, L. A. (2006). *Bioarchaeology: The contextual analysis of human remains*. Burlington: Academic.

Buikstra, J. E., & Cook, D. C. (1980). Palaeopathology: An American account. *Annual Review of Anthropology, 9*, 433–470.

Cohen, M. N., & Armelagos, G. J. (1984). *Paleopathology at the origins of agriculture*. Orlando: Academic.

Cohen, M. N., & Crane-Kramer, G. M. M. (2007). Ancient health: Skeletal indicators of agricultural and economic intensification. In C. S. Larsen (Ed.), *Bioarchaeological interpretations of the human past: Local, regional, and global perspectives*. Gainesville: University Press of Florida.

Ferguson, T. J., Dongoske, K. E., & Kuwanwisiwma, L. J. (2001). Hopi perspectives on southwestern mortuary studies. In D. R. Mitchell & J. L. Brunson-Hadley (Eds.), *Ancient burial practices in the American Southwest* (pp. 9–26). Albuquerque: University of New Mexico Press.

Goodman, A. H., & Armelagos, G. J. (1989). Infant and childhood mortality and mortality risks in archaeological populations. *World Archaeology, 21*, 225–243.

Goodman, A. H., & Leatherman, T. L. (1998). Traversing the Chasm between biology and culture: An introduction. In A. H. Goodman & T. L. Leatherman (Eds.), *Building a new biocultural synthesis: Political-economic perspectives on human biology* (pp. 3–41). Ann Arbor: University of Michigan Press.

Goodman, A. H., Brooke Thomas, R., Swedlund, A. C., & Armelagos, G. J. (1988). Biocultural perspectives on stress in prehistoric, historical, and contemporary population research. *American Journal of Physical Anthropology, 31*(S9), 169–202.

Gravlee, C. C. (2009). How race becomes biology: Embodiment of social inequality. *American Journal of Physical Anthropology, 139*(1), 47–57.

Kuckelman, K. A., Lightfoot, R. R., & Martin, D. L. (2002). The bioarchaeology and taphonomy of violence at Castle Rock and Sand Canyon Pueblos, Southwestern Colorado. *American Antiquity, 67*, 486–513.

Larsen, C. S. (1987). Bioarchaeological interpretations of subsistence economy and behavior from human skeletal remains. In M. B. Schiffer (Ed.), *Advances in archaeological method and theory* (Vol. 10, pp. 339–445). San Diego: Academic.

Larsen, C. S. (1997). *Bioarchaeology: Interpreting behavior from the human skeleton*. Cambridge: Cambridge University Press.

Martin, D. L. (1998). Owning the sins of the past: Historical trends in the study of Southwest human remains. In A. H. Goodman & T. L. Leatherman (Eds.), *Building a new biocultural synthesis: Political-economic perspectives on human biology* (pp. 171–190). Ann Arbor: University of Michigan Press.

Martin, D. L., Goodman, A. H., Armelagos, G. J., & Magennis, A. L. (1991). *Black Mesa Anasazi health: Reconstructing life from patterns of death and disease*. Carbondale: Southern Illinois University Press.

Martin, D. L., & Harrod, R. P. (2012). Special forum: New directions in bioarchaeology. *SAA Archaeological Record, 12*(2), 31.

Ogilvie, M. D., & Hilton, C. E. (2000). Ritualized violence in the prehistoric American Southwest. *International Journal of Osteoarchaeology, 10*, 27–48.

Rankin-Hill, L. M., & Blakey, M. L. (1994). W. Montague Cobb (1904–1990): Physical anthropologist, anatomist, and activist. *American Anthropologist, 96*(1), 74–96.

Sheridan, S. G. (2000). 'New Life the Dead Receive': The relationship between human remains and the cultural record for Byzantine St. Stephen's. *Revue Biblique, 106*(4), 574–611.

Sofaer, J. R. (2006). *The body as material culture: A theoretical osteoarchaeology*. Cambridge: Cambridge University Press.

Spurr, K. (1993). *NAGPRA and archaeology on Black Mesa, Arizona*. Window Rock: Navajo Nation Papers in Anthropology, No. 30, Navajo Nation Archaeological Department.

Swedlund, A. C. (1994). Issues in demography and health. In G. J. Gumerman & M. Gell-Mann (Eds.), *Understanding complexity in the prehistoric Southwest. Santa Fe Institute Studies in the Sciences of Complexity Proceedings* (Vol. 16). Reading: Addison-Wesley Publishing Company.

Walker, P. L. (2001). A bioarchaeological perspective on the history of violence. *Annual Review of Anthropology, 30*, 573–596.

White, T. D., Folkens, P. A., & Black, M. T. (2012). *Human osteology* (3rd ed.). Burlington: Academic.

Wood, J. J., & Powell, S. (1993). An ethos for archaeological practice. *Human Organization, 52*(4), 405–413.

Zuckerman, M. K., & Armelagos, G. J. (2011). The origins of biocultural dimensions in bioarchaeology. In S. C. Agarwal & B. A. Glencross (Eds.), *Social bioarchaeology* (pp. 15–43). Malden: Wiley-Blackwell.

Chapter 2
An Ethos for Bioarchaeologists

There is no doubt that working with ancient and historic human remains is fraught with legal, ethical, and moral implications. For the young scholar the issues raised by restrictive legislation and outcry from indigenous people who say it is wrong to study human remains may seem daunting. However, instead of seeing these alternative perspectives as roadblocks and challenges, many bioarchaeologists are embracing the issues being raised by transforming how bioarchaeology is taught and how research is conducted. Instead of framing the issues as what must be done as responsible scientists, bioarchaeologists have the potential to rewrite their agendas and to frame a more encompassing worldview on ways of working with archaeological resources. Wood and Powell (1993) present an essential piece of scholarship powerfully relevant for bioarchaeology. In their presentation, they provide a compelling set of reasons for shifting the ethos of how archaeology is practiced. Bioarchaeology can also be transformed by changing the underlying ethos regarding how research is done. Ethos implies a fundamental set of beliefs that shape daily practice. This chapter suggests that a basic tenant for research involving human remains must embrace an engagement at every level with the larger context within which the human remains and artifacts are connected. This includes descendant populations, local communities, county, state and national legislation, government and local statutes, and repositories and museums that house related materials.

Bioarchaeologists working in this complex and intermeshed context increasingly need to convey and demonstrate the importance of their research. They must be able to convey why they should be permitted access to human remains and other artifacts for their research. Their research cannot be seen as esoteric because the individuals under study are part of a larger sociopolitical context that extends far beyond the bones. The modern political and legal arena within which bioarchaeology must also operate and comply continues to permeate the study of the human remains. As mentioned in Chap. 1, laws such as the Native American Graves Protection and Repatriation Act (NAGPRA) passed in 1990 and more recent amendments to it, as well as the National Museum of the American Indian Act

D.L. Martin et al., *Bioarchaeology: An Integrated Approach to Working with Human Remains*, Manuals in Archaeological Method, Theory and Technique, DOI 10.1007/978-1-4614-6378-8_2, © Springer Science+Business Media New York 2013

(NMAIA) passed in 1989, continue to exert limitations and challenges to research focused on human remains.

To appreciate and understand where modern bioarchaeology is today, it is essential to provide an overview of the historical trends in the study of human remains. First is a discussion of the ethos of the early physical anthropologists that pioneered the scientific analysis of ancient human remains using a variety of approaches. Because human remains were analyzed largely without context prior to the 1980s and without the permission or collaboration of descendant tribal groups prior to the 1990s, there were many missed opportunities to make the case that working with human remains was of broad relevance to the modern world. The historical focus on descriptive morphology and typology dominated analyses and thus precluded integrative studies that could have linked the past to the present (and the dead to the living) in valuable ways.

Second, it is important to trace the impact that NAGPRA legislation had on the field of bioarchaeology. There were a number of different responses from the bioarchaeological community in the 1990s, but few of these involved self-critique or reflexive assessment of their research on a grand scale. And, the ethos of bioarchaeology was very slow to shift from the traditional worldview that scientists should have access to all human remains for study to the more contemporary notion that descendant groups have the right to say when and if scientific studies should be conducted on their ancestors. In the 1990s, some bioarchaeologists shifted into neutral research areas such as forensic anthropology (see a discussion of this field in Chap. 3), and the discipline did not undergo any major transition or change its ethos at that time. Yet simply problematizing the position of physical anthropologists and bioarchaeologists vis-à-vis NAGPRA and NAGPRA-like legislation would have made clear the need for change.

Recent scholarship by many bioarchaeologists is now showing a fundamental shift in ethos. The emerging protocols for conducting bioarchaeological research have been enriched and codified by a new generation working closely with tribal representatives and legislative bodies. Regardless of county, state, and national laws, before beginning any project involving human remains, bioarchaeologists must consider the following: (1) What are the full implications of conducting the research? (2) How might the research impact the descendant and local communities? (3) Are there potentially negative ways that the information being collected and disseminated might be utilized by people outside the field of anthropology? The last consideration is the one that is often hardest for researchers to judge but is arguably the most critical to consider. For any scientist who generates data, it is difficult to assess and track how the data will be utilized in the future. Even cultural anthropologists struggle with this issue. Chacon and Mendoza (2012) present compelling case studies on the ways that cultural anthropologists grapple with the many ethical ramifications of publishing (and not publishing) sensitive ethnographic data on indigenous groups. Case studies in this chapter help illuminate the complexities of working with human remains both in the USA and in international settings.

2.1 Historical Trends and Missed Opportunities for Integration and Engagement in the USA

Bioarchaeology straddles both the natural and social sciences and as such generates a great deal of both quantitative and qualitative data about the past. Most bioarchaeologists believe fervently that understanding the past is as important as conducting cancer research or research on global warming. Many would agree with the historian and moralist of his time, Lord Acton, when he stated "If the past has been an obstacle and a burden, knowledge of the past is the safest and surest emancipation" (Weaver 1960:22). Yet bioarchaeologists have been very bad at explaining exactly how their studies from the past can have a beneficial impact on solving today's problems. In the critical examination of physical anthropology (and bioarchaeology as one of its sub-disciplinary foci), many scholars have suggested that it was due to a failure to frame research by posing questions that connect the past to the present (Walker 2000; Martin 1998; Alfonso and Powell 2007; Larsen and Walker 2005; Walsh-Haney and Lieberman 2005; Kakaliouras 2008; Turner and Andrushko 2011).

In every science there are studies which are purely descriptive and that do not seek to make broader conclusions. This is certainly true for bioarchaeology and its twin subdiscipline, paleopathology (discussed in Chaps. 6 and 7). It was once very easy for archaeologists and physical anthropologists to excavate ancient skeletons or cemeteries, measure the bones, and then examine the bone surfaces for age at death, sex, and pathology. Early publications abound with studies that simply document all the measurements and any finding of pathology. While there is value to some of these purely case study-based publications, they are limited. Sometimes other researchers can take several case studies and begin to parse out patterns across a temporal or spatial dimension, but this is usually difficult to do. Rarely do descriptive studies utilize a set of standard methods for collection and reporting of data.

Thus, descriptive studies can add to a general growing body of observations and to the understanding of how pathology is expressed on human bone tissue, but these are very difficult to connect to broader themes relating to the human condition. Descriptive studies based on quantitative measures may be building blocks that can be used to construct larger notions about human fragility and resilience, but often they are simply too particularistic and narrowly focused to use in this way.

NAGPRA played an indirect role in forcing bioarchaeologists to become more engaged with larger questions and with linking their research to contemporary problems. In the 1990s, it became increasingly necessary that bioarchaeologists wishing to have access to human remains for study would need to articulate in clear and nontechnical language why it was so important for them to study indigenous skeletons. In doing so, it became clear that any answer that implied that it was the right of all scientists to have access to human remains was insufficient. In an attempt to rectify and repair relationships between bioarchaeologists and indigenous communities, research programs that were responsive to the concerns raised by living descendants became models for the new post-NAGPRA bioarchaeology. These approaches laid the foundation for what is now the norm in bioarchaeology.

Bioarchaeologists have worked hard in recent years to remediate that disconnect between their work and the public perception of their work by making the questions they ask and the answers they seek more relevant and applicable to the modern world. Collaborations between indigenous groups and bioarchaeologists are the way of the future, as it is at the heart of the new "best practices" for the field. For example, in the USA, researchers in recent decades have been increasingly consulting and working in conjunction with Native American groups (Harrod 2011; Dongoske 1996; Stapp and Longnecker 2008; Miller 1995). We argue that this collaboration is not simply the consequence of NAGPRA, as bioarchaeologists working in countries that do not have these laws are following these same best practices (Turner and Andrushko 2011; Pérez 2010). Cultural anthropologists and archaeologists also have found that collaboration can be extremely productive (Chacon and Dye 2007).

Not every Native American in the USA or indigenous person in another country will be convinced that human ancestral bones should end up on the cold, hard tables in bioarchaeology laboratories undergoing scientific examination. Popular literature and media are full of examples of this sentiment, perhaps best expressed by Leslie Marmon Silko, "The interpretation of our reality through patterns not our own serves only to make us ever more unknown, ever less free, ever more solitary" (1987:93). Bioarchaeologists need to be prepared to empathize and respond to those who do not see the value of measuring and analyzing bones. As discussed in Chap. 1, bioarchaeologists are anthropologists. Being trained in anthropology helps with this issue because clearly there are many ideological differences about what the dead means to the living. It is imperative that practicing bioarchaeologists be well versed in these highly varied ideologies regarding death and the afterlife (this is discussed in great depth in Chap. 5).

2.1.1 The History of Physical Anthropology

Physical anthropology was born out of the principle of morphological comparison developed by Linnaeus, and divergence from a common ancestor proposed by Darwin and Wallace. The result of this origin is that throughout history, researchers have focused on the qualities that distinguish different groups (usually referred to as "types") as a means of creating categories or typologies. While much of this work was primarily focused on identifying humans and their relationship to fossil hominids, some researchers applied these principles to living humans. The result of trying to classify modern humans into types and distinct groups was the origin of the concept of "race" (Brace 2005). The term "race" is used to describe distinct groups of people who differ based on perceived physical characteristics. The belief in this approach is bolstered by the assumption that "pure" types existed at one time, and although there has been interbreeding, measurement of physical features can capture what those original types were.

This concept is problematic because in the past there was a tendency to associate different "races" with different abilities such as intelligence, capabilities, political leanings, and personality. For example, the association of "race" with certain social and political movements in the USA contributed to the justification for a number of atrocities that include slavery and subjugation, segregation, and forced sterilization (i.e., eugenics). These were enacted upon Africans exported to the USA, Native Americans, and various other immigrant groups.

Within the field of physical anthropology, "race" has been a concept that has divided researchers (Stocking 1982). Some have argued that physical variations among human populations are so slight and meaningless that separation into well-defined, exclusive groups is unachievable (Brace et al. 1993; Brace 1964, 2005; Armelagos and Salzmann 1976; Van Gerven et al. 1973; Goodman 1994). Others have argued that the observable differences between groups are meaningful and do define discrete "races" (Birkby 1966; Gill 1998; Ousley et al. 2009; Sauer 1992, 1993). The scholarly research on both sides of this debate attempts to deal with the underlying causes and significance of variation seen at the phenotypic (anatomical) level.

Traditionally, researchers favoring a typological approach to human variation have separated populations into groups primarily using external characteristics referred to as "classical traits" that included things like skin color, hair type, nose form, and body structure. Other researchers have demonstrated that these kinds of physical traits are the result of environmental adaptations, and thus they simply are a reflection of regional variation in phenotypes. More current methods using genetics and DNA have been utilized in these debates to demonstrate the importance of individual variation over the utility of typing traits. Regardless of whether anatomical or genetic traits are used, there is overwhelming evidence to suggest that the differences among the so-called races are more reflective of regional adaptation and phenotypic similarities (Brace 2005). Human variation is fluid *between* populations and it is complex *within* populations.

Early researchers who laid the groundwork for the foundation of physical anthropology tended to view human variation among populations hierarchically. They were influenced by the now outdated notion that evolution was progressive. The problem is that the differences they noted among human populations were due to regional specialization similar to what Darwin found in his evaluation of variation among finches. Even though the differences were recognized to be adaptations to climatic conditions, some physical anthropologists used the variation as evidence of the existence of different human species (Nott and Gliddon 1854). Not only were these so-called races viewed as separate species, but these researchers argued that each could be ranked by its affinity to God in a hierarchical system known as the Great Chain of Being (Lovejoy 1936; Hoeveler 2007).

Constructing humans as separate species is problematic on many levels. These alleged different human species violate the means of distinguishing different species as defined in biological terms. In biology, different species are any two organisms that cannot breed and produce viable offspring. With regard to human populations, it is obvious that any member of any human group can successfully mate with any other person of the opposite sex and have a child that can do the same. Furthermore, the species designation is implausible because, aside from being scientifically unsound, it is logically false. Even before an awareness of what genetic information reveals about "race," the physical differences or "classical traits" that were used to separate humans into distinct populations were so slight that even the most distant people resembled one another in more ways than they differed.

With the formal development of physical anthropology in America, science became a tool that played a significant role in the "race" ideology as it institutionalized the concept and promoted its use as a means for directing social and political policies such as restrictions on marriage, segregation, eugenics, and eventually the holocaust in Europe. It is critical to understand this legacy because in many ways it influenced how human remains were retrieved and analyzed by physical anthropologists. This in turn directly led to increasing outcry by Native Americans that because of their "race" their ancestors' bones were being treated differently than European American's ancestors' bones. It was common prior to 1990 to excavate and retrieve historic and precontact Native American human remains, even though it was (and still is) illegal to disturb human burials. In Nebraska, for example, in an analysis of the effect of NAGPRA on anthropology, the author states that by law it is illegal for any set of human remains to be disturbed (Brown 1995–1996), and yet a federal law had to be passed to protect Native American burials. In 1979, a trial was held (*Sequoyah v. TVA*, 620F.2d 1159) where the Cherokee tribe was attempting to prevent the Tennessee Valley Authority from the removal and analysis of indigenous burials. The importance of this case is that the graves of Native Americans were curated and studied, while the burials of individuals of both Euro-American and African-American decent were immediately reinterred (Ferris 2003:161). The fight led by Native Americans for NAGPRA legislation is a direct result of this typological and ultimately racist approach to human remains.

2.1.1.1 The Role of Measurement in "Scientific Racism"

The reason for the intimate link between physical anthropology and "race" (and bones and types) is that the focus of early research in the USA was to maintain the focus on explaining variations in phenotype (physical appearances). In the field of physical anthropology, this concept was made more quantifiable with the implementation of measurements and statistics through the technique called anthropometry. Anthropometry is the study of the physical structure of the human body in an attempt to differentiate populations. In his book, Bass (2005) identifies four categories of anthropometry, which include somatometry, cephalometry, osteometry, and craniometry. The two techniques generally associated with anthropometry in the traditional sense of being the study of living humans are somatometry, which is defined as the "measurement of the body of the living and of cadavers," and cephalometry, "measurement of the head and face" (Bass 2005:62). Physical anthropologists, for the most part, are no longer actively employing the techniques of somatometry and cephalometry. In contrast, craniometry and osteometry still form the basis of quantitative research in the study of human remains, so it is important to discuss some of the problems historically associated with the application of these techniques.

The first of these methods, craniometry, or the use of measurement to determine the size and shape of the skull and facial bones, was developed in 1775 by Johann Friedrich Blumenbach. Craniometry is and has always been a controversial method of analysis. According to Armelagos et al. (1982:308), as far back as 1896 Rudolf Virchow was "... extremely critical of the use of crania for assessing biological affinity." Yet its use persists even today.

Perhaps one of the most prolific practitioners of craniometrics was the physician Samuel George Morton who El-Najjar and McWilliams (1978) acknowledge as the founder of physical anthropology in the United States. Morton was known for his extensive research that compared the cranial capacity or average size of the brain of a multitude of different populations around the world. Using these data and other research he conducted, Morton created a hierarchy of human groups in the mid-1800s. Other researchers built on this research and continued to perpetuate a hierarchy of "races" often based on supposedly differing levels of intelligence.

Like Linnaeus and other scientists of the time, it is no surprise that the top of this hierarchy was the lightest-skinned populations located in Europe, especially England, while the lowest of the populations were the dark-skinned people of Africa. The problem with Morton's study and others like it is that there is no evidence that these minute variations in cranial capacity have any consistency among supposed racial populations or correlation with intelligence. This lack of a correlation was first exposed by Gould (1981) although there have been arguments that Gould's reanalysis of Morton's work was flawed (Lewis et al. 2011). Despite the criticisms of Gould, other research has supported his original assertion that there is no correlation between cranial or brain size and intelligence (Jackson 2010; Gravlee et al. 2003a, b; Boas 1912; Carey 2007).

The second method used by physical anthropologists to differentiate populations is osteometrics, which is generally defined as the measurement of bones, but some researchers use the term to refer to measures of the postcranial features of the body only (Bass 2005:62). This technique has been less controversial than craniometrics because intelligence has not been correlated with body size, and ". . . the racial assessment of postcranial remains never captured the interest of physical anthropologists to the degree that cranial studies had . . ." (Armelagos et al. 1982:318).

The purpose of exploring the history of skeletal analysis is not to discredit the field of physical anthropology but to acknowledge that some of the ways bioarchaeologists today approach the analysis of human remains are largely based on the same methods used by others in research that led to the production of racist pronouncements backed up with scientific data. This is not unique to physical anthropology, as it has been shown that nearly all scientists at times are influenced by ideological and political agendas. However, there has been much progress in scientists acknowledging their biases and reflecting on the meaning of their research.

The misuse of scientific data can lead to very negative outcomes. The extreme examples of this include the use of scientific racism whereby political motivations to maintain slavery and to justify the displacement of Native or indigenous peoples were upheld by anthropological and medical research (Johanson 1971; Brooks 1955; Horsman 1975). To understand how these policies could come to pass, this brief review of the historical development of the term "race" in the field of physical anthropology forms a backdrop to the subsequent discussion of formulating a new ethos for bioarchaeology in the modern moment. Without this historical contextualization of the field, it may seem less urgent or relevant to promote an ethos of integration, engagement, and ethical consideration around the analysis of human remains.

2.1.2 Developing a Biocultural Approach and the Origin of Modern Physical Anthropology

Throughout the history of physical anthropology, there were researchers arguing against racial classification and promoting more nuanced methods of understanding human variation (Washburn 1951; Boas 1912; Cobb 1939; Montagu 1942). Over the last several decades, the field has actively begun to move away from asking questions about superficial physical differences between groups precisely because that approach fails to answer any broad theoretical or practical questions. Instead, there has been a push for framing questions about the ways that human variation is a product of biological, cultural, ecological, and geographical variation and how these variations change among populations.

The motivation behind this shift in emphasis was to salvage a valuable comparative tool of science. The revitalization of anthropometrics is a good example of this. In anthropology, it is used in a wide number of studies relating to growth and development and the problems of undernutrition and disease (Bogin and Keep 1999; Vaughan et al. 1997; Komlos 1989, 1995).

Outside of anthropology there exists a field known as ergonomics, which is defined as ". . . the science of work: of the people who do it and the ways it is done, of the tools and equipment they use, the places they work in, and the psychosocial aspects of the working situation . . ." (Pheasant and Haslegrave 2006:4). It is a multidisciplinary field that comprises researchers from engineering, biomechanics, psychology, and, increasingly, anthropology. The goal of ergonomics is to design products, workspace, and occupational activities so that they function as efficiently as possible with the human user, typically in a business (Sagot et al. 2003; Chaiklieng et al. 2010; Hendrick 2003) and military (Gordon 1994; Rogers 2011; Huishu and Damin 2011) setting. With potential to have an impact on the success of a product on the market and a reduction of significant costs related to workplace injury, this field is highly applicable to the modern world. A second and perhaps even more valuable way in which anthropometrics is being utilized by physical anthropologists today is in the study of metabolic disease and nutrition (Vaughan et al. 1997; Komlos 1995; Hsieh and Muto 2005). Thus, even though anthropometry was a method used in typological projects, it does have a role to play in more important areas of research.

Looking more specifically at the analysis of the skeletal remains, there has also been a shift in focus. At its worst, the analyses of skeletal remains were focused on craniometry and pathology; at its best, analyses provided population estimates on a range of mortuary, demographic, and health factors. Combined with a more integrative approach (discussed in Chap. 1), the biocultural model has facilitated using metric measures to address the effects of environment and culture on skeletal growth and development. It is important to note that some of the earliest researchers examining human remains were physicians or anatomists and not anthropologists. In the 1970s there was a major shift in anthropologists being interested in more holistic approaches to understanding human change of time and space (Larsen 1987; Cohen and Armelagos 1984; Buikstra 1977; Gilbert and Mielke 1985; Huss-Ashmore et al. 1982).

Bioarchaeology as a subdiscipline within physical anthropology grew out of this attempt at integration and holism. This approach to the study of human remains was much more inclusive, with a focus on understanding not only the remains but the contextual information in which they were situated (e.g., type of interment, grave goods associated with the burial, and mortuary landscape). The goal of bioarchaeology was not to analyze skeletal remains, but to understand the life histories of individuals. Measurements from human remains are now more likely to be used to examine differences between populations as a function of the interaction between genetics, culture, and the environment. The result of this shift was the development and employment of biocultural models (Goodman and Leatherman 1998; Buikstra 1977; Blakely 1977) and human behavioral ecology (Smith and Winterhalder 1992; Cronk 1991).

The term physical anthropology is interchangeable with biological anthropology, but there is an interesting history to the current shift toward using biological anthropology instead of the former. The term "physical" within anthropology harks back to a reference about the physical body but also as it pertains to the physical laws of nature. The body and morphology were used to understand the divergence from a common ancestor as proposed by Darwin and Wallace. These early studies focused on comparison of extinct and extant populations of humans and our ancestors in an attempt to understand our development. Some of this research took a dangerous turn when morphology and differences became reduced to "types" and typological model. This focus on physical variation between types often failed to offer a means for obtaining answers to questions about what accounts for variability in the first place.

In contrast to the term physical, "biological" is generally accepted to mean the study of life, which includes the organisms and their life histories. When biology is coupled with anthropology, it seems to expand the topics and possibilities of what can be addressed, beyond physical characteristics of the body. Biological anthropology is actually more inclusive in an approach to research that looks at all aspects of life and what it means to be human. More so than physical anthropology, biological anthropology does cover topics that include nutrition, trauma, disease, metabolic disease, hormones, cognition, behavior, and more.

From physical anthropology of the past to biological anthropology of the present, the terms are used interchangeably although new textbooks and almost all bioarchaeologists practicing today would consider themselves biological anthropologists. The shift from physical to biological is both a symbol of the expansion of the questions that researchers are striving to answer and a real marker of more integrative (i.e., biocultural) questions.

Bioarchaeological research using the biocultural approach is crucial for the study of human adaptation and disease in the past (Brickley and Ives 2008; Martin 1994; Merbs and Miller 1985; Powell 2000; Roberts and Manchester 2005; Ortner 2003). This research is crucial for understanding the spread of epidemics and the prevention of future disease outbreaks (Armelagos and Barnes 1999; Barrett et al. 1998; Roberts and Buikstra 2003; Roberts 2010). Additionally, this research is useful in understanding how environmental conditions affect the long-term health of a population over a short period of time. The utilization of modern clinical literature requires relying on medical records that were often discontinuous and incomplete and long-term studies that lasted decades but still did not reveal disease pattern. Bioarchaeological research can link social processes to health conditions, and in this sense it has much wider applicability to solving human problems today.

2.2 The Rise of Legislation and Its Effect on Bioarchaeology

To understand the profound impact that emerging legislation and statutes have had on bioarchaeology in the USA, it needs to be discussed within the broader context of Native American sovereignty. Most people, including some of our most influential leaders, are unfamiliar with the concept of tribal sovereignty (Cobb 2005:119). Sovereignty is defined as ". . . The supreme, absolute, and uncontrollable power by which any independent state is governed; . . . the international independence of a state, combined with the right and power of regulating its internal affairs without foreign dictation . . ." (Black 1968:1568; Garner 2009). Clearly, tribes were sovereign nations prior to 1492 and for a few centuries following depending on the exact time of contact. The colonies and ruling powers across the Atlantic recognized individual tribal nations, dealing with them via treaties and other legal documents (Pevar 2012:1–6). When the Founding Fathers drafted a Constitution for the new United States of America, they referenced Indians twice (Pevar 2012:56–59). The Commerce Clause, located in Article II, section 2, clause 2, provides that "Congress shall have the Power . . . to regulate Commerce with foreign Nations, and among the several States, and with the Indian Tribes" (Pevar 2012:57). The Treaty Clause, located in Article II, section 2, clause 2, gives the president and the Senate the power to make treaties. This includes treaties with Indian tribes (Pevar 2012:57).

While recognizing Indian nations within the wording of the Constitution seems to put them on the same footing as foreign nations, the Supreme Court has interpreted this inclusion differently. In 1832, the Supreme Court held in Worcester v. Georgia (31 U.S. 515) that these two constitutional provisions give Congress "all that is required" to have plenary power over Indians and tribes. In the previous decade, the Supreme Court held in Johnson v. M'Intosh (21 U.S. 543) (1823), that because of the "discovery" of North America and the "conquest" of its inhabitants, all persons and property within the USA are subject to its laws.

The reference to Indians and tribes within the Constitution, and subsequent Supreme Court interpretations, has severely limited tribal sovereignty within the United States. For example, lacking complete sovereignty means that tribes may not declare war against foreign nations. However, the federal government does recognize tribal self-government. Indian tribes have the inherent right to govern themselves. As noted by a federal appellate court as recently as 2002, "Indian tribes are neither states, nor part of the federal government, nor subdivisions of either. Rather, they are sovereign political entities neither possessed by sovereign authority nor derived from the United States, which they predate. [Indian tribes are] qualified to exercise powers of self-government . . . by reason of their original tribal sovereignty" (Pevar 2012:82).

This original tribal sovereignty has been severely limited by the federal government over the years. The Supreme Court has interpreted federal documents, including the Constitution, as giving Congress plenary power over Indians and tribes. This plenary power has been exercised in numerous pieces of federal legislation to limit or eliminate numerous tribal powers (Pevar 2012:82–83). Express limitations include prohibiting tribes from selling tribal lands without the federal government's permission.

As mentioned earlier, tribes may also not declare war against foreign countries. The most severe limitation on tribal powers occurs when the federal government either implicitly does not recognize or expressly terminates the government-to-government relationship with a tribe. A terminated tribe is considered not federally acknowledged (Pevar 2012:271–274). Tribes that are not federally acknowledged may continue to exist as tribal entities but without recognition from the US government, they are ineligible for government programs established for Indian tribes, and their tribal members are not considered American Indian for most governmental purposes. Additionally, such tribes are not covered under NAGPRA.

While limits on tribal sovereignty continue to be upheld by the Supreme Court, there are several areas in which tribes may continue to self-govern. Tribes may form their own governments (Pevar 2012:84–85), choose their own leaders (Pevar 2012:87), maintain their own court systems (Pevar 2012:88), and determine tribal membership without interference from the federal government (Pevar 2012:90–93). These are all exceptionally important aspects of tribal sovereignty and tribal self-government. While tribes do not exercise the same level of sovereignty they did prior to the formation of the US government, it is important to note that Congress, with its self-appointed plenary power over Indians and tribes, has not gone so far as to eliminate all aspects of tribal self-government.

This very brief introduction to the postcolonial arrangement of tribal groups within the USA is important in making sense of both the benefits and the challenges that came with the passage of NAGPRA and other legislation. Too often, there is only the briefest explanation of the profound impact of these kinds of laws on the development of bioarchaeology. If bioarchaeology is truly to be transformed by a shift in the ethos of practicing bioarchaeologists, it must begin with a more full appreciation of the complexities of what it means to do ethical research in these arenas.

Bioarchaeologists need to be familiar with NAGPRA and its vast outreach programs that aim to link tribes with bioarchaeologists and archaeologists. Everything one needs to know about NAGPRA can be found on at http://www.nps.gov/nagpra/. However, it is important to acknowledge that NAGPRA (and any legally mandated rules and regulations that will come along in the future) are not perfect and laws will never cover every ancient burial, bone or archaeological site. That is why this text advocates for bioarchaeologists to have an ethos, that is, an everyday practice that is built on scientific responsibility, a moral code that goes beyond laws and regulations, and a commitment to social justice and inclusivity.

2.2.1   Who Owns the Past?

How tribal sovereignty has most effected bioarchaeology is in the establishment of laws like NAGPRA that grant the right to reclaim and protect their ancestor's skeletal remains. Under NAGPRA, only federally recognized Native American tribes and Native Hawaiian organizations may claim burials and cultural items. Prior to these laws, any one could dig up or remove bodies and artifacts from their place of interment without much fear of consequence. As a result, nonacademic people often looted unmarked graves looking for items of value or souvenirs. Early archaeologists were not exactly looters or grave robbers, but there were many cases of archaeological excavation of burials that took place that were unethical by any standard understanding of the term ethical. These activities ranged from not considering the descent group that identifies the remains as their ancestors and how they feel about these activities, to actively stealing remains from graves (Riding In 1992; Cole 1985; Thomas 2000).

The argument for digging up the graves was that the findings would benefit society. It was explained that excavating human remains and examining them produced a more scientific understanding of the continent's original inhabitants. These reasons and the methods of obtaining the remains of the original inhabitants are now viewed as suspect.

It may be argued that studying skeletal remains is not the same as medical experimentation since bioarchaeological research does not result in the death or physical mutilation of living people. However, according to Echo-Hawk (1988:2) "Regardless of the motive for expropriating Indian graves, the impact of this activity upon the affected Indians is always the same: emotional trauma and spiritual distress." As Ferguson and colleagues state "One thing archaeologists should keep in mind is that the disturbance of human remains is agonizing for Hopi people . . . In presenting results of mortuary studies, archaeologists need to understand that for the Hopis, the heartfelt spiritual concerns about the disruption of graves far outweighs any [of the] scientific studies . . . what archaeologists find to be interesting results and findings are colored by the desecration of the graves that led to those results" (2001:22).

In shifting the ethos concerning bioarchaeological research, it is incumbent upon bioarchaeologists to use information that they collect in a way that does not trivialize or diminish the lives of the living descendants. Collaboration and consultation with representatives from indigenous groups regarding data derived from the excavation of the burial and the analysis of the remains is one way to adhere to these ethical considerations. If bioarchaeologists fail to follow this simple tenet, and instead push for the rights of scientists over consideration of the impact of the research on living people, the consequences will be a continued fracturing of the discipline. According to Watkins (2003) this approach has resulted in Native Americans often distrusting anthropologists and archaeologists.

This is apparent in the contrasting ethical doctrines of the Vermillion Accord on Human Remains (passed in 1989) and the SAA Code of Ethics (as it was written in 2003). Though both guide anthropologists and archaeologists on how to appropriately conduct research that involves human remains, the difference between the two is the goal they are working toward. The SAA Code of Ethics presumes the past belongs to everyone, while the Vermillion Accord argues for a recognition that the cultural materials of the past are related to the cultures living today so anthropologists must work with the native groups to understand the past.

Watkins (2003) challenges that we need to move beyond arguing over who owns the past because what is of more concern is which group should get to represent the past. Using the myth of the "Mound Builders" as an example he illustrates how science is not objective and as such it is possible that indigenous history can be distorted by the worldview of the people analyzing the material and remains (Watkins 2003:132). This is not to say that science is inherently flawed, but just that "true" objectivity is impossible as "...knowledge is necessarily embodied, partial, and situated and, further, that its construction, claiming, and enacting are activities with moral and political ramifications" (Lang 2011:75). As bioarchaeologists, we work with the dead who cannot provide us with all the details of their daily lives, so any interpretation we make from the data we collect is devoid of a significant portion of the context. While archaeological reconstructions and ethnographic records can provide a large portion of the context, often times the indigenous groups can play a crucial role in fleshing out the scientific interpretation.

2.2.2   The Impact of NAGPRA on Bioarchaeology

The responses of physical anthropologists working with human remains to NAGPRA were varied and idiosyncratic. Some left the profession altogether, relocated to medical schools, or shifted to working in the private sector for cultural resource management operations. In many states, burials encountered in the course of excavation can still be analyzed in situ, but cannot be fully analyzed in a lab, a practice that needs to be thoughtfully considered (see Chap. 4). Others have simply stopped working with US collections. There are also physical anthropologists that still lament the loss of academic freedom and are working to reverse or diminish NAGPRA.

However, many bioarchaeologists stayed their course and helped museums and other repositories with skeletal remains comply with specific NAGPRA requirements. In the course of complying with the legal mandates of NAGPRA, many bioarchaeologists discovered hundreds of collections in museums and other repositories that had never been studied. Compliance activities involved a thorough and systematic listing of every burial and human bone being held in state or federal repositories. This generated much work and employment for bioarchaeologists. One of the major activities for compliance involved consultation meetings between museum and tribal representatives. For many bioarchaeologists, these consultations were the first time they had ever talked to, or worked with, Native Americans. Because the experience was often educational and positive, a decade of compliance activities and consultations helped to lay the groundwork for a new ethos in bioarchaeology. This is just one example of how NAGPRA and similar legal mandates have helped to change bioarchaeologists.

Today, bioarchaeology is a subdiscipline that is thriving. Over the last several years, the breadth, depth and amount of bioarchaeological scholarship has increased dramatically. This growth is evident in the development and expansion of university programs with faculty who specialize in bioarchaeology, forensic anthropology and paleopathology, as well as the explosion of researchers trained in these overlapping fields. According to one source, there was a potential increase of 28% in employment for bioarchaeologists between 2008 and 2010 (Huds 2011). The increase in the number of people specializing in bioarchaeology is seen also in the increase in the number of anthropology departments advertising for bioarchaeologists, with 12 positions for hire in 2013 compared with 3 positions in 2012 and 2 in 2010 (AAA website).

Thus, the impact of NAGPRA can now be seen as an important corrective. It has forced researchers to formulate meaningful research questions that can be answered with data derived from the archaeological context. It has also shaped the discipline toward being a more inclusive enterprise and one that engages with real world problems.

2.2.2.1   Case Study: American Southwest

Martin (1998) captured the above trends for one region of the USA in the article *Owning the Sins of the Past: Historical Trends in the Study of Southwest Human Remains*. The Southwest was an early field site for many archaeologists and physical anthropologists. The Southwest was essentially a training ground and laboratory to some of anthropology's most prominent scholars beginning in the late 1800s. As early as 1908, Ales Hrdlicka, one of the founding fathers of physical anthropology, conducted studies in the region. Focusing primarily on craniometrics (measurements of the skull), he developed an approach to skeletal analysis that relied on comparative morphology to place individual into typologies.

In the 1920s, Alfred Kidder was excavating a very large ancestral Pueblo site outside of Santa Fe, New Mexico called Pecos Pueblo. The human remains were sent directly to Earnest Hooten at the Peabody Museum at Harvard. In his analysis of over one thousand burials, he provided information on age, sex and disease. However, the major focus of his study was on the determination of morphological types based on metric data from the adult crania. His results were not surprising; at the completion of his extensive analysis, he concluded that the people inhabiting the pueblo were likely Native American. "Of the eight morphological skull types distinguished by me in the Pecos collection, all except the long-headed Basket-makers, pseudo-australoids, and pseudo-negroids show clear evidence of Mongoloid admixture and they are in fact predominatingly Mongoloid in features" (Hooten 1930:344–363).

This finding was not very new or interesting as the site of Pecos was populated by Native Americans, a fact well documented in numerous historic records. Pecos was a bustling town of over 2,000 people when the Spanish raided it for food throughout the 1500s, before finally establishing a brutal and coercive mission in 1619. For the next 200 years, the people were literally beat into submission and forced to labor for Spanish conquistadors and the missionaries. The people at Pecos stuck it out until 1838 when the last members of the community abandoned their village and joined another historic pueblo.

The focus on morphology and typology dominated studies using human remains for several decades. By the 1980s, there were only a handful of studies conducted on Southwestern remains that focused on other biological indicators, such as pathology. Though many studies still adhered rigidly to descriptive analysis, some did move beyond to conduct biocultural analyses that looked at adaptation, demography and subgroups at risk. However, this trend took place a little too late, and the result is that many of the Native Americans in the Southwest still associate physical anthropology with grave robbing, skull doctoring, and looting.

Instead of working on research that was of some interest to Native Americans prior to the 1980s, bioarchaeologists continued to emphasize their own goals and interests even after the establishment of NAGPRA. Since the 1940s there has been demographic, epidemiological and medical evidence that the contemporary people still living in the Southwest were at risk of early death and that they carried a higher illness burden than white counterparts. As Native Americans were beginning the long fight for improvement of life on the reservations and for social justice, physical anthropologists continued to study skeletal material in their laboratories, oblivious to these facts. Native American life on reservations is fraught with racism, poverty, disease and early death. The ancestors of these people were being studied with little regard for the ancestor-descendant relationship that existed.

Indeed, reports from the medical journals demonstrated that American Indian infant mortality and adult morbidity were alarmingly high and disproportionate to the rates for the general US population. Concerning health effects of environmental pollutants, Native Americans have borne the brunt of doses of radiation due to their proximity to major areas of nuclear testing such as the Nevada Test Site and Los

Alamos. Levels of lead in HUD housing may have caused and continues to cause high levels of lead in native children. Traditional subsistence activities have exposed native people to toxic waste in rivers and oceans. Armed with hard data from bone and teeth that demonstrate "before" and "after" levels of these toxic mineral and trace elements, activists working for better monitoring could use these data to improve lives. This linking of political processes and biological effects demands a broadly historical perspective and a multidimensional approach.

By sidestepping issues of importance to native people, scientific data generated by bioarchaeologists has been slow to be considered of value to Native Americans in the American Southwest. Some of the data have been used in ways that aid in the continued tyranny of native people today. For example, Deloria argues that elite, largely eastern scientists, plied their trade at the expense of Indians. In one example, he writes, "[In the 1930s] the idea that human cranial capacity demonstrated the intelligence of the different races [was] a piously proclaimed scientific truth. Indians were hardly on their reservations before government employees began robbing graves at night to sever skulls from freshly buried bodies for eastern scientists to measure" (1997:6). These kinds of activities, where skulls are used for the "progress of anthropological study" are the ones that Native Americans most associate with bioarchaeology.

During the 1980s and continuing to the present, some bioarchaeologists have been doing what can be considered state-of-the-art research in the American Southwest. For example, Stodder (1990) examined a range of demographic and epidemiological factors in the adaptation of two ancestral Pueblo groups during the protohistoric period (circa 1400s). Her careful analysis of context and the interplay of various biocultural factors demonstrated that adaptation to marginal desert environments by these early farmers presented challenges for some segments of the population. This kind of information is crucial in today's discussion of droughts and starvation in groups living in marginal environmental conditions undergoing desertification.

Martin (1994) worked on several large archaeological projects such as the Black Mesa Archaeological Project in northern Arizona and the La Plata Highway Project in northern New Mexico. The resulting analyses of the human remains relied on integrating a range of data sets using a biocultural model (discussed in Chaps. 1 and 5). Some of these health problems are related to the challenges of being desert farmers in areas where rainfall was unpredictable and growing seasons were short. Other underlying reasons for poor health, such as chronic middle ear infections in children, are problems Navajo and Hopi children still suffer from today. The desert winds and general environmental conditions are part of what keeps ear infections endemic even today with all of the modern interventions. The persistence of ear infections also points a general inaccessibility to health care for individuals living in the Southwest today.

Many of the larger skeletal collections from the American Southwest have been repatriated. For example, in the summer of 1999 the largest repatriation of human remains in the USA took place. The Pecos Pueblo burials numbering over 2,000 were returned to the Jemez Pueblo in New Mexico by the administrators at Harvard

University. The remains were stored for at Harvard for about 70 years. Included with the human remains were also hundreds of grave offerings. The Jemez Pueblo Indians were extremely gratified to have the return of their ancestors, and they reburied them in a process called "reverse archaeology" at the site of Pecos Pueblo, now a National Park (Archaeology, Volume 52 Number 4, July/August 1999). Recently, Morgan (2010) edited a volume entitled *Pecos Pueblo Revisited, the Social and Biological Dimensions* which represents the current scholarship based on the original archaeological excavations. While this volume does include some specialized studies carried out before the remains were repatriated, they are not fully realized bioarchaeological studies.

2.2.2.2 Case Study: Columbia Plateau Region

The Columbia Plateau presents a different history of pre- and post-NAGPRA work by bioarchaeologists. Analysis of the history of archaeological research in the Columbia Plateau reveals that there is now a movement toward opening dialogue and promoting cooperation among researcher studying the human remains. However, this was not always the case. This shift is a direct consequence of NAGPRA legislation.

The Columbia Plateau lacks the long and illustrious history of research that characterizes the American Southwest. However, this area is particularly interesting for a discussion of the impact of NAGPRA on bioarchaeology as it was the focus of one of the most famous NAGPRA cases in the USA. The case involves the dispute over who owns the bones of an adult male of great antiquity whose skeletonized remains were found in 1996. He has come to be known as Kennewick Man or "The Ancient One" (Mason 2000).

This individual was of interest because it was dated to approximately 8,340–9,200 calibrated years ago (Chatters 2000:299), and he was found to have been a victim of violence. There is a projectile point in the right ilium as well as evidence of several other nonlethal traumatic injuries (Chatters 2000, 2002). In addition to the extremely old age of Kennewick Man and the fact that he shows evidence of violence, the initial analysis of the cranial size and shape suggested to some researchers that he was not Native American but of an entirely different "race" altogether.

A dispute followed that has not yet been completely resolved between the Army Corp of Engineers who wanted to repatriate the remains to the tribes in the region and the scientists who wanted to study him. The local tribes (the Confederated Tribes of the Umatilla Reservation, the Consolidated Tribes and Bands of the Yakama Nation, the Nez Perce, and the Colville Confederated Tribes) all claimed that this individual was one of their ancestors. A group of bioarchaeologists contested this claim and argued that it was of historic scientific importance and thus it is imperative that the remains be thoroughly analyzed. The case lasted nearly 6 years before a judge finally decided in favor of the scientists and denied the repatriation of the remains to the tribes. The importance of this case stretches beyond reburial because it was the first case to present a fundamental challenge to the notion of tribal sovereignty and the right to scientific inquiry. Although the court made a decision, it has been appealed and contested numerous times, and new contestations are forthcoming almost yearly.

Recently, Doug Owsley, a bioarchaeologists from the Smithsonian who was one of the scientists who fought to have the remains studied by scientists and not repatriated, revealed new data that suggested to him that Kennewick Man was not even from the inland region of the Columbia Plateau where he was found. Owsley, in an interview, suggested that isotopic data from the bones revealed that he consumed marine animals and so was a coastal dweller (Mapes 2012). Based on craniometry, Owsley also stated that he did not think that this individual was even related to Native Americans, but rather was of Asian-Polynesian ancestry. Other bioarchaeologists who have analyzed the remains have demonstrated that Kennewick Man contains a mix of features seen in modern groups, including East Asians, American Indians, and Europeans (Powell and Rose 1999). Additionally, based on research by Boas (1912) that has been independently confirmed by Gravlee et al. (2003a, b), craniometrics are fluid generation to generation. Given over ~8,000 years and dramatic shifts in the climate that affected available flora and fauna (Chatters 1998), there is no way to know exactly where to place him in terms of present-day cultures. This example highlights the challenges of NAGPRA because of the gray areas presented by human remains that are over 5,000 years old. Proving ancestral affiliation is almost impossible.

Today, there is very little bioarchaeology conducted in the Columbia Plateau on the USA side. The Canadian portion of the Columbian Plateau is quite active with bioarchaeological research. For example, the Canoe Creek, Soda Creek, and Dog Creek bands of the Northern Shuswap Tribal Council have recently granted bioarchaeologist Malhi et al. (2007) permission to analyze DNA from two burials from the mid-Holocene (circa 5,000 years old) recovered in British Columbia. This project is significant because the mtDNA analyzed from the two burials revealed a new haplogroup, which is a group of haplotypes or combination of alleles that allow researchers to identify genetic populations. The implication of this is that it is possible that early populations possessed more genetic diversity than those found today. This represents an integrative, engaged, and ethical bioarchaeological project that involves collaboration between nonnative bioarchaeologists and Native American representatives. Had relations not been cultivated between the anthropologists and the tribe, this study would not have been completed.

2.2.3 NAGPRA and Bioarchaeology Can Coexist

A common misconception about NAGPRA is that it impedes the research of physical anthropologists. The truth is that for the most part, the goals of anthropologists and Native Americans are not irreconcilable. Research suggests that NAGPRA may have actually had a positive effect on the analysis of human remains in physical anthropology. "The repatriation movement and most recently NAGPRA have made significant positive contributions to osteology as a research enterprise and to the bioarchaeology of North America" (Rose et al. 1996:99). Two of the main advances that have been revealed are improvements to the inventory process, as well as the evaluation of unexamined human remains previously held in storage. Prior to the passage of the NAGPRA in 1990, much of the analysis of ancient skeletal material was done without permission of, input by, or accountability to Native Americans. In the past, skeletal remains were often sent to labs for analysis by physical anthropologists. This divorced the interpretation of biology from its historical, cultural, and environmental setting.

NAGPRA was especially beneficial to museums because it required the collections of human remains be inventoried (Ousley et al. 2005). Inventories require funding, which allows for the addition of more staff if only temporally. Overall, the benefit of NAGPRA is that it not only promotes the reevaluation and inventory of human remains in collections but also, if excavations are necessary, because of accidental exposure through construction projects or natural erosion, pressures anthropologists to conduct analysis as rapidly and efficiently as possible. "One bioarchaeology overview shows that 64% of 20,947 excavated skeletons have not been studied at all. These skeletons remain unstudied not because osteologists were not interested in them, but because there was never enough time or funds to study them all" (Rose and Green 2002:215–216).

A final, often overlooked, benefit of NAGPRA is that it has promoted communication between various academic organizations and researchers involved in the field of physical anthropology, as well as promoted the diffusion of information outside of academia. This latter trend is perhaps the greatest development to arise from the establishment of NAGPRA. The implications for future research are vast as bioarchaeologists are increasingly sharing data with other researchers, a development that is expanding the information uncovered about the past.

2.2.4   Beyond Legislation: Bioarchaeology Outside of the USA

While this critical analysis of the history and future of working with human remains arguably grew out of bioarchaeological research among the indigenous populations in North American, these same considerations should be applied to populations throughout the world. Although laws like NAGPRA do not apply to the study of skeletal remains outside of the United States, other countries are establishing their own laws that protect human remains. As of 2004, several countries had created such laws, such as South Africa with the National Heritage Resources Act (NHRA), Australia with the Aboriginal and Torres Strait Islander Heritage Protection Act of 1984 (ATSIHPA), and New Zealand with the Historic Places Act (HPA) (Seidemann 2004). Additionally other countries have established repatriation movements that lack a formal law protecting burials (e.g., Canada, Denmark, and Scotland) (Curtis 2010; Thorleifsen 2009; Simpson 2009).

Finally, there has been an increase in museums repatriating remains to the descendant culture directly without use of government laws (Ferri 2009; Pérez 2010). The idea behind this movement is that even in countries where there are no formal laws in place to protect burials, researchers have a responsibility to acknowledge the indigenous populations the remains represent and when possible involve the descendants in the research (Singleton and Orser 2003; Martin 1998:171). The message one should get from the establishment of these laws and the increasing cooperation between anthropologists and indigenous groups is that the ethical treatment of the bones as individuals is one of the foremost concerns of any research involving human remains. The importance of cooperation was of great concern for the research in the southern portion of the Columbia Plateau (Harrod 2011). Evidenced by the fact that even though laws like NAGPRA did not apply because no actual remains were handled or disturbed, and the data being analyzed was from Native American remains that are or are in the process of being repatriated, the tribal bodies were contacted and permission to conduct the study was attainted. Although it may seem unnecessary, this cooperation led to communication that in the end greatly enhanced the research.

2.2.4.1 Case Study: Yaqui People

From 2007 to 2009, Ventura Pérez was part of team that helped facilitate an international repatriation between the American Museum of Natural History (AMNH) and the Yaqui people. Everyone involved was pleased with the eventual outcome of returning these warriors to their homeland and families. On the day the tribe was to take custody of their brethren, a tribal elder told Pérez "this was meant to happen now." Her words, as they are for most tribal elders, were profound (Fig. 2.1). Everyone involved needed to be at a point where they could offer a meaningful contribution because solutions would ultimately lie beyond the scope of NAGPRA.

It is often said that history becomes meaningful when seen through the lens of personal experience. This is the story of Los Guerreros (the warrior) Yaqui and their social interaction with the decedent Yaqui community and the global impact of this repatriation. There was a delicate and complex dance that took place between sorrow and joy with this repatriation. For the Yaqui people this repatriation had a profound impact on the community and reopened old wounds and traumatic memories. The social reality of the Yaqui people was affected by the lives, deaths, and prolonged burial and grieving process for Los Guerreros. Their repatriation and the stories stirred memories of violence that had a profound impact on generations of people who had not directly experienced the violence but whose mothers, fathers and grandparents had. The repatriation of the remains was not only emotionally powerful for the descendants but for the bioarchaeologist who through his interactions with the Yaqui learned more about these men, women and children that analysis of the bones alone could ever reveal.

To understand the poetics of revolt, particularly those in native communities which, in the face of economic, political and social pressures brought to bear by European societies, have sought to retain their own identities and social structures, it is imperative to recognize how dominate cultures impose and define minority cultures through the legitimized acts of structural violence. This is illustrated by examining the June 8, 1902, massacre of 124 Yaqui men, women and children

Fig. 2.1 Dr. Ventura Pérez showing how he analyzed their ancestor's human remains to Yaqui school children from the pueblo of Vicam, Sonora, Mexico

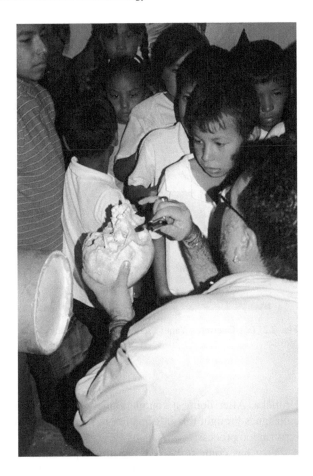

by Mexican troops; the subsequent collection of human remains and material culture by Ales Hrdlička and their transport to the AMNH in New York; and the successful efforts of the Yaqui to have their brethren repatriated to the Yaqui Zona Indigena while exploring the social impact of that process on the Yaqui descendant community the larger global community.

The USA papers of the time reported this as just another minor battle in the ongoing Yaqui war, and it may have simply remained a footnote in history had it not been for the actions of an American Physical Anthropologist who was traveling in Mexico at the time. Ales Hrdlička was in Mexico conducting research for the AMNH, which had been financed as part of the Hyde expedition. Three weeks after the Yaqui massacre, Hrdlička was taken to site with blessing of General Torres. While at the site Hrdlička collected skeletal remains from 12 individuals and items of material culture from the bodies of others (Fig. 2.2).

In 2007, Pérez was conducting research at the AMNH in New York and it was brought to his attention that the Yaqui remains that had been brought there by

Fig. 2.2 Los Guerreros Yaqui (Yaqui warriors) on the day of their reburial in Vicam, Sonora, Mexico

Hrdlička. After doing a complete analysis of the human remains, Pérez and his colleagues brought this information to the Pascua Yaqui tribe, and this started the repatriation process.

Pérez has come to believe that this repatriation would not have happened as quickly as it did, if at all, had Yaqui pursued it under NAGPRA. Pérez firmly believes that the Pascua Yaqui, being a federally recognized tribe, had the right to claim the remains and material culture of their ancestors from AMNH under NAGPRA. However, it was made clear to the tribe very early on that this was not going to happen. The argument was simple: The Pascua Yaqui is a United States federally recognized tribe and NAGPRA is a USA law. The human remains and material culture were collected in Mexico; a USA tribe cannot use a USA law to facilitate their repatriation. This, of course, makes no sense given that the Yaqui, like most tribes along the US-Mexico border, have always occupied and traveled both sides of this imposed international border. The remains were clearly the ancestors of the Pascua Yaqui. They were in a USA museum, and yet it was argued that they were not subject to NAGPRA. Instead, AMNH insisted that the remains be returned to the Instituto Nacional de Antropología e Historia (INAH, National Institute of Anthropology and History). INAH was established in 1939 as a federal bureau of the Mexican government to protect and advocate the research, preservation, and promotion of the precontact, archaeological, anthropological, historical, and paleontological heritage of Mexico.

Fig. 2.3 Dr. Ventura Pérez with Yaqui girls the day their ancestors were returned to Vicam Pueblo, Sonora Mexico, October, 2009

It is not our intention to debate the wisdom of this decision but rather to consider the implications of the statement. Consider for a moment AMNH and INAH's positions. What kind of Pandora's Box would have been opened had the Pascua Yaqui successfully repatriated this collection under NAGPRA? How many archaeological collections exist in this country that were collected in Mexico but are culturally affiliated to federally recognized tribes here in the USA? It is important to emphasize one of the principle reasons this repatriation was allowed to go forward. The collection was not considered archaeological material by AMNH or INAH but rather it was seen as a historic massacre site and thus a human rights issue. AMNH agreed that it would return the human remains and material culture to the Yaqui if INAH would permit it. AMNH returned the material to INAH and then INAH returned the material to the Yaqui. This is an incredibly important point to recognize. AMNH did not give material collected in Mexico to a culturally affiliated US federally recognized tribe, and INAH did not return archaeological material to a Mexican tribe (Fig. 2.3).

This begs the question—can and should NAGPRA apply if similar situations arise? Right now, there is no clear answer to this question. However, it is an important one to consider.

2.3 Indigenous Archaeology

Indigenous archaeology is a term used to describe archaeology that is carried out and supervised by, or done in conjunction with, indigenous groups of that particular area or who are the descendants of the groups under archaeological study. This represents an important step in empowering indigenous scholars and others to be part of shaping the way research is conducted. Indigenous groups are taking control of their own cultural heritage and sometimes even utilizing the scientific knowledge archaeology can provide. Rather than compromising scientific inquiry, collaboration with other knowledge traditions has challenged scholarly epistemologies and has led to "... substantial contributions to the intellectual growth of our discipline" (Colwell-Chanthaphonh 2010).

Setting up indigenous archaeology as an academic subdiscipline within archaeology is an important thing to do. It is similar to the civil rights movement which prompted the creation of fields of study such as African-American studies, women's studies, ethnic studies, and Native American studies. It is important to open up a space in academia where there can be a reorganization of resources and a way to shine a light on a previously unknown field of study.

Beyond the benefits to researchers, projects that fall under the goals of indigenous archaeology have been aimed at addressing the ongoing estrangement between specific indigenous groups and their cultural heritage. This is a result of a history of oppression that is often not acknowledged by academic entities. Prior to the development of indigenous archaeology, groups lacked a voice in the argument for where the material and human remains associated with their ancestors should be handled, curated, stored and analyzed (Lippert 2006:431).

A recent special theme issue in American Indian Quarterly called *Decolonizing Archaeology* presents numerous research projects conducted on groups from all around the world who are actively engaging in taking control of their own heritage. It is titled Decolonizing Archaeology because they feel that contemporary archaeology is just another form of colonialism in that it is the study of the "other" from a Western perspective with little to no input from the people that are the subject of analysis (Atalay 2006). "If colonialism has meant Indigenous peoples living within a framework of non-Indigenous control, the decolonization of archaeology has to involve archaeologists working within a framework of Indigenous control, a framework in which research process, outcomes, and benefits are genuinely negotiated between researcher and community"(Smith and Jackson 2006:341). This quote captures the broader intellectual themes opened up by situating archaeology as part of the problem and suggesting ways that it might transform itself so that it is part of the solution.

Following an indigenous archaeology paradigm does not require one to be indigenous or to be conducting archaeology related to indigenous history. The intent is for theory and practice to intersect with indigenous values while being attentive and responsible to indigenous communities in order to redress real inequality while enriching the archaeological record (Atalay 2006). The goals then of indigenous archaeology are actually the same goals that this text is advocating for with a discussion that guides shifting the ethos of bioarchaeology. For example, when consulting or collaborating with indigenous and tribal representatives, the rules of engagement include the understanding that everyone at the table will be allowed to express their point of view freely (see Mador et al. 1995: 481).

Central to the mission and ethical views of the Society for American Archaeology is the conservation and protection of the archaeological record. The primary value in archaeological resources is derived from the information archaeologists are able to discern through excavation (Lynott and Wylie 1995). Archaeologists are positioned as "[...] one group specifically qualified to study the archaeological record" (Lynott and Wylie 1995:29).

This raises the question, who else is qualified? If one subscribes to the idea that there are many ways of knowing the past (i.e., archaeological and ethnographic techniques, oral history), then it would seem that many stakeholders would be "qualified" to help in the interpretation of the archaeological record. Archaeology should be engaged in as a human endeavor not limited to the study of material culture (McGuire 1997:86). As such, "[...] a diversity of archaeologies should arise from our relationships with different communities" (McGuire 1997:86). The idea of multiple archaeologies and stakeholders is perhaps where indigenous archaeology makes one of its greatest contributions to our discipline. The focus on stakeholders is amplified in the argument that "[...] we must recognize that [scientific and scholarly] interests are not the only legitimate ones at stake" (McGuire 1997:86).

Not all archaeologists are in favor of the creation of indigenous archaeology, arguing that it simply represents a movement toward old models of "Aboriginalism" or ideas of the "Noble Savage" (McGhee 2008). Many of these researchers believe that instead of conducting objective science that takes a critical approach to understanding the past, researchers are allowing indigenous groups to dictate how we interpret the past. In response to this, Croes (2010:215) argues that indigenous archaeology cannot be viewed as a sacrifice for scientists but has to be seen for what it is, a mutually beneficial and equal partnership. There are numerous projects in many parts of the world where indigenous groups are in control of the research being conducted and presented about their ancestors and these groups have very successful and long-term relationships with nonindigenous archaeologist. Indigenous archaeology promotes collaboration, but as Conklin (2003:5) argues this in no way means that the research should be modified to satisfy a particular group.

There are now many examples of projects where indigenous groups and archaeologists have established collaborative relationships that have produced research that is relevant to both parties. In Virginia the Chickahominy, Mattaponi, Nansemond, Pamunkey, Rappahannock, and Upper Mattaponi formed the Virginia Indian Advisory Board (VIAB) that worked with local archaeologists in the Werowocomoco Research Group (WRG). The goal of this collaboration was to develop a better understanding of the history of site of Werowocomoco where the Virginia Company encountered the Powhatan chiefdom (Gallivan and Moretti-Langholtz 2007). In Arizona, collaboration between the White Mountain Apache Tribe and archaeologists provided a great example of the power of collaboration. By listening to and working with the Apache, the indigenous people and the researchers were able to transform Fort Apache, a place that had symbolized the loss of their traditional way of life, into a heritage center that both revealed the negative history of the fort and celebrated the Apache culture both in the past and today (Colwell-Chanthaphonh 2007).

One of the more important outcomes of indigenous archaeology becoming more visible within the broader communities is that it provides a means for achieving justice for past wrongs. This is especially true in the case of indigenous groups that have suffered, been exploited, and been subjugated by colonial powers (Colwell-Chanthaphonh 2007). By justice Colwell-Chanthaphonh (2007) does not mean retribution as retributive justice is often damaging instead of helpful. Instead he suggests that they type of justice that can be achieved through collaboration is restorative. The difference between these two concepts is illustrated by how justice can be sought for the genocide of millions American Indian people in the United States. Retributive justice would be to punish those responsible, but since the genocide happened many generations ago, there is no one to punish. Even if there were, the satisfaction of punishing a few for an atrocity that destroyed millions of people's lives and devastated whole cultures, this punishment would not be satisfying for many of the descendant communities. Instead, he argues that justice needs to be restorative and seek to reconcile the past with the present. Justice is an ongoing process of revealing the truth of the past, and while there can be multiple ways of interpreting data, all perspectives/conclusions should be considered. However, it is important to point out that not all perspectives/conclusions are equally valid. "Restorative justice" is thus an important form of justice a way for individuals and communities to seek healing when violence has suffused an entire community (Colwell-Chanthaphonh 2007:37). It is in deciding which conclusion has more validity that collaboration between the indigenous group and the archaeologists is the most productive. Chacon and Mendoza (2012b:489–490) acknowledge the potential pitfalls of assuming an unbridled advocacy position and provide a cautionary case study.

Although there is little doubt that indigenous archaeology is a valuable new approach in archaeology, Silliman (2008:4) makes a provocative point in stating that, though some are shining a light on what he terms "collaborative indigenous archaeology," others are setting it apart as different from the rest of archaeology. He argues that instead there needs to be a change in archaeology as a whole so it is more like this approach: "… more methodologically rich, theoretically interesting, culturally sensitive, community responsive, ethically aware, and socially just" (Silliman 2008:4–5). These calls for collaboration are appearing in increasingly more studies that include researchers from many different areas and examples of this are provided by Chacon and Mendoza (2012).

Chacon and Mendoza (2012) have participated in a number of collaborations, and one of their recommendations is a very specific protocol for what to do when the tribal representative disagrees with the conclusions of a study. They suggest that a private meeting be arranged with tribal elders or representatives to hear an explanation of the findings. This should normally occur before publication. Tribal representatives should be encouraged to present alternative conclusions or other ways to interpret the data. Sufficient time should be given to this process. If the researcher still believes that their original conclusions are correct, any publications that result should provide an accurate synopsis of the counterarguments put forth by the tribe. The tribe should have an opportunity to see the publication with their alternative viewpoints summarized. This protocol provides readers with the opportunity to decide for themselves which version to endorse based on the merits of both ways of explaining the data (Richard Chacon, personal communication, 2012).

It is this fundamental shift in ethos that we are suggesting for the future of bioarchaeology as well. Instead of having students of bioarchaeology learn the various county, state, and national laws as a way of teaching them how to practice ethically, there should be a broader agenda to cultivate an ethical approach that goes beyond what is legally mandated. Instead of students taking one course on indigenous archaeology, they should be exposed to a broad range of other ways of knowing and thinking about the past that integrates the past with the present in ways that offer new pathways to collaborative research.

2.4 Summary

NAGPRA and NAGPRA-like legislation and mandates are here to stay. This will mean operating often in a bit of a gray area, with no clear guidelines for what it means to do the right thing ethically and morally. Bioarchaeologists of the world will need to be nimble and flexible in figuring out what it means to be responsible to the living descendants. This might take very different shapes depending on the groups involved, the timing of the research, and the place. Developing an ethos that is embedded in how bioarchaeologists do their work will take time to develop precisely because there will be no one right way to proceed. Modern bioarchaeology must attend to understanding what responsible scientific research looks like in any given situation. While it may include collaboration with indigenous groups, in other situations it may simply mean filing the appropriate forms with the tribal representatives. Chacon and Mendoza (Chacon and Mendoza 2012) argue that ethical guidelines should be crafted on a case-by-case, region-by-region basis.

The term "politically correct" is sometimes used to characterize the kind of approach that this text is advocating. The response to NAGPRA and NAGPRA-like legislation, rules, and regulations has not been embraced by the bioarchaeologists who feel that academic freedom and scientific integrity are compromised by such laws. The phrase "political correctness" is a smoke screen that dismisses everything that is distasteful to some researchers, especially if it raises the possibility that some scientific research is more ethical than others. Students and practitioners of bioarchaeology can generally ignore these kinds of name-calling.

Since an ethical approach to bioarchaeology will never be proscriptive or follow a predetermined set of steps, it is crucial to develop an ethos that encompasses how to practice an ethical bioarchaeology. If the worldview of all bioarchaeologists can shift to include practices that enhance the operationalization of responsible and ethical scientific research, it will ensure its role as an integrative, engaged and ethical enterprise. Green (1984:22) cited in Wood and Powell (1993:409) provides a succinct rejoinder to what it means to do ethical archaeology: "Be sensitive to, and respect the legitimate concerns of, groups whose culture histories are the subjects of archaeological investigation." The careful wording here by stating *legitimate* concerns is important because it suggests that each researcher will have to decide for themselves how to interpret and deal with issues raised by various groups.

References

Alfonso, M. P., & Powell, J. (2007). Ethics of flesh and bone, or ethics in the practice of paleopathology, osteology, and bioarchaeology. In V. Cassman, N. Odegaard, & J. Powell (Eds.), *Human remains: Guide for museums and academic institutions* (pp. 5–20). Lanham: AltaMira Press.

Armelagos, G. J., & Barnes, K. (1999). The evolution of human disease and the rise of allergy: Epidemiological transitions. *Medical Anthropology, 18*(2), 187–213.

Armelagos, G. J., Carlson, D. S., & Van Gerven, D. P. (1982). The theoretical foundations and development of skeletal biology. In F. Spencer (Ed.), *A history of American physical anthropology, 1930–1980* (pp. 305–328). New York: Academic.

Armelagos, G. J., & Salzmann, Z. (1976). Problems of racial classification. *Acta Facultatis Naturalium Universitatis Comenianae Anthropologia, XXII,* 11–13.

Atalay, S. (2006). Indigenous archaeology as decolonizing practice. *American Indian Quarterly, 30*(3/4), 280–310.

Barrett, R., Kuzawa, C. W., McDade, T. W., & Armelagos, G. J. (1998). Emerging and re-emerging infectious diseases: The third epidemologic transition. *Annual Review of Anthropology, 27,* 247–271.

Bass, W. M. (2005). *Human osteology: A laboratory and field manual* (5th ed.). Columbia: Missouri Archaeological Society.

Birkby, W. H. (1966). An evaluation of race and sex identification from cranial measurements. *American Journal of Physical Anthropology, 24*(1), 21–27.

Black, H. C. (1968). *Black's Law Dictionary: Definitions of the terms and phrases of American and English juriprudence, ancient and modern.* St. Paul: West Publication, Co.

Blakely, R. L. (1977). *Biocultural adaptation in prehistoric America.* Athens: Southern Anthropological Society Proceedings, No. 11, University of Georgia Press.

Boas, F. (1912). Changes in the bodily form of descendants of immigrants. *American Anthropologist, 14*(3), 530–562.

Bogin, B., & Keep, R. (1999). Eight thousand years of economic and political history in Latin America revealed by Anthropometry. *Annals of Human Biology, 26,* 333–351.

Brace, C. L. (1964). A non-racial approach toward the understanding of human diversity. In A. Montagu (Ed.), *The concept of race* (pp. 103–152). London: Collier Books.

Brace, C. L. (2005). *'Race' is a four-letter word: The genesis of the concept.* Oxford: Oxford University Press.

Brace, C. L., Tracer, D. P., Yaroch, L. A., Robb, J. E., Brandt, K., & Nelson, R. (1993). Clines and clusters versus "Race:" A test in ancient Egypt and the case of a death on the Nile. *American Journal of Physical Anthropology, 36,* 1–31.

Brickley, M., & Ives, R. (2008). *The bioarchaeology of metabolic bone disease.* London: Academic.

Brooks, S. T. (1955). Skeletal age at death: The reliability of cranial and pubic age indicators. *American Journal of Physical Anthropology, 26*(1), 67–77.

Brown, T. F. (1995–1996). The Native American Grave Protection and Repatriation Act: A necessary but costly measure. *The Nebraska Anthropologist, 12*(1), 89–98.

Buikstra, J. E. (1977). Biocultural Dimensions of Archaeological Study: A regional perspective. In R. L. Blakely (Ed.), *Biocultural adaptation in prehistoric America* (pp. 67–84). Athens: Southern Anthropological Society Proceedings, No. 11, University of Georgia Press.

Carey, D. P. (2007). Is bigger really better? The search for brain size and intelligence in the twenty-first century. In S. D. Sala (Ed.), *Tall tales about the mind and brain: Separating fact from fiction* (pp. 105–122). Oxford: Oxford University Press.

Chacon, R. J., & Dye, D. H. (2007). *The taking and displaying of human body parts as trophies by Amerindians*. New York: Springer Science and Business Media.

Chacon, R. J., & Mendoza, R. G. (2012). *The ethics of anthropology and Amerindian research: Reporting on environmental degradation and warfare*. New York: Springer.

Chaiklieng, S., Suggaravetsiri, P., & Boonprakob, Y. (2010). Work ergonomic hazards for musculoskeletal pain among University Office Workers. *Walailak Journal of Science and Technology, 7*(2), 169–176.

Chatters, J. C. (1998). Environment. In D. E. Walker Jr. (Ed.), *Handbook of North American Indians, Volume 12: Plateau* (pp. 29–48). Washington, DC: Smithsonian Institution Press.

Chatters, J. C. (2000). The recovery and first analysis of an early Holocene human skeleton from Kennewick, Washington. *American Antiquity, 65*(2), 291–316.

Chatters, J. C. (2002). *Ancient encounters: Kennewick Man and the first Americans*. New York: Touchstone.

Cobb, A. J. (2005). Understanding tribal sovereignty: Definitions, conceptualizations, and interpretations. *American Studies, 46*(3–4), 115–132.

Cobb, W. M. (1939). Race and runners. *The Journal of Health and Physical Education, 7*(1), 1–9.

Cohen, M. N., & Armelagos, G. J. (1984). *Paleopathology at the origins of agriculture*. Orlando: Academic.

Cole, D. (1985). *Captured heritage: The scramble for Northwest Coast artifacts*. Vancouver: UBC Press.

Colwell-Chanthaphonh, C. (2007). History, justice, and reconciliation. In J. L. Barbara & P. A. Shackel (Eds.), *Archaeology as a tool of civic engagement* (pp. 23–46). Lanham: AltaMira Press.

Colwell-Chanthaphonh, C. (2010). Remains unknown: Repatriating culturally unaffiliated human remains. *Anthropology News, 51*(3), 4–8.

Conklin, B. (2003). Speaking truth to power. *Anthropology News, 44*(7), 3.

Croes, D. R. (2010). Courage and thoughtful scholarship: Indigenous archaeology partnerships. *American Antiquity, 75*(2), 211–216.

Cronk, L. (1991). Human behavioral ecology. *Annual Review of Anthropology, 20*, 25–53.

Curtis, N. G. W. (2010). Repatriation from Scottish Museums: Learning from NAGPRA. *Museum Anthropology, 33*(2), 234–248.

Deloria, V., Jr. (1997). *Red Earth, White Lies: Native Americans and the myth of scientific fact*. Golden: Fulcrum Publishing.

Dongoske, K. E. (1996). The Native American Graves Protection and Repatriation Act: A new beginning, not the end, for osteological analysis—A Hopi perspective. *American Indian Quarterly, 20*(2), 287–297.

Echo-Hawk, W. R. (1988). Tribal efforts to protect against mistreatment of Indian Dead: The quest for equal protection of the laws. *Native American Rights Fund Legal Review, 14*(1), 1–5.

El-Najjar, M. Y., & McWilliams, K. R. (1978). *Forensic anthropology: The structure, morphology, and variation of human bone and dentition*. Springfield: Thomas.

Ferguson, T. J., Dongoske, K. E., & Kuwanwisiwma, L. J. (2001). Hopi perspectives on southwestern mortuary studies. In D. R. Mitchell & J. L. Brunson-Hadley (Eds.), *Ancient burial practices in the American Southwest* (pp. 9–26). Albuquerque: University of New Mexico Press.

Ferri, P. G. (2009). New types of cooperation between museums and countries of origin. *Museum International, 61*(1–2), 91–94.

Ferris, N. (2003). Between colonial and indigenous archaeologies: Legal and extra-legal ownership of the archaeological past in North America. *Canadian Journal of Archaeology, 27*, 154–190.

Gallivan, M. D., & Moretti-Langholtz, D. (2007). Civic engagement at Werowocomoco: Reasserting Native narratives from a Powhatan place of power. In J. L. Barbara & P. A. Shackel (Eds.), *Archaeology as a tool of civic engagement* (pp. 47–66). Lanham: AltaMira Press.

Garner, B. A. (2009). *Black's law dictionary digital* (9th ed). Westlaw BLACKS: West Group.

Gilbert, R. I., & Mielke, J. H. (1985). *The analysis of prehistoric diets*. Orlando: Academic.

Gill, G. W. (1998). The beauty of race and races. *Anthropology News, 39*(3), 1–5.

Goodman, A. H. (1994). Problematics of "Race" in contemporary biological anthropology. In N. T. Boaz & L. D. Woke (Eds.), *Biological anthropology: The state of the science* (pp. 221–243). Bend: International Institute for Human Evolutionary Research.

Goodman, A. H., & Leatherman, T. L. (1998). *Building a new biocultural synthesis: Political-economic perspectives on human biology*. Ann Arbor: University of Michigan Press.

Gordon, C. C. (1994). Anthropometry in the U.S. Armed Forces. In S. J. Ulijaszek & C. G. N. Mascie-Taylor (Eds.), *Anthropometry. The individual and the population* (pp. 178–210). Cambridge: Cambridge University Press.

Gould, S. J. (1981). *The mismeasure of man*. New York: W. W. Norton and Company.

Gravlee, C. C., Russell Bernard, H., & Leonard, W. R. (2003a). Boas's changes in bodily form: The immigrant study, cranial plasticity, and Boas's physical anthropology. *American Anthropologist, 105*(2), 326–332.

Gravlee, C. C., Russell Bernard, H., & Leonard, W. R. (2003b). Heredity, environment, and cranial form: A re-analysis of Boas's immigrant data. *American Anthropologist, 105*(1), 125–138.

Green, E. L. (1984). *Ethics and values in archaeology*. New York: The Free Press.

Harrod, R. P. (2011). Phylogeny of the southern Plateau–An osteometric evaluation of inter-tribal relations. *HOMO—Journal of Comparative Human Biology, 62*(3), 182–201. doi:10.1016/j.jchb.2011.01.005.

Hendrick, H. W. (2003). Determining the cost-benefits of ergonomics projects and factors that lead to their success. *Applied Ergonomics, 34*(5), 419–427.

Hoeveler, D. J. (2007). The measure of mind. *Reviews in American History, 35*(4), 573–579.

Hooten, E. A. (1930). *Indians of Pecos Pueblo: A study of their skeletal remains*. New Haven: Yale University Press.

Horsman, R. (1975). Scientific Racism and the American Indian in the mid-nineteenth century. *American Quarterly, 27*(2), 152–168.

Hsieh, S. D., & Muto, T. (2005). The superiority of waist-to-height ratio as an anthropometric index to evaluate clustering of coronary risk factors among non-obese men and women. *Preventive Medicine, 40*(2), 216–220.

Huds, D. (2011). Salaries of Bioarchaeologists. eHow. Accessed October 15, 2012.

Huishu, Z., & Damin, Z. (2011). Simulation and ergonomics analysis of pilot visual information flow intensity. *Journal of Beijing University of Aeronautics and Astronautics, 5*, Article 5.

Huss-Ashmore, R., Goodman, A. H., & Armelagos, G. J. (1982). Nutritional inference from paleopathology. *Advances in Archaeological Method and Theory, 5*, 395–474.

Jackson, J. P., Jr. (2010). Whatever happened to the cephalic index? The reality of race and the burden of proof. *Rhetoric Society Quarterly, 40*(5), 438–458.

Johanson, G. (1971). Age determination from human teeth. *Odontologisk Revy, 22*(Suppl 21), 1–126.

Kakaliouras, A. M. (2008). Toward a new and different osteology: A reflexive critique of physical anthropology in the United States since the passage of NAGPRA. In T. W. Killion (Ed.), *Opening archaeology: Repatriation's impact on contemporary research and practice* (pp. 109–129). Santa Fe: School of Advanced Research Press.

Komlos, J. (1989). *Nutrition and economic development in the eighteenth-century Habsburg Monarchy: An anthropometric history*. Princeton: Princeton University Press.

Komlos, J. (1995). *The biological standard of living on three continents: Further explorations in anthropometric history*. Boulder: Westview.

Lang, J. C. (2011). Epistemologies of situated knowledges: "Troubling" knowledge in philosophy of education. *Educational Theory, 61*(1), 75–96.

Larsen, C. S. (1987). Bioarchaeological interpretations of subsistence economy and behavior from human skeletal remains. In M. B. Schiffer (Ed.), *Advances in archaeological method and theory* (Vol. 10, pp. 339–445). San Diego: Academic.

Larsen, C. S., & Walker, P. L. (2005). The ethics of bioarchaeology. In T. R. Turner (Ed.), *Biological anthropology and ethics: From repatriation to genetic identity* (pp. 111–120). Albany: State University of New York Press.

Lewis, J. E., DeGusta, D., Meyer, M. R., Monge, J. M., Mann, A. E., & Holloway., R. L. (2011). The mismeasure of science: Stephen Jay Gould versus Samuel George Morton on skulls and bias. *PLoS Biology, 9*(6), 1–6.

Lippert, D. (2006). Building a bridge to cross a thousand years. *American Indian Quarterly, 30*(3/4), 431–440.

Lovejoy, A. O. (1936). *The great chain of being: A study of the history of an idea.* Cambridge: Harvard University Press.

Lynott, M. J., & Wylie, A. (1995). *Ethics in American archaeology: Challenges for the 1990s, revised* (2nd ed.). Washington, DC: Society for American Archaeology.

Mador, M. J., Rodis, A., & Magalang, U. J. (1995). Reproducibility of Borg scale measurements of dyspnea during exercise in patients with COPD. *Chest, 107*(6), 1590–1597.

Malhi, R. S., Kemp, B. M., Eshleman, J. A., Cybulski, J. S., Smith, D. G., Cousins, S., et al. (2007). Haplogroup M discovered in prehistoric North America. *Journal of Archaeological Science, 34*, 642–648.

Mapes, L. V. (2012). Kennewick Man Bones not from Columbia Valley, scientist tells tribes. Seattle Times. Accessed October 15, 2012.

Maresh, M. M. (1955). Linear growth of the long bone of extremities from infancy through adolescence. *American Journal of Diseases of Children, 89*, 725–742.

Martin, D. L. (1994). Patterns of health and disease: Health profiles for the prehistoric Southwest. In G. J. Gumerman (Ed.), *Themes in Southwest prehistory* (pp. 87–108). Santa Fe: School of American Research Press.

Martin, D. L. (1998). Owning the sins of the past: Historical trends in the study of Southwest human remains. In A. H. Goodman & T. L. Leatherman (Eds.), *Building a new biocultural synthesis: Political-economic perspectives on human biology* (pp. 171–190). Ann Arbor: University of Michigan Press.

Mason, R. J. (2000). Archaeology and Native American oral tradition. *American Antiquity, 65*(2), 239–266.

McGhee, R. (2008). Aboriginalism and the problems of indigenous archaeology. *American Antiquity, 73*(4), 579–597.

McGuire, R. H. (1997). Why have archaeologists thought that the real Indians are dead and what can we do about it. In T. Biolsi & L. J. Zimmerman (Eds.), *Indians and anthropologists in post colonial America* (pp. 63–91). Tucson: University of Arizona Press.

Merbs, C. F., & Miller, R. J. (1985). *Health and disease in the prehistoric Southwest.* Tempe: Arizona State University.

Miller, E. (1995). *Refuse to be Ill: European contact and aboriginal health in northeastern Nebraska.* Unpublished Ph.D. dissertation, Arizona State University, Tempe.

Montagu, A. (1942). *Man's most dangerous myth: The fallacy of race.* New York: Columbia University Press.

Morgan, M. E. (2010). *Pecos Pueblo revisited: The biological and social context.* Cambridge: Peabody Museum of Archaeology and Ethnology Harvard University.

Nott, J. C., & Gliddon, G. R. (1854). *Types of Mankind: Or Ethnological researches, based upon the ancient monuments, paintings, sculptures, and crania of races, and upon their natural, geographical, philological and Biblical history.* Philadelphia: Contributions by Louis Agassiz, William Usher, and Henry Stuart Patterson. Lippincott, Grambo & Co.

Ortner, D. J. (2003). *Indentification of pathological conditions in human skeletal remains.* London: Academic.

Ousley, S. D., Billeck, W. T., & Hollinger, R. F. (2005). Federal repatriation legislation and the role of physical anthropology in repatriation. *American Journal of Physical Anthropology, 128*(S41), 11.

Ousley, S. D., Jantz, R. L., & Freid, D. (2009). Understanding race and human variation: Why forensic anthropologists are good at identifying race. *American Journal of Physical Anthropology, 139*(1), 68–76.

Pérez, V. R. (2010). From the singing tree to the hanging tree: Structural violence and death with the Yaqui Landscape. *Landscapes of Violence, 1*(1), Article 4.

Pevar, S. L. (2012). *The rights of Indians and tribes*. Oxford: Oxford University Press.

Pheasant, S., & Haslegrave, C. M. (2006). *Body space: Anthropometry, ergonomics, and the design of work* (3rd ed.). Boca Raton: CRC Press.

Powell, J. F., & Rose, J. C. (1999). Chapter 2. Report on the Osteological Assessment of the "Kennewick Man" Skeleton (CENWW.97.Kennewick). In *Report on the Nondestructive Examination, Description, and Analysis of the Human Remains from Columbia Park, Kennewick, Washington [October 9]*. Washington, DC: National Park Service. http://www.nps. gov/archeology/kennewick/.

Powell, M. L. (2000). Ancient diseases, modern perspective: Treponematosis and tuberculosis in the age of agriculture. In P. M. Lambert (Ed.), *Bioarchaeological studies in the age of agriculture: A view from the Southeast* (pp. 6–34). Tuscaloosa: The University of Alabama Press.

Riding In, J. (1992). Without ethics and morality: A historical overview of imperial archaeology and American Indians. *Arizona State Law Journal, 24*, 11–34.

Roberts, C. A. (2010). Adaptation of populations to changing environments: Bioarchaeological perspectives on health for the past, present and future. *Bulletins et Mémoires de la Société d'anthropologie de Paris, 22*(1–2), 38–46.

Roberts, C. A., & Buikstra, J. E. (2003). *The bioarchaeology of tuberculosis: A global perspective on a re-emerging disease*. Gainesville: University Press of Florida.

Roberts, C. A., & Manchester, K. (2005). *The archaeology of disease* (3rd ed.). Ithaca: Cornell University Press.

Rogers, R. M. (2011). *Design of military vehicles with the soldier in mind: Functionality and safety combined*. Warren: US Army RDECOM-TARDEC, No. 21658.

Rose, J. C., & Green, T. J. (2002). NAGPRA and the future of skeletal research. In M. A. Park (Ed.), *Biological anthropology: An introductory reader* (pp. 214–217). Boston: Reprint *General Anthropology, 4*(1), 8–10 (1997). McGraw-Hill.

Rose, J. C., Green, T. J., & Green, V. D. (1996). NAGPRA is forever: Osteology and the repatriation of skeletons. *Annual Review of Anthropology, 25*, 81–103.

Sagot, J.-C., Gouin, V., & Gomes, S. (2003). Ergonomics in product design: Safety factor. *Safety Science, 41*(2–3), 137–154.

Sauer, N. J. (1992). Forensic anthropology and the concept of race: If races don't exist, why are forensic anthropologists so good at identifying them? *Social Science & Medicine, 34*(2), 107–111.

Sauer, N. J. (1993). Applied anthropology and the concept of race: A legacy of Linneaus. *National Association for the Practice of Anthropology, Bulletin, 13*, 79–84.

Seidemann, R. M. (2004). Bones of contention: A comprehensive examination of law governing human remains from archaeological contexts in formerly colonial countries. *Louisiana Law Review, 64*(3), 545–588.

Silko, L. M. (1987). Landscape, history, and the pueblo imagination. In D. Halpren (Ed.), *On nature* (pp. 83–94). San Francisco: North Point.

Silliman, S. W. (2008). Collaborative indigenous archaeology: Troweling at the edges, eyeing the center. In S. W. Silliman (Ed.), *Collaborating at the Trowel's edge: Teaching and learning in indigenous archaeology* (pp. 1–21). Tucson: University of Arizona Press.

Simpson, M. (2009). Museums and restorative justice: Heritage, repatriation and cultural education. *Museum International, 61*(1–2), 121–129.

Singleton, T. A., & Orser, C. E., Jr. (2003). Descendant communities: Linking people in the present to the past. In L. J. Zimmerman, K. D. Vitelli, & J. Hollowell-Zimmer (Eds.), *Ethical issues in archaeology* (pp. 143–152). Walnut Creek: AltaMira Press.

Smith, C., & Jackson, G. (2006). Decolonizing indigenous archaeology: Developments from down under. *American Indian Quarterly, 30*(3/4), 311–349.

Smith, E. A., & Winterhalder, B. (1992). *Evolutionary ecology and human behavior*. New York: Aldine de Gruyter.

Stapp, D. C., & Longnecker, J. (2008). *Avoiding archaeological disasters: Risk management for heritage professionals*. Walnut Creek: Left Coast Press, Inc.

Stocking, G. W., Jr. (1982). *Race, culture and evolution: Essays in the history of anthropology.* Chicago: The University of Chicago.

Stodder, A. L. W. (1990). *Paleoepidemiology of Eastern and Western Pueblo Communities.* Unpublished PhD Dissertation, University of Colorado, Boulder, Boulder.

Thomas, D. H. (2000). *The Skull Wars: Kennewick man, archaeology, and the battle for Native American identity.* New York: Basic Books.

Thorleifsen, D. (2009). The repatriation of Greenland's cultural heritage. *Museum International, 61*(1–2), 25–29.

Turner, B. L., & Andrushko, V. A. (2011). Partnerships, pitfalls, and ethical concerns in international bioarchaeology. In S. C. Agarwal & B. A. Glencross (Eds.), *Social bioarchaeology* (pp. 44–67). Malden: Wiley-Blackwell.

Gerven, V., Dennis, P., Carlson, D. S., & Armelagos, G. J. (1973). Racial history and bio-cultural adaptation of Nubian archaeological populations. *The Journal of African History, 14*(4), 555–564.

Vaughan, L. A., Benyshek, D. C., & Martin, J. F. (1997). Food acquisition habits, nutrient intakes, and anthropometric data of Havasupai adults. *Journal of the American Dietetic Association, 97*(1), 1275–1282.

Walker, P. L. (2000). Bioarchaeological ethics: A historical perspective on the value of human remains. In M. Anne Katzenberg & R. S. Shelley (Eds.), *Biological anthropology of the human skeleton* (pp. 3–39). Hoboken: Wiley.

Walsh-Haney, H., & Lieberman, L. S. (2005). Ethical concerns in forensic anthropology. In T. R. Turner (Ed.), *Biological anthropology and ethics: From repatriation to genetic identity* (pp. 121–132). Albany: State University of New York Press.

Washburn, S. L. (1951). The new physical anthropology. *Yearbook of Physical Anthropology, 7,* 124–130.

Watkins, J. (2003). Archaeological ethics and American Indians. In L. J. Zimmerman, K. D. Vitelli, & J. Hollowell-Zimmer (Eds.), *Ethical issues in archaeology* (pp. 129–142). Walnut Creek: AltaMira Press.

Weaver, R. M. (1960). Lord Acton: The historian as thinker. *Modern Age, 4,* 13–22.

Wood, J. J., & Powell, S. (1993). An ethos for archaeological practice. *Human Organization, 52*(4), 405–413.

Chapter 3
Formulating Research Projects Involving Human Remains

Given the historical trajectories of research with human remains discussed in Chap. 2, it is now mandatory that bioarchaeologists take seriously the legal and ethical issues at every step of the process involving research and human remains. The legal aspects will involve seeking official permission to carry out a study. Permission could be anything from writing a proposal that a tribal entity will review to filling out a form and having a museum representative approve it. The moral and ethical issues raised by formulating research projects involving human remains will be more complicated to attend to than the legal issues, and these are just as important.

Research projects need to be carefully designed with as much integration and engagement with legal and ethical considerations as possible. Prior to NAGPRA in the USA and other kinds of legislation worldwide, designing a study was, in some cases, something that the researcher could do without necessitating very much discussion with anyone, but this is no longer the case. Those carrying out a project involving human remains need to consider and build in the ethical dimension as the project is being conceived. Even the legal aspects need to be investigated prior to designing the research to see if there are any caveats or conditions under which some research might be deemed inappropriate or unacceptable. For example, a research project that involves destruction of a small piece of bone or tooth enamel in order to study ancestry or diet through isotopic analyses (see Chap. 8) might not be something that is legally permitted. Furthermore, there may be tribal sanctions against certain kinds of research. On the other hand, techniques requiring destruction of a small portion of human remains have been carried out with state and local authority's approval (see, e.g., Dongoske 1996; O'Rourke et al. 2005).

D.L. Martin et al., *Bioarchaeology: An Integrated Approach to Working with Human Remains*, Manuals in Archaeological Method, Theory and Technique, DOI 10.1007/978-1-4614-6378-8_3, © Springer Science+Business Media New York 2013

3.1   Bioarchaeology and Forensic Anthropology: Complementary Methods but Different Approaches

Forensic anthropology is closely allied to bioarchaeology. Like bioarchaeology, forensic anthropology is also a subdiscipline of physical anthropology. It uses many of the same methods as bioarchaeology. The major differences between these approaches to working with human remains are that forensic anthropologists work primarily in today's modern context on fairly recent deaths and they work in conjunction with law enforcement, human rights groups, or criminal justice institutions. The term forensic means "pertaining to courts of law" (McCaffrey et al. 1997:ix), so a forensic anthropologist identifies and analyzes human remains in a legal context. Although it shares methods with bioarchaeology, forensic anthropology also is closely allied with the field of forensic science. Forensic archaeology is also part of this legal arena in which anthropologists are increasingly employed (Hunter and Cox 2005; Dupras et al. 2011). Another area of growing expertise has been labeled "disaster archaeology" (Gould 2007).

As Nicholas (2004) points out, archaeologists and physical anthologists have had a long and illustrious history of working in the field of forensics and human rights abuses. The field of forensic archaeology is a recognized component of biological anthropology with regard to medicolegal issues in the Americas (Hunter 2002). Many of the earliest anthropologists working with human skeletal remains were trained in anatomy within the field of medicine. Immediately after the development of a formal field of study, early physical anthropologists such as Hooton and Hrdlička began conducting forensic cases (Thompson 1982). In 1936, Hrdlička became the first anthropologist to assist the FBI on criminal cases and according to the FBI he is said to have been ". . . the best informed man in the United States on anthropology" (Ubelaker 1999:728).

Standard archaeological techniques have been used with great success in the careful in situ exposure of bodies and recording and photographing of material culture relevant to the crime scene (Skinner et al. 2003; Sigler-Eisenberg 1985; Saul and Saul 2002; Sorg and Haglund 2002). In addition, Gargett (1999) has focused on human agency in buried remains as another area of cross-pollination between the two fields. Drawing on the work of Micozzi (1997) and Galloway et al. (1989), Gargett explores how decomposition rates impact disarticulation and what they can say about intentional versus natural burial. Finally, Gould (2005) has been instrumental in organizing and training a special archaeological response team (Forensic Archaeology Recovery) for recovering and interpreting human remains from disaster sites.

Forensic work is a methods-driven approach focused upon the identification of victims and the adjudication of criminal and civil cases and includes criminalistics, forensic pathology, forensic anthropology, forensic odontology, forensic engineering, toxicology, behavioral sciences, and questioned documents. The work is made more complicated when it is in the context of terrorist actions (World Trade Center, Oklahoma City), sectarian civil wars (Bosnia), ethnic genocides (Rwanda), massacres (Guatemala), or warfare (Vietnam, Afghanistan). This kind of humanitarian work put anthropologists trained in forensics in contact with the local communities and with people who are searching for the identification of their loved ones who may have perished (Steele 2008). The nature of extreme violence permeates their work because they are often asked to piece together evidence of what occurred between perpetrator and victims (see many examples of this in the edited volume by Ferllini 2007).

Forensic anthropology is done largely in consort with police at state and federal levels in the USA or with humanitarian and international governments beyond the USA. Virtually every state in the USA now employs former physical anthropologists and bioarchaeologists within their crime agencies. There are also private organizations, such as NecroSearch, that employ physical anthropologists and archaeologists to work with police to locate and retrieve buried and hidden bodies. The recent interest in working with human remains in the last decade has been driven by a public interest in forensic anthropology due to the highly popularized books by authors such as Kathy Reichs, Beverly Connor, William Bass and his collaborator Jon Jefferson, and Jan Burke. The proliferation of television crime shows such as CSI and Bones also has spurred an interest in the ways that human remains are used to solve cases.

The demographic and identifying characteristics that forensic anthropologists evaluate to aid in the identification of unidentified victims include age, sex, height, and ancestry. The history of forensic anthropology runs parallel to that of physical anthropology in that it has more recently adopted the notion of "ancestral groups" to discuss variation among human populations in lieu of "race." This is largely because the variation among and between groups is difficult to quantify and the older typological categories (i.e., *Caucasoid*, *Negroid*, and *Mongoloid*) do not capture existing variation in phenotypic expression (Steadman 2009).

Though many social scientists would adamantly deny the existence of race and ancestral groups, many forensic anthropologists argue that people tend to welcome any tactic that helps to identify missing or unidentified loved ones (e.g., soldiers missing in action [MIAs], victims of domestic terrorism [World Trade Center], and victims of homicide). The argument for discernible human variability rests on the assumption that the deep past human populations were to some degree geographically and genetically isolated from one another and adaptive physical traits evolved in different regions. Providing information from skeletons on ancestry involves providing anatomically specific traits that can be confirmed with known physical and genetic markers (Gill 1998; Ousley et al. 2009; Sauer 1993, 1992).

The role of most forensic anthropologists centers on identification, postmortem changes, and trauma (Rhine 1998; Maples and Browning 1994). One of the issues facing the relatively young discipline of forensic anthropology is the need to create a balance between specialized knowledge embedded in the forensic sciences and making sure practitioners are trained in the other subfields of anthropology (Buikstra et al. 2003). New and previously unimaginable roles are becoming increasingly more common as anthropologists engage in fieldwork associated with death and violence. It is apparent that anthropologists are fulfilling important roles in helping to locate, identify, and examine dead and missing bodies. Through interdisciplinary inquiry and engagement across the subfields, anthropology provides a way to view the diversity of opinions about violence, warfare, and human rights issues that often result in death, trauma, and social upheaval.

A variety of forms of violence, covert and vicious types of warfare, and a broad range of activities regarding surveillance or the elimination of human rights have created work for anthropologists that could not have been imagined 100 years ago. Thus, the following section is not just for bioarchaeologists but also for those interested in forensic anthropology. Future forensic anthropologists need training not just in fundamental skeletal analysis. They also need training in all four subfields of anthropology in order to broaden their theoretical and practical approach to dealing with everyday violence. In this way forensic anthropology can engage in an interdisciplinary inquiry of the theoretical and empirical issues within the study of violence, warfare, surveillance, and human rights. The strength anthropology brings to the study of violence is a critical, self-reflective, and non-reductionist perspective that allows for a holistic examination of the dynamics that have led to the wide array of human atrocities committed throughout the world. It is this strength that is needed to keep forensic anthropology from falling into a trap of creating highly specialized technicians who are disconnected from the rest of anthropology.

3.2 The Research Question: Putting Human Remains into Context

The context within which human remains are discovered is as important as the bones themselves. Chapter 5 provides an overview on the mortuary contexts where human remains are often located. These include formal cemeteries and necropolises, tombs, shallow pits in midden areas, burials underneath the floors of habitation areas, and unusual or unique places external to where people lived. Burials may be complete, partially complete, disarticulated, commingled, or high fragmentary. The data derived from these contexts are crucial to add to the information from the human remains. But this is only one part of a larger and multilayered notion of what context may include. Context also includes the larger archaeological site that often provides information on such things as habitation and work areas, ceremonial architecture, food storage, material objects used in everyday life (such as ceramics and lithics), and domesticated plants and animals. Beyond the physical setting of the mortuary spaces and the archaeological site, context also includes ideology and culture as it relates to the humans who lived at the site. Thus, context can be stretched to include almost anything about the humans under consideration that can be reconstructed from archaeological and other sources of information.

But some contextual information is more important than other aspects, and so it is crucial that the method to integrating data uses a systematic and scientific approach. Bioarchaeologists have a deep commitment to scientific inquiry (White et al. 2012; Larsen 1997). Formulating testable hypotheses along with using multiple methodologies and interdisciplinary approaches is a major departure from the more descriptive techniques of earlier approaches (Martin 1998). Bioarchaeology has moved skeletal analysis beyond description, and it blends the methods and data from skeletal biology with other perspectives. Related disciplinary approaches from ethnography, taphonomy, forensics, medicine, history, and geology also present potential complementary and revealing data sets that can be used in conjunction with skeletal analyses and interpretation.

An example of this integration of context and data is provided by Walker (2001) who presented a review of bioarchaeological method, theory, and data regarding the study of violence in past populations. He presents a useful diagram (2001:577) that aids in using strong inference in the interpretation of injury and trauma from skeletal remains. While some studies that document trauma on skeletal remains stop once the data on injury from bones have been collected, Walker feels that this is actually the starting point for a much more integrated study that brings in other aspects of context. Starting with evidence from the bones, his flow chart diagram begins the more integrative work by asking a series of questions relating to taphonomy (see Chap. 4) and the archaeological reconstruction of the mortuary component (see Chap. 5) and the ritual/ceremonial (see Chap. 9) aspects. His approach demonstrates the necessity of integration of contextual data in order to produce an interpretation that extends far beyond description. This is possible because reconstructing context is part of the research design. Also because of the systematic way the data are collected and analyzed, hypotheses can be tested.

Bioarchaeology can and should be used to independently test hypotheses (Armelagos 2003). The difference between descriptive skeletal analysis and bioarchaeology is that the latter employs an interdisciplinary and cross-cultural research tool that aids in the analysis of a wide range of data on human biology that is useful in hypothesis testing. One of the goals of bioarchaeology is to interpret the biological data in relationship to social and ecological contexts such as changes in diet, increases in population size and density, shifts in power and stratification, and differential access to resources. By using multiple working hypotheses, scientific methodology, and strong inference, it provides a means for utilizing multiple lines of evidence in interpretations. Most importantly, variability is not factored out; rather, it is considered and weighed against all other available lines of evidence so that it is accounted for (instead of being discounted). Formulating research questions using human skeletal remains is something that needs to take many factors into consideration. Failing to do so will result in projects that may later be called into question by legal or tribal authorities, peer reviewers, or others who evaluate the final product and find it limited in scope.

Nearly all projects dealing with human remains deal with some aspect of the question: Why do humans die before they are really old? Almost all of the answers to this question lie in three realms. They get sick (paleopathology), they have accidents (premortem and perimortem signs of trauma plus circumstantial evidence), or they are killed (perimortem signs of trauma plus circumstantial evidence). At the population and community level, deaths are often patterned by factors that include age, sex, occupation, social status, economic status, and access to resources such as food, water, and shelter. Just as the Centers for Disease Control (CDC) in the USA and the World Health Organization (WHO) at the international level track patterns and trends in disease, poor health, disability, and death for contemporary populations, analysis of ancient and historic remains permits similar kinds of analyses for past populations.

The compelling questions regarding patterns in poor health and early death can be framed even more broadly to investigate which environmental, cultural, physiological, and biological factors best explain those patterns. Studies that compare data either cross-culturally or through time provide a broader perspective which is useful in formulating interpretations. This highlights the fact that some groups experience less disease, trauma, or early death than other groups, and explanations can then be theorized and offered. Ultimately bioarchaeology seeks to be able to explain human behavior. It provides a very unique slant on human life and death that no other disciplinary science can provide. Crumley (2006) published a very compelling set of reasons for the importance of archaeological research in her essay entitled "Archaeology in the New World Order: What We Can Offer the Planet." She provides a multitude of concrete ways that archaeological (and bioarchaeological) data underscore that while humans have contributed to the crisis of global warming and environmental degradation, there is also an abundance of scientific data to show the ways that some human groups were innovative and creative in building communities that were sustainable and nondestructive to the planet. Armelagos (2003:34) emphasizes this point as well when he states that bioarchaeology ". . . can provide insights that are essential for understanding our relationship to our environment, how we interacted with it throughout history, and how we are interacting with it now."

In conducting research, bioarchaeologists generally start with an interest that they have in answering some question or explaining some phenomenon. Usually an interest in something sparks some ideas or notions about a particular topic. Also at this stage bioarchaeologists tend to do a lot of reading in an effort to both learn more about the topic as well as to see what kinds of findings already exist in the literature (DiGangi 2012). This is an important stage because research almost always builds on prior studies. One of the outcomes of reading broadly about a topic is that there is an opportunity to discover how other researchers problematize their research on the topic you are interested in understanding. How has the topic already been or is being approached and from what angle or lens are they framing their questions? Are there new ways for collecting data? These are the kinds of questions that need to be thought about in the process of developing a research project.

3.2.1 The Need for Empirical Data

It has only been from empirical data on human adaptation from the archaeological record that anthropologists and historians have come to understand how changes over time in environmental, political, and economic structure, subsistence and diet, and settlement patterns can and do have profound effects on population structure and rates of morbidity and mortality. A now classic and exceedingly comprehensive set of examples for this can be found in the volume *Paleopathology at the Origins of Agriculture*, which focuses very systematically on the changes in health related to shifts in subsistence economy in many different locales around the world (Cohen and Armelagos 1984).

One reason to focus analysis on human remains from the past, even though there are legal and ethical challenges to doing so, is that they represent literally the only available information on human biology in prestate and precolonial groups. Even groups that have left written records about births and deaths sometime lack the specificity and accuracy that can be obtained through a thorough analysis of human skeletal remains (see Chaps. 6, 7, and 8). Disease has profoundly affected the course of human history, yet few models of culture change integrate the effects of endemic disease on decisions to migrate, have more children, have fewer children, go to war, or to abandon a region. Empirical demographic and disease data analyzed on regional levels can suggest the role of poor health, dietary inadequacies, and differential mortality during periods of stability and centralization versus instability and population movements (see good examples of this in Steckel and Rose 2002).

Early scholars, in the absence of empirical data from skeletal remains, visualized the precontact past using observations of historic and contemporary indigenous lifestyles. For example, in the American Southwest there was a concentrated effort by archaeologists to reconstruct life prior to colonization. Expeditions were common as early as the 1930s and were underwritten by museums such as the American Museum of Natural History and the National Museum of Natural History at the Smithsonian (Stodder 2012). One resulting publication in the journal *Science* by the archaeologist Colton (1936) was heavily influenced by his stay in a Hopi village where he witnessed communities with a high density of people living in close proximity to refuse and stagnant water supplies. Using ethnographic analogy, he suggested that settled village life in precolonial times must have been fraught with disease and sickness. Likewise, Titiev, a cultural anthropologist living in a Hopi village in 1933, repeatedly mentions the unsanitary conditions and general poor health of many of the inhabitants, but he often related contemporary attitudes about health and sickness to earlier ancestral conditioning to such a lifestyle (Titiev 1972). Thus, Colton felt that because Pueblos are sick today, they must have been sick in the past, while Titiev thought that it is because they were sick in the past that they are sick today as well.

Neither of these hypotheses about the chronology of health and disease over time was tested using empirical data. Both Colton's and Titiev's observations may in fact be relevant for our understanding of the Hopi experience as well as for ancestral Pueblo peoples, but these hypotheses need to be tested against all available evidence of disease and death. A more comprehensive understanding of the health and disease dimensions was not established until more integrated approaches that combined bioarchaeological investigations, ethnohistoric data, and collaborations with others were utilized (Martin et al. 1991; Stodder 2012).

The bioarchaeologist is in a unique position to monitor the dynamics between changes in the ecological and cultural environment and changes in human response. The documentation of patterns of disease in ancient times can be channeled back into the discussion of contemporary health problems. In modern society, health of infants and children is directly linked to the function of mothers, families, and communities. Understanding physiological disruption and the impact of stress on any population feeds directly back into the understanding of cultural buffering and environmental constraints. It is extremely important to understand how disease and early death affect the functional and adaptive consequences for any community. For example, poor health can reduce work capacity of adults without necessarily causing death. Decreased reproductive capacity may occur if maternal morbidity and mortality is high in the youngest adult females. Individuals experiencing debilitating or chronic health problems may disrupt the patterning of social interactions and social unity and may strain the system of social support. Similar dynamics can be assumed for all human groups, and these interrelated issues must be explored for early communities because it supplies a much needed time depth to understanding the origin and history of diseases.

The result is that future researchers interested in addressing these hypotheses concerning health dynamics can measure the demographic and biological impact of stress by evaluating skeletal indicators of growth disruption, disease, and death (see Chap. 6). Disease and illness as quantified by pathological alterations on bone are assessed primarily through the systematic description of lesions. Stress is revealed by differential patterns of growth and development. Demographically, a great majority of the human remains recovered from many archaeological sites are under the age of 18, and growth and development of children using dental and skeletal data from critical stages could be compared to contemporary groups living in similarly marginal areas (see Chap. 7). Identifiable, age-specific disruptions in growth yield important information on patterns of childhood developmental disturbances and physiological disruption. The distribution and frequency of specific diseases (nutritional, infectious, degenerative) are also an essential part of the health profile. The patterning and frequencies of nutritional diseases such as iron deficiency anemia are documented for many precontact populations and have obvious implications for understanding adequacy of diet. Infectious diseases, also well documented for many skeletal series, provide an indicator of demographic patterning, population density, and degree of sedentism.

3.2.2 Documenting Patterns and Processes in Life and in Death

Human behavior is generally highly patterned in ways that facilitate adaptation to environmental challenges. Adaptation can be at many different levels, from genetic adaptations over many generations to physiological adaptation and acclimatization in the course of a lifetime. The biocultural model discussed in Chap. 1 has proven to be a productive tool for teasing out the challenges in any given environment that humans must adapt to. Bioarchaeological studies are important additions to understanding the limits of human adaptation. It provides information on individuals most at risk for succumbing to stressors such as undernutrition, extremely hot or cold weather, exposure to insects or predators, or interpersonal violence.

The linking of demographic (Chap. 7), biological (Chap. 6), and cultural processes within an ecological context is essential for dealing with the kinds of questions that are of crucial interest to scientists and others. Empirical data from bioarchaeology has a predictive quality that can contribute to understanding the age, sex, and physiological composition of individuals who are most vulnerable within a given population. For example, understanding the relationship between political centralization and illness, the impact of population reorganization or collapse on mortality, and the relationship between social stratification, differential access to resources, and health are all useful for understanding the groups who are most at risk for poor health and early death. These kinds of problems demand a multidimensional approach to collecting data because they cross over numerous disciplinary boundaries.

While patterns reveal particular kinds of information about subgroups at risk within populations, there is also much to be gained from the analysis of single individuals. Questions having to do with the origin and evolution of certain diseases often hinge on locating an individual within a context that can be dated who shows signs of diseases such as leprosy, cancer, tuberculosis, or other diseases that still plague humans today. Stodder and Palkovich (2012) have edited a collection of studies focused on single individuals. This collection demonstrates the ways that unusual or unique individuals within a burial population or a skeletal collection can reveal important data about the group. Each of these studies placed the individuals within their larger biocultural and environmental context so that the interpretations made about their lives prior to death were well integrated with other kinds of data.

Bioarchaeologists who have obtained access to human remains for study find that it is fairly straightforward to collect raw data and produce a description of basic findings. These kinds of studies are very common in the literature. But descriptive studies do not attempt to establish patterns, interpretations, or pathways to understanding human behavior on a more grand scale. In fact, it is much harder to produce a theory-driven and critically analytical study that involves an analysis at the population level in order to understand processes of adaptation for whole groups. However, this latter type of study is much more informative and valuable to the future of research than a simple descriptive analysis.

Schug (2011) presents an example of the utility of documenting patterns and processes of health and disease using a temporal and spatial analysis. Her analysis focuses on using data on health, growth, and development from children's remains from the Chalcolithic period in India (circa 1400–700 bc). Integrating the biological data, archaeological reconstruction, and paleoclimate data, she investigates the effects of climate change and monsoon variability on human adaptation. Having temporal depth (skeletal remains from various time periods), she could chart the significant changes that occurred from periods of agricultural abundance through collapse of many of these communities. Using a biocultural approach to human adaptation over time, Schug's bioarchaeological study provides new perspectives on the complexity of human-environment transformations. Additionally, her study has relevance and important implications for the modern world as it attempts to understand the impact of climate change, which is a topic under intense scrutiny today by scientists the world over.

3.3   Social Theory in Bioarchaeology

As discussed in Chap. 1, models are a simplified version of how something works. They are heuristic tools that can help bioarchaeologists keep track of variables that are likely to be important in the interpretation of human behaviors. It is simply impossible to view any phenomenon in its entirety. Models reflect limited but crucially important aspects of the phenomenon under study. Models reveal the reality of how something works from a particular perspective. Thus each researcher may focus on different aspects of something under study, but it is unusual for one researcher to be able to model something in its totality. The use of models provides a way to think through what the context of the human remains comprises. This will include everything of importance from the reconstruction of the environment to analysis of the mortuary component. Most bioarchaeologists heavily rely on using the biocultural perspective and the model because it has proven so useful (see Zuckerman and Armelagos 2011; DiGangi and Moore 2012).

However models are not theories. There may be some underlying theoretical ideas about why certain cultural behaviors are adaptive or why some ecological variables are more important than others, but the biocultural perspective or the biocultural model is simply a tool for identifying those variables that seem most correlated with, or predictive of, human adaptation. The most important thing about a biocultural perspective is that the research does not privilege culture over biology or biology over culture. The two domains are given equal weight in the analysis.

Theories tend to include general principles that explain a phenomenon and that hold up over time as new data comes to light. But the explanatory nature of theories sometimes comes under scrutiny and reevaluation. Robust theories are those that hold explanatory power across broad applications. Theories that are brought into an analysis are what help frame and shape a particular interpretation of the raw data. Parsons (1938) presented a thoughtful piece on the role of theory in social research (although it is now almost 80 years old). He reminds the reader of something that he had heard Max Weber state that went something like this: "To understand Caesar, you do not need to have been Caesar" (1938:13). For bioarchaeologists, to understand past behaviors, one need not have lived in the past. Parsons demonstrated the intimate relationship between theory and empirical research showing that each is enhanced by the other and each is essentially meaningless without the other.

The use of some kind of theory is often done without researchers even being aware and consciously acknowledging that they are drawing on social theory. For example, in order to make sense out of bioarchaeological data that demonstrate that infants die in higher frequencies in agricultural populations, or that males sustain head wounds during periods of warfare, or that captives are tortured and killed during ceremonies, it is relatively easy to draw on well-known ideas about why these phenomena are occurring in the times and places they are documented. As Schiffer declared in the opening introduction to an edited volume entitled *Social Theory in Archaeology*, "... (this book) is intended, obviously, to reach archaeologists interested in social theory; in practice this means all archaeologists, for ***everyone employs social theory, explicitly or implicitly*** [emphasis added], in explanations of past human phenomena" (2000:vii). In other words, human motivations, ideologies, and other facets of behavior are drawn on every time a researcher simply tries to make sense of the data that they have collected.

There are many different kinds of theories that can be utilized to aid in the interpretation of data derived from human remains. The graduate students at the University of Alabama have put together a wonderful menu of general theories used in anthropology, and this is a really good starting point [go to http://anthropology.ua.edu/cultures/cultures.php]. The thing that needs to be decided is which theory should be used. Theory aids in anchoring a study and narrowing its focus to something manageable, and theory also provides a framework within which to expand the interpretive power of the findings. Theories concerning human behavior and adaptation have been formulated within many different intellectual traditions including the natural and social sciences as well as the humanities.

As well, anthropologists have generated a great deal of theory about human behavior, and those theories often draw on a wide range of evolutionary and behavioral studies. There is really no one way to identify the theory one might use in a bioarchaeological study. The important thing is to find a theory that is exciting and that provides possibilities for thinking about human behavior in innovative or boundary-pushing ways. Any theory can be appropriate if it is relevant to what the researcher is interested in finding out. Provided here are brief overviews of theories that have been used productively by bioarchaeologists. Theories also tend to overlap with each other, and some scholars have productively blended theories together.

3.3.1 Theory: Evolution of Human Behavior

Evolutionary theory in anthropology has a long history. Originally the primary theoretical perspective of the earliest physical anthropologists, it lost favor in the early twentieth century as it was argued that evolution could not explain complex human behaviors or the existence of culture. The problem was that early evolutionary models being utilized by anthropologists were too simple. It was not until the emergence of neoevolutionism under White (1943) that more advanced models of understanding how evolutionary processes could contribute to understanding human behavior were developed. However, even these theories had severe limitations as they failed to understand the complexity of the interaction between cultural and evolutionary factors. While there are data to support the existence of some biological predispositions, humans are not genetically hardwired for any behavior. Genes plus environment produce particular phenotypes, and this is generally accepted by all physical anthropologists.

The role of evolutionary forces should be understood as adaptations that have more or less influence in different contexts. Given that humans have possessed complex brains capable of producing cultural innovations that have allowed us to modify the environment to meet our own needs for at least two millions years, it is more accurate to say that humans are the product of "biocultural" evolution.

Using biocultural evolution as a theoretical framework, a growing number of researchers in bioarchaeology have been trying to provide a better understanding of topics from nutrition and disease to violence. In a recent article Zuckerman et al. (2012) argued for the importance of utilizing evolutionary theory to understand human health and pathology. It is critical to understand that diseases and humans have evolved together both in response to one another and to the environmental context as well (Woolhouse et al. 2002). Malaria and the development of sickle-cell anemia is the most often cited example of how humans and diseases have coevolved and continue to evolve together (Etkins 2003).

Evolutionary perspectives are also very useful for understanding the origin and continuation of violence among humans (Martin et al. 2012). Taking an evolutionary approach, it is apparent that violence is not some unexplained aberrant act but is adaptive in many situations. For example, looking at coalitional violence, there is some evidence to suggest humans have been engaging in this behavior for a long part of our evolutionary history. Researchers suggest that the reason we have coalitional violence, especially among males, is that it is a product of cooperation. LeBlanc (1999) suggests that it is through cooperation that we are able to enhance our evolutionary success as we eliminate competition. This argument relies on the theory of "parochial altruism" that states that there is a selective advantage for males to cooperate with males in their own group and attack males outside of their group (Durrant 2011; Choi and Bowles 2007).

Durrant (2011: 429) states that coalitional or "collective" violence is at the heart of raiding, ambushes, feuding, and warfare. Looking in the bioarchaeological record, researchers have found that violence is not uncommon or atypical and that it has been around for a long time (Frayer 1997; Walker 2001; Zollikofer et al. 2002). Kelly (2005) argues that there are three stages or eras of violence in human history: (1) the development of coalitional violence; (2) a period of warlessness due to the development of a defensive advantage; and (3) warfare (2005:15298).

The importance of incorporating evolutionary theory into research is not to disprove or downplay the role of culture. Instead, it should be viewed as a means of allowing one more line of inquiry to enable researchers to ask more nuanced questions about human behavior that takes into consideration a longer time span. In looking for the origins and evolution of behaviors such as sexual division of labor or violence, or of particular diseases, or on the impact that major cultural innovations have on the body, evolutionary theory can provide a strong analytical framework for the interpretation of data from human remains.

3.3.2 Theory: Human Ecology

Originating as part of the revivalism of evolutionary theory in anthropology, ecologically based theories argued that to understand the evolution of human behavior, we had to put it back into context (i.e., the ecological setting). It differed from other evolutionary theories, however, in that it understood that humans were unique. Unlike the evolution and adaptation of other organisms, human beings are more difficult to understand because they not only adapted to their environment but also caused significant changes to it. The existence of culture gave humans the power to shape and alter their environments at many levels.

The first ecological theory, cultural ecology, was developed by Steward (1955) and is best articulated as ". . . the examination of the cultural adaptations formulated by human beings to meet the challenges posed by their environments" (McGee and Warms 1996:221–222). This does not mean, however, that culture for Steward was strictly determined by the environment. Instead, he suggests that the environment offered certain constraints that humans adapted their culture to (e.g., technoeconomic factors) in order to survive. Thus, Steward believed that cultural phenomena was the product of adaptations to the environment as limited by subsistence strategies and technology (Bennett 1976). Despite arguing for a dynamic relationship between humans and the environment that was reciprocal in nature, cultural ecology was often equated with environmental determinism or ecological reductionism.

Ecological reductionism is itself a type of adaptation model that developed out of cultural ecology. According to Gnecco (2003:13), this perspective in archaeology implies that there is a relationship between culture and the environment but the relationship is unequal. Environment places limits on what culture or adaptations can develop. The problem with an ecological reductionism model is that the dichotomy of environment and culture is treated like any other simple correlational relationship. The focus on two variables inevitably leaves out a multitude of other variables. The most important factor not considered in this relationship is the role that people play. Human beings are often not factored into the equation when research is designed around a theoretical perspective like ecological reductionism.

In response to ecological reductionism, a revised ecological theory was developed called human ecology. Similar to cultural ecology, human ecology acknowledges that the evolution and adaptation of humans is different from other organisms. Humans not only physically adapt to their environment but develop complex technoeconomic adaptations. Unlike cultural ecology, Bennett (1993) argues that people adapt not only to the natural environment but also to the social environment, adopting particular cultural patterns to cope with the environment. This added layer implies that ". . . while we may be a part of Nature, our fate is in *our* hands, not Nature's" (Bennett 1977:215). The social environment includes the institutional systems in place, as well as the needs of each individual within the society. The function of the institutional systems is that they offer a means of controlling resources, population size, and individual needs. As such, these institutional systems fluctuate among societies, which is why there are variations among cultures inhabiting similar environments.

A second major point of Bennett's concept of human ecology is that unlike Steward he perceives a system of input and feedback in the human-environment interaction, which implies that humans are not only affected by the environment but in turn cause significant changes to it. Essentially, the existence of culture gave humans the power to not only survive in but also redesign the natural world. This notion of humans having an impact on the environment is an important aspect of Bennett's work on human ecology and humanity's ongoing ecological transition (Bennett 1976, 1993) as ". . . adaptive behavior in one context may be maladaptive in another" (Bennett 1993:49). The theory helps to understand how populations in similar environments may use very different strategies in their adaptation.

The next stage of ecological theory to develop involved a greater emphasis on the dual influence of both ecology and evolution. Called human behavioral ecology (HBE) (Cronk 1991; Smith and Winterhalder 1992), this theory focuses on understanding the ". . . link between ecological factors and adaptive behavior" (Smith 2000:29). The difference between this approach and earlier ecological theories is that adaptation is no longer driven by culture but by individuals. Individuals within the society are viewed as active agents involved in how the overall group adapts to and modifies the environment. However it is crucial to realize that in this theory the decisions, actions, and behaviors that the individuals make in order to adapt are not necessarily conscious and they are often a product of cost-benefit analysis by individuals as a means of surviving in a particular environment (Sutton and Anderson 2010).

The importance of any one of these three ecological theories to bioarchaeology is that they facilitate asking questions about human variation at the population level. Van Gerven et al. (1973) provide a case study for illustrating this point. Earlier researchers assumed that changes in size and shape of the cranium of ancient Nubians over time (from the Mesolithic to the Christian periods) were due to an influx of people from other regions who came in and genetically replaced the existing population. By carefully documenting the environmental and cultural context, Van Gerven and colleagues demonstrated that the biological changes were driven purely by changes in subsistence from hunting and gathering in the earliest populations to an increasing reliance on agricultural products such as millet and sorghum. Other bioarchaeological research projects have also incorporated an ecological perspective in order to show that variation within a particular region was not the product of biology but a result of adaptation, both cultural and biological, to a particular environment (Harrod 2011; Buikstra et al. 1988; Larsen 2001; Ruff 1987). This is also evident in the volumes that have been produced that look at the effect of the adoption of agriculture on the health and nutrition of populations (Cohen and Armelagos 1984; Cohen and Crane-Kramer 2007; Steckel and Rose 2002; Pinhasi and Stock 2011).

3.3.3   Theory: Human Body and Identity

Cultural anthropologists have often thought of the body as both corporeal but also as the location for a variety of cultural processes to be expressed and symbolized.

Social identity often weaves in ideas about the social body. For example, Scheper-Hughes and Lock (1987:7–8) provide a theoretical framework for viewing the body from a number of different perspectives, each providing a unique insight. The three bodies that individual's possess include the individual or biological body, the cultural body, and the political body or body politic.

Cartesian (or Western) notions tend to separate the body from the mind and decouple the body from the contexts within which humans live. The importance of understanding this conception of the body is that in the study of skeletal remains there has been a long history of simply describing the characteristics of age and sex without consideration of how these physical characteristics would have affected a person's individual life experiences within a particular culture (e.g., being a male or female in ancient Rome). This is where the importance of understanding the social body is emphasized. "The human organism and its natural products of blood, milk, tears, semen, and excreta may be used as a cognitive map to represent other natural, supernatural, social, and even spatial relations… Insofar as the body is both physical and cultural artifact, it is not always possible to see where nature ends and culture begins in the symbolic equations" (Scheper-Hughes and Lock 1987:18–19). Finally, the body politic is the notion that the body is not only a representation of nature and culture but a medium through which relationships of power and control are played out (Scheper-Hughes and Lock 1987:24). As such, it can reveal information about how the person was doing and the quality of life they had within a particular community (e.g., a captive female compared to an elite male).

How might bioarchaeologists use these theories about identity and the body? The biological body or biological identity can be approximated in skeletonized human remains through standardized methods for assignment of age, sex, stature, health status, and other biological variables.

The cultural body or cultural identity may be assessed for skeletonized remains by examining the archaeological reconstruction of their lived experience. This would include site descriptions, including the location, layout, and size, as well as mortuary context on burial location (intramural or extramural) and the type of burial goods (presence of absence of grave goods). These contextual elements that can be related back to human remains may provide information on the position of individuals within the society.

Many studies based on human remains have been able to establish a range of notions about cultural identity. For example, if the burial is unaccompanied by grave goods and is in a more haphazard position, it can suggest a more expedient and less ceremonial burial. The amount and type of grave goods (e.g., artifacts, precious stones, ornaments, tools, etc.) can reveal the social persona of the person while alive. Prior work has illustrated that mortuary context can depict a great deal of information about social identity, gender and class differences of individuals (Neitzel 2000), their social ranking (Akins 1986), and the social organization of their society (Palkovich 1980). Thus, by looking at the archaeological and mortuary context, a person's cultural identity (or cultural body) can be approximated.

The body politic may be more challenging to assess through human remains and the archaeological context, but it is also the important one to attempt. The body politic as discussed by Scheper-Hughes and Locke attempts to understand the ways that politics and institutionalized forms of social control and domination (and violence) have an impact on bodies (Scheper-Hughes and Lock 1987:7). Human remains can reveal the effects of political oppression, structural violence, and other forms of coercion and domination in the form of healed and unhealed trauma, pathologies relating to beatings and torture, massacres and warfare, and diseases that may come with lack of proper diet and starvation (e.g., Watkins 2012; Erdal 2012; Osterholtz 2012; Shuler 2011). Isotopic data can also reveal the presence of locals and nonlocals within burial populations, and this data can be very useful in determining how migration, captivity, and other factors play into political actions (see chapters in Knudson and Stojanowski 2009).

Interrogating the "three bodies" through skeletal analyses is likely to produce a much more multidimensional interpretation of the data (see Chap. 6). Trauma, pathologies, and isotopic data can be theorized in a more complex and nuanced way using the notion of the three bodies than if a more standard descriptive analysis is employed. There are now many bioarchaeological studies that seek to theorize about social identity and identity politics and the ways that the body becomes both a real and symbolic vehicle (e.g., Knudson and Stojanowski 2009; Agarwal and Glencross 2011).

3.3.4 Theory: Sex and Gender

Nearly half a century ago researchers began to develop new theoretical approaches focused on understanding the roles that women play in society. Known as the feminist movement, this theoretical shift toward considering the various roles that women hold within society eventually took hold in archaeology in the 1980s and early 1990s (Conkey and Spector 1984; Gero and Conkey 1991; Dahlberg 1981). A reaction to over a century of focus being placed on the role of males epitomized in *Man the Hunter* (Lee and Devore 1968), this new movement emphasized the importance of understanding females in the past beyond just being the mothers and wives of men.

Gender theory is a paradigm that had a major impact on the field of anthropology as a whole. The consideration of sex and gender in bioarchaeology more specifically changed both the questions that were being asked and the way that the bodies were being analyzed. While differences between the traditional male-female binary have been noted by early bioarchaeologists looking at population-level changes, the development of gender theory led to researchers critically analyzing why those differences existed. It is a result of this critical analysis that some very important ideas about health and longevity discrepancies between those individuals assigned male versus those assigned female first came to light. In the edited volume *Exploring Differences: Sex and Gender in Paleopathological Perspective* (Grauer and Stuart-Macadam 1998), there are numerous chapters that take skeletal pathologies such as iron deficiency, osteoporosis, osteoarthritis, trauma, and infections and present frequencies for age-matched males and females. What this research and the studies that followed it argued for was that instead of just noting when differences are statistically significant, researchers *must* incorporate as much contextual information as possible to try to understand why these differences may be occurring.

The push for a better understanding of context resulted in the development of an understanding that different patterns of sexual divisions of labor and cultural ideology had a tremendous impact on the levels of differential access to resources, exposure to pathogens, and reproductive and occupational stressors that women in the society faced. Prior to this understanding, the disproportionate mortality and morbidity that existed between males and females was often thought of a consequence of biology. Cultural ideology, sociopolitical organization, and patterns of subsistence and labor production were all brought in as possibilities for explanation. Thus theorizing about male–female differences that go beyond biology had provided new ways to interpret differences in pathology and longevity between males and females as being part of the larger cultural sphere of influence.

Sex differences in pathology and age at death have been standardized in the bioarchaeological literature, and there are very few population-level analyses that do not present the data on age and pathologies by sex. But this kind of analysis that focuses on differences between males and females in indicators of stress is not de facto part of what it means to use sex and gender theory in a study. Scholarship in gender theory is not only concerned with asymmetries in various biological indicators. Theorizing the social roles of males and females also includes an examination of difference within the categories of female and male. Gender theory emphasizes interrogating a number of categories where power may be differentially held by various women in the society depending on an array of factors, such as status, ethnicity, and kinship (Geller and Stockett 2006). Gender theory provides a means for looking closely at subgroups within any population and locating which individuals have power over others and better access to resources. This necessitates a study that integrates the human remains within much broader cultural and ideological contexts and using new methodologies that reveal the complexity of social relations within a society, such as the three bodies (Scheper-Hughes and Lock 1987).

In terms of methodology, for over a century researchers had without question assigned the labels of sex to skeletal remains. What some bioarchaeologists are arguing is that sex is not a binary category with distinct divisions of male and female (Johnson and Repta 2012). Instead, sex should be viewed on a continuum or spectrum. The medical community has for decades recognized that there are individuals who are born neither male nor female but as intersexed or transgendered (Fausto-Sterling 1993). Yet the majority of studies continue to place human remains into one of the two categories. This is despite the fact that the technique used to estimate the sex of a set of skeletal remains is itself not binary. Following the standards of osteological analysis (Buikstra and Douglas 1994), researchers must decide if a particular feature on the bones is female (1), probable female (2), indeterminate (3), probable male (4), or male (5). Despite this, most individuals are reported as either male or female.

Even more problematic than this lack of flexibility in assigning sex is that sex is then translated into gender. Sex and gender as typically used by social scientists are distinctively different terms, but they are often used interchangeably in some bioarchaeological studies. Walker and Cook (1998) were the first to make the plea that bioarchaeologists not conflate these two terms. "Sex refers to the anatomical or chromosomal categories of male and female. Gender refers to socially constructed roles that are related to sex distinctions" (1998:255). This is an important issue to both theorize and seek to understand because cultures in the past had more gender categories than male and female, and these were not tied to biological sex. Throughout North America, there have been ethnographic reports and archaeological evidence that supports the presence of people considered third gender (i.e., *berdache* or two spirit) beyond male and female (Hollimon 2011). This concept of third gender is important because it allows biological males an opportunity to identify as gender females, which would have an impact on how the remains are analyzed and interpreted in bioarchaeological studies. Both sex and gender have an effect on a person's access to resources and daily activities. Hollimon (2011) and Sofaer (2006) both suggest that bioarchaeologists may be uniquely suited to expanding on nonbinary gender and its role in human social systems.

More recently Agarwal and Glencross (2011) edited a volume of studies that encouraged authors to draw heavily on feminist theory, gender theory, and social theory to aid in the interpretation of a wide range of studies based on human remains. Hollimon (2011) suggests that bioarchaeologists may be uniquely suited to locating third-sex individuals and to theorizing about nonbinary genders. Bioarchaeologists are playing a crucial role in this movement, as researchers like Hollimon (2011) and Geller (2005, 2008) present arguments that researchers must question many of the long-standing assumptions about the nature of human remains in terms of sex and gender.

Bioarchaeologists utilize gender theory in a multitude of ways to answer an array of different research questions. For example, a researcher analyzing human remains can use gender theory to ask about differential labor between and within the categories of male and female. However, they could also ask more nuanced questions such as how often women compete for status (analyzing nonlethal trauma, cranial trauma) and if there are size differences on some women that coincide with age-matched males suggesting they were engaged in a similar level of mobility (i.e., women as traders or hunters). Using gender theory offers a way to break down the normative narratives about males and females and to frame questions that are more dimensional and dynamic in terms of the complexities of what sex and gender mean within human social systems.

3.3.5   Theory: Human Violence

All violence is predicated on systems of societal norms. Cultural norms play a large part in regulating how violence is used and maintained for social control through the creation of fear and chaos. The study of violence requires researchers to understand the transformative powers of its use in social relations and cultural practices.

To accomplish this, researchers must understand that they are not simply studying a punctuated event but rather a transformative process within a historical trajectory. Understanding violence in the archaeological past helps to contextualize violence in the present.

Perspectives on violence have been too narrowly conceived, and it is time for theoretical paradigms to be broadened. This is why it is essential that anthropologists try to understand and explain the cultural mediation of real world conditions that foster the use of violence. Arguably the most influential current definition of violence is from Riches's *The Anthropology of Violence*, which classifies violence as ". . . an act of physical hurt deemed legitimate by the performer and illegitimate by (some) witness" (Riches 1986:8). The study of violence has often been conducted with little or no consideration for the specific and often unique cultural meanings associated with it. Warfare and violence are not merely reactions to a set of external variables but rather are encoded with intricate cultural meaning. To ignore these cultural expressions or, worse yet, suggest they do not exist, minimizes our understanding of violence as a complex expression of cultural performance. Violence should never be reduced to its physicality when trying to understand its use (Scheper-Hughes and Bourgois 2004). This is because violent acts often exemplify intricate social and cultural dimensions and are frequently themselves defined by these same social contexts.

The symbolic aspects of violence have the potential to create order and disorder depending on the specific social context within which the violence is expressed (Galtung 1990; Sluka 1992). This is the apparent paradox of violence studies. Most cultures feel that their safety lies in their ability to control violence with violence. While people fear and abhor violent acts that they see as senseless, they are more than willing to condone the "legitimate" use of violence to promote social control and economic stability (Turpin and Kurtz 1997). Sluka (1992:28) refers to this apparent paradox as the dual nature of conflict. Violence and conflict often have the ability to unite, create stability, and be progressive while at the same time generating the antithesis of these positive forces. This is why Whitehead (2005:23) argues that acts of violence and warfare should be viewed as cultural performances that may be unfamiliar to "Western cultural experience." The consequence of not doing so is that violence may become seen as a "natural" component of human behavior rather than contingent upon historical consequences.

Most of the violence practiced in human societies is not considered deviant behavior. In fact, it is often seen as honorable when committed in the service of conventional social, economic, and political norms. These social and cultural contexts are what give violence its power and meaning. To see violence as only an aberrant behavior committed solely by deviants blinds us to the role that violence has and continues to play in the foundation of many human societies.

Anthropologists have long used the human body in its corporeal state as a lens through which to examine cultural processes. How dead bodies are discussed, hidden, and displayed can be used as a point of departure for examining the forms of violence that produced the deaths. We can observe the way that the deaths are perceived and further used as people try to make sense of violent acts. Dead bodies are far more than just decaying matter. To quote Douglas (1966/1992:115), "The body is a model which can stand for any kind of bound system." To do even minimal justice to the symbolic complexity of the human body requires consideration of its political symbolism, cultural death rituals, analysis of the type of corpse manipulation within the wider regional cultural dynamics, and how the manipulation of the corpse will impact local histories and create spatial memory (Verdery 1999:3). The body is often seen as a "natural symbol" through which the social world can be ordered (Douglas 1973). When the body is viewed in this way, it holds a particular view of society and the cosmos. Meaning flows from the body to the cosmos or through the body and society. When this flow is interrupted, the consequence can be profoundly damaging for the grieving family and their community (Martin and Pérez 2001). This is because the mourners' transition from grief to the affirmation of their own lives, their own certain deaths, and journey to the afterlife is echoed in the fate of that corpse.

The meaning associated with the corpse has a great deal to do with the condition and location in which it is found. If the goal is creating what Taussig (1984:467–497) has appropriately termed a "culture of terror," then mutilation, destruction, and/or disappearance of this powerful symbol are effective mechanisms in achieving the desired effect. Examples of this can be observed throughout human history. During much of American history, Africans and African-Americans were routinely lynched to instill fear and maintain the control of the population. Regardless of the form of the lynching (rope or fire), dismemberment and distribution of body parts were regularly carried out. These body parts, which often included teeth, ears, toes, fingers, nails, kneecaps, bits of charred skin, and bone, were turned into watch fobs or displayed for public viewing (Litwack 2000).

3.3.5.1   Massive Trauma

This type of intercommunity or intergroup violence can lead to a concept known as "massive trauma." Krystal (1968) first coined the term "massive trauma" to mean extreme circumstances of traumatization such as natural disasters, technological catastrophes, and social, political, cultural, gender, ethnic, or religious persecution that leaves any society, ethnic group, social category, or class fearful. This, in turn, can destroy culturally constructed webs of trust, based on social norms, worldviews, and moral convictions that may then produce overwhelming feelings of terror and anxiety. This is because social violence does not simply affect the psyche of the victims but the coexisting psychic spaces of the person's outer world. Extreme violence "unmakes" the "normal" everyday world, changing it from a nurturing and loving space to one filled with the reminders of horrors and atrocities.

People who survive massive traumas are left with the daunting task of having to cope with unbelievable and previously unimaginable horrors that are incompatible with their previous lives. This concept is known as the "uncanny," a term first coined by Freud. Gampel's (1996) definition of the term includes frightening experiences that cannot be expressed in words. Many of the forms of social violence discussed above would be produced by unthinkable behavior and thus be unrepresentable or unspeakable. Furthermore, these atrocities are encoded in the everyday objects and environments where they took place. Often, ordinary items become used as instruments of torture and murder. Public space becomes a battleground where maimed and dead are gathered. Religious centers, public parks, community centers, schoolyards, and markets all become places, the original meanings of which are lost to the imprint of war and violence. Because this type of extreme social violence transforms reality into ambiguity, the victim never feels safe. Thus, the uncanny is an emotional/psychological state that falls between anxiety and terror (Gampel 2000:51).

What makes this form of violence so potent is that it does not stay confined to its victim. The pains and fears of massive traumas transcend generations as they are conveyed to family members and loved ones (Suárez-Orozco and Robben 2000:44). The transmission of this fear often creates a sense of hatred and violence as a way to cope with traumatic wounds (Apfel and Simon 2000:102). Children who have been negatively affected by sociocultural violence live with the memory of physical and emotional suffering. This often manifests itself in a lack of trust and a fear of the unknown (Quesada 1998). While it is true that, given enough time, humans can and will adapt to terror and fear, the low-level anxiety becomes a constant companion in their day-to-day affairs (Green 1999).

The concept of "massive trauma" supports Ember and Ember's (1997) idea that it is fear that motivates people to go to war and maintain the cycle of violence. When a cultural group is socialized for aggression and violence, it takes minimal external stimuli to trigger a violent response (Apfel and Simon 2000; Ember and Ember 1997; Ferguson 1997; Sluka 1992; Knauft 1991; Whitehead 2004b). Many non-Western, non-industrial societies that have had to deal with massive trauma do so through healing rituals, religious ceremonies, communal dances, revitalization movements, restored symbolic places, and community centers (deVries 1996). However, if the massive trauma is endemic, these mechanisms do not always restore the community to a state of equilibrium. The concept of agency (Bourdieu 1977; Giddens 1979) and culture (Sahlins 1981:7) stresses that the actions of individuals and groups create the system in which they operate. The fundamental question facing archaeology regarding the production of violence should be "How does the reproduction of a [social] structure become its transformation?" (Sahlins 1981:8).

Nordstrom (2009:63–64) wrote that "in the case of war and institutionalized inequality, fault lines do not reside within landmasses, but in certain political, economic, and ethical relations that span the world's countries. Fault lines are flows— often unrecorded—of goods, services, money, and people that precipitate unstable inequalities, uneven access to power, and unevenly distributed resources. They represent fissures in humanity."

The bioarchaeology of violence utilizes a range of theories from theoreticians (many discussed above) as a way to interpret the signs of trauma on the bones. However, data from the human remains pose one way that bioarchaeologists can begin to unravel the ways that violence infiltrates people's lives and the ways it is patterned across time and space. Without using theory about how violence operates and is part of the complex web of social order, trauma cannot be really understood. Bioarchaeologists are embracing a more nuanced theoretical approach that permits an understanding of violence in its broadest context (see, e.g., the chapters in Martin et al. 2012).

Pérez (2012) are bioarchaeologists who have published on violence utilizing the theoretical distinctions made between physical violence and structural (or cultural) violence. Physical violence is interpersonal violence among individuals. Structural violence involves all of the cultural, political, and social institutions that legitimize and sanction certain kinds of violence. Culturally sanctioned violence includes such activities as raiding or warfare. Social structures that create inequality may render some portions of the population malnourished and dying of diseases. These are also considered part of structural violence.

3.3.6 Theory: Inequality

Theories that account for inequality in human social systems have come out of many disciplinary traditions. Some have been variously referred to as Marxist theory, economic theory, or political-economic theory. Theories about inequality, social hierarchies, and class stratification examine people's lived experience, social relations, and historical contingencies as a way of clarifying the underlying political and economic factors that produce and maintain inequality in human groups

There are theories about social inequality that specifically relate to class and hierarchy within social institutions and theories that relate to sexual division or gender-based divisions that create inequality within social systems. There are articles, websites, and texts dedicated to the origin and maintenance of social inequality. The important feature of these kinds of processes is that they attempt to expose the political-economic underpinnings of inequality and how societies maintain social stratification.

Within the bioarchaeology, the study of inequality often starts with raw data collected from skeletal remains that show patterning and asymmetries among subgroups within a population. Using theories about the origin, production, and maintenance of inequality provides a way to interpret the raw data within a set of principles that have been well studied for many different populations. Measuring the impact of inequality on the lives of individuals requires unraveling complex political and economic relationships that are interwoven into the social and economic relations within the culture. Careful analyses of skeletal indicators of growth, development, activity, and pathology can highlight the degree to which certain groups across the life span may have been subordinated and targeted. There is a growing body of data from human remains documenting the ways that inequality creates marginalized peoples. Previous studies have used patterns of skeletal pathology in archaeological contexts (e.g., burial position or presence and type of grave goods) to examine how social stratification could have affected past human health and well-being (Martin et al. 2001).

Recently, there has been a shift in the way entheses or musculoskeletal stress markers (MSMs) are utilized (see Chap. 6). Development of sites on the bone where muscles, ligaments, and tendons attach (Benjamin et al. 2006) can reveal social inequality (Robb et al. 2001; Martin et al. 2010; Perry 2008). The value of looking at entheses is that if analyzed in context of the culture as a whole, these changes to the skeleton can provide clues to the nature of the political and economic structure within the society (Robb 1998; Stefanović and Porčić 2011).

Evidence of osteological trauma can provide insights into the development of sociopolitical power relations. High-status individuals may be using violence and domination to keep lower-status individuals under their charge through aggressive means (Whitehead 2004a). Other scenarios might include nonlethal interpersonal violence being used as a means of competing for status (Tung 2007; Powell 1991; Harrod 2012; Walker 1989). Questions that arise surrounding this association include the following: Can bioarchaeology identify patterns of inequality and violence that result from the development of social stratification? What can be understood about the co-occurrence of increasing poor health and increasing social hierarchy?

Raiding and warfare in the past resulted not only in death for some but in captive-taking and enslavement. Ethnographic, historic, and ethnohistoric data reveal that warfare and captive-taking was and still is a form of exploiting a subgroup of individuals who do not have equal access to resources and who are kept in subordinate positions (Harrod et al. 2012). Captives were mostly women and children who were incorporated into their captor's societies in a variety of ways—as wives, drudge wives, concubines, and slaves (Cameron 2011). These people often made up a substantial proportion of the population of non-state and state-level societies, but these marginalized people are often difficult to see in the archaeological record.

Realizing the need for more nuanced approaches to interpreting inequality in the past requires an understanding that as inequality becomes culturally embedded, the roles of those involved in maintenance of a subclass of people are intimately linked with power. Bioarchaeologists have only begun the task of identifying the material remains of marginalized people in the past. Goodman (1998) provides an excellent model for integrating theories about inequality for populations in the past. He was able to link sociopolitical systems and health for several different regions of the world as examples of the utility of linking data from the skeletal remains with the archaeological reconstruction of systems of power. However the interpretive power of these integrated data sets was magnified by the utilization of theories about differential access to resources. He draws on a theoretical perspective developed by Roseberry (1998) who provides ways of theorizing power within human groups. Roseberry's theory about power is that it is located in "fields" and these fields of power can shift and change within lifetimes and over generations. Goodman also draws on a theory from Krieger (1994) about the ways that power is part of an extended web, and to understand it one needs to look at both proximate forms of power and the ways that power extends out into other domains within societies. One can see how utilizing theory to extend the biological data from human remains is crucial for interpreting it within some broader set of ideas about human behavior, in this case around domination and the creation of subordinate groups.

In general, the cultural construction of inequality and stratification of human societies has been coupled with emerging social theory about the development and maintenance of social control (in other words, power) related to subsistence intensification, specialization, and the rise of social complexity. Armelagos and Brown (2002) propose that social stratification evolved where elite members of society attempted to maintain their health at the expense of subordinate social groups. To refine our understanding of the role of hierarchy, more nuanced studies of osteological data attempt to identify differential patterns of trauma and how these patterns may relate to the sociopolitical realm.

Bioarchaeological analyses combined with theories about the ways that social dynamics and power relations impact the material and biological lives of people is a productive way to interpret asymmetries found in frequencies and patterning of disease at the population level. It can further lead to a broader contextualization of populations by examining demographic shifts (i.e., population growth and migration), changing subsistence strategies, and distribution of resources. By creating empirical links between inequality, pathology, and culture, bioarchaeology provides insights into the human propensity for constructing and legitimizing force and control by some over others.

3.3.7   Theory: Colonization and Imperialism

Related to theories about violence and theories about inequality, there are many theories in the literature about the colonial process. Bioarchaeologists have been publishing a great deal on the impact of the colonial encounter between Europeans and Native Americans in North America (see, e.g., Larsen 2001). The focus of literature on the epidemics and genocides that resulted from colonial encounters places great specificity on recording the name, timing, and place of the events. A Deleuzian reinterpretation (Finzsch 2008; Deleuze 1990) argues that an "event" such as an epidemic or genocide is really only a moment in time that must be understood in a much larger context. Using the model of a rhizome (Deleuze and Guattari 1987) or braided river (Moore 1994) helps put these events into perspective as simply temporary surface effects. Each effect or event is caused by a multitude of underlying, hard-to-define processes which are intertwined and complex, defying being named or being fixed in time and space. In reference to the atrocities that took place as part of colonialism, what underlies these events is a progression from dehumanization and small-scale violence against the other to the removal of land and resources and eventually domination or extermination of an entire people and their culture. Genocide and epidemics are actually forces both visible and invisible that become normalized and part of daily encounters. In violence theory, this is referred to as structural violence (discussed above) that is the invisible ways that violence becomes part of the daily reality, even when it affects large numbers of people.

Bioarchaeology is in a unique position to be able to explore the underlying structural violence behind epidemics and genocide by digging below the surface of these seemingly discrete events and examining the connected causal factors that both precede and continue long after the event. A bioarchaeological approach can contribute to a better understanding of the colonial encounter precisely because it provides a richly nuanced spatial and temporal approach to reconstructing human behavior from multiple perspectives.

3.4 Research Design

Bioarchaeological studies that seek to connect with bigger ideas and theories about the nature of human behaviors and adaptations generally will include the following:

- *Studies need a thesis.* A thesis needs to be developed that clarifies what is being investigated and why. It can be framed as a series of questions that can be answered with the data that will be collected. It can also be seen as a point of view that helps organize how the study will be done.
- *Methods used need to be tailored to the thesis.* There are many different methods for analysis of human remains (see Chaps. 6, 7, and 8). No one study can use every method and collect every piece of data. For example, methods might be utilized that focus on age, sex, and activity patterns by looking at the gross morphology of the bones. Other levels of analysis might be to utilize radiographic or histological techniques to obtain data about growth and development across age categories. Some studies may only need age at death and sex to reconstruct demographic profiles.
- *Studies need to problematize issues.* This aspect of designing a study differentiates simple descriptive studies from broader more engaged and integrative studies. As the thesis is developed, tensions and conflicts in the literature may become apparent. Gaps in knowledge may be uncovered. It is important for the bioarchaeologists to think about what might be at stake in doing the project. Articulating clearly the thesis of the project and the problem areas that the research is likely to need to attend to helps to narrow the scope of the project.
- *Studies need to theorize about human behavior and human adaptation.* Theories are ideas about how things work and can come from many different perspectives. They are the glue between raw data and interpretations.
- *Studies need to be engaging and compelling.* Without a thesis, appropriate use of methods, and using theory to frame the interpretations, studies will not be widely read or used because they will be narrow and limited in scope.

The general ways that bioarchaeological studies accomplish the above is to use a variety of approaches to frame the project. There may be a focus on comparing and contrasting data derived from the human skeleton by regions. A single disease process or a pattern in mortality by age or sex present in a population may be analyzed within a matrix of biocultural factors (see Chap. 1 for examples of this approach).

Table 3.1 A model for how to connect research focus, skeletal data, methods, and theory

Focus of analysis	Data	Skeletal correlates	Theory	Methodology
Biological identity	Age	Assessment of growth/development related changes in the pelvis, cranium and long bones	Crowder and Austin (2005), Potter (2010)	Bass (2005), Buikstra and Ubelaker (1994), Scheuer and Black (2000)
	Sex	Assessment of sexually dimorphic features in the pelvis, cranium and long bones	Geller (2005, 2008), Hollimon (2011)	Bass (2005), Buikstra and Ubelaker (1994), Scheuer and Black (2000)
Diet and nutrition	Stature	Reconstruction of achieved height from adult long bones; Reconstruction of growth and development from subadult long bones evaluated with dental age	Auerbach (2011), Mummert et al. (2011), Steckel (1995)	Genoves (1967), Trotter (1970), Trotter and Gleser (1952), Ousley and Jantz (2005)
	Nutritional deficiencies	Porotic hyperostosis/Cribra orbitalia assessed from the cranium	Vercellotti et al. (2010), Wapler et al. (2004), Walker et al. (2009)	Ortner (2003), Aufderheide and Rodríguez-Martin (2003)
	Dental health	Prevalence of caries, dental wear and enamel hypoplasias	Goodman and Rose (1991), Klaus and Tam (2010), Larsen et al. (1991)	Hillson (2008), Irish and Nelson (2008), Scott and Tuner (1988)
Health	Infectious disease	Periosteal reactions on the long bone and osteomyletis	Farmer (1999), Barrett et al. (1998)	Weston (2008), Ortner (2003), Aufderheide and Rodríguez-Martin (2003)
	Dental disease	Periodontal disease and antemortem tooth loss	DeWitte and Bekvalac (2011), Clarke et al. (1986)	Hillson (2008), Irish and Nelson (2008), Scott and Tuner (1988)
Activity	Robusticity	Measurement of the size and shape of the long bones assessed by age and sex	Marchi and Shaw (2011), Mummert et al. (2011), Ruff (2008)	Bass (2005), Cole (1994), Stock and Shaw (2007)
	Entheses (MSMs)	Analysis of bone deposition at the muscle insertion sites of the bone assessed by age and sex	Robb et al. (2001), Stefanović and Porčić (2013)	Capasso et al. (1999), Mariotti et al. (2007), Robb (1998)
Traumatic injury	Violence-related	Lethal and non-lethal fractures of the cranium, ribs and long bones	Martin et al. (2010), Walker (2001), Durrant (2011)	Brink (2009), Kremer et al. (2008), Walker (1989)
	Accidental and occupational	Lethal and non-lethal fractures of the postcranial skeleton	Harrod et al. (2012), Walker-Bone and Palmer (2002)	Finegan (2008), Guyomarc'h et al. (2010), Wakely (1997)

Evidence for disease or violence may be examined within a broader context of life history and cultural ideology. Analyses may provide cross-cultural comparisons and/or changes over time. There may be a focus on adaptation to extreme environmental conditions. Studies may examine long chronologies of disease by geographic setting or cultural location to understand the processes underlying the patterns (Table 3.1).

Bioarchaeologists carrying out studies must adhere to the same principles as other anthropologists which includes operationalizing the idea of cultural relativity. This is the idea that all cultures are equally worthy of respect and that in studying another culture, anthropologists should suspend judgment, empathize, and try and understand the way that particular behaviors and motivations are ideologically driven. Also, anthropology research generally seeks to avoid stereotyping other cultures by focusing on nuance, difference, and variability within cultures, not just between cultures. In organizing theory, method, and data at the beginning of a project, it is helpful to think about the data that will be most important to the questions being asked.

3.5 Summary

Bioarchaeology projects are best served when there is a proactive approach that coordinates the research question with the research design. There needs to be planning for how integration across data sets will be implemented and for the ways that the research question may engage with broader issues of importance to descendant populations and other stakeholders. Finally, researchers need to consider the ethical implications of their particular study. Bioarchaeology and forensic anthropology and forensic archaeology are complementary in terms of sharing methodologies for identification of unknown skeletonized human remains, but the legal aspects of the forensics work make it a distinctively different enterprise.

It is incumbent upon bioarchaeologists to use information that has been collected in a way that does not trivialize or diminish the lives of the living descendants. For example, collaboration and consultation with representatives from Pueblo groups regarding data derived from the excavation of ancestral Pueblo sites is one way to adhere to these ethical considerations. Tribal collaboration in research and in review of scholarly work *prior to publication* has not been the norm, but a (very small) trend in doing so among bioarchaeologists has resulted in a much richer interpretation because of the inclusive nature of the enterprise (Spurr 1993; Martin 1998; Ogilvie and Hilton 2000; Ferguson et al. 2001; Kuckelman et al. 2002). Without collaboration and the nuanced layering of additional knowledge from those most closely related to the people studied, the interpretations scientists formulate may be grounded in expert technique and state-of-the-art methodology but utterly wrong or incomplete. Bioarchaeologists must also accept that Native Americans (and indigenous people the world over) may sometimes refuse to participate or may withhold information deemed esoteric and inappropriate for publication. Bioarchaeology within these kinds of frameworks is transforming the physical anthropology of the past into a more dialogical process, similar to Wood and Powell's (1993) proposal for archaeology. This will necessarily include decision-making about the kinds of research that is undertaken.

References

Agarwal, S. C., & Glencross, B. A. (2011). *Social bioarchaeology*. Malden: Wiley-Blackwell.

Akins, N. J. (1986). *A biocultural approach to human burials from Chaco Canyon, New Mexico, Reports of the Chaco Center, No. 9*. Santa Fe: National Park Service.

Apfel, R. J., & Simon, B. (2000). Mitigating discontents with children in war: An ongoing psychoanalytical analysis. In A. C. G. M. Robben & M. M. Suárez-Orozco (Eds.), *Cultures under Siege: Collective violence and trauma* (pp. 102–131). Cambridge: Cambridge University Press.

Armelagos, G. J. (2003). Bioarchaeology as anthropology. In S. D. Gillespie & D. L. Nichols (Eds.), *Archaeology is anthropology*. Washington, DC: Archaeological Papers of the American Anthropological Association, No. 13.

Armelagos, G. J., & Brown, P. J. (2002). The body as evidence; The body of evidence. In R. H. Steckel & J. C. Rose (Eds.), *The backbone of history: Health and nutrition in the western hemisphere* (pp. 593–602). Cambridge: Cambridge University Press.

Auerbach, B. M. (2011). Reaching Great Heights: Changes in indigenous stature, body size and body shape with agricultural intensification in North America. In R. Pinhasi & J. T. Stock (Eds.), *Human Bioarchaeology of the Transition to Agriculture* (pp. 203–233). Chichester: Wiley-Blackwell.

Aufderheide, A. C., & Rodríguez-Martin, C. (2003). *The Cambridge Encyclopedia of Human Paleopathology, Reprint edition*. Cambridge: Cambridge University Press.

Barrett, R., Kuzawa, C. W., McDade, T. W., & Armelagos, G. J. (1998). Emerging and Re-emerging Infectious Diseases: The third epidemiologic transition. *Annual Review of Anthropology, 27*, 247–271.

Bass, W. M. (2005). *Human Osteology: A laboratory and field manual, fifth edition*. Columbia: Missouri Archaeological Society.

Benjamin, M., Toumi, H., Ralphs, J. R., Bydder, G., Best, T. M., & Milz, S. (2006). Where tendons and ligaments meet bone: Attachment sites ('entheses') in relation to exercise and/or mechanical load. *Journal of Anatomy, 208*, 471–490.

Bennett, J. W. (1976). *The ecological transition: Cultural anthropology and human adaptation*. New York: Pergamon.

Bennett, J. W. (1977). Bennett on Johnson's and Hardesty's reviews of *The Ecological Transition*. *Reviews in Anthropology, Commentary, 4*(2), 211–219.

Bennett, J. W. (1993). *Human ecology as human behavior: Essays in environmental and development anthropology*. New Brunswick: Transaction.

Bourdieu, P. (1977). *Outline of a theory of practice* (R. Nice, Trans.). Cambridge: Cambridge University Press.

Brink, O. (2009). When Violence Strikes the Head, Neck, and Face. *The Journal of TRAUMA: Injury, Infection, and Critical Care, 67*(1), 147–151.

Buikstra, J. E., Autry, W., Breitburg, E., Eisenberg, L. E., & van der Merwe, N. (1988). Diet and health in the Nashville Basin: Human adaptation and maize agriculture in middle Tennessee. In B. V Kennedy & G. M. LeMoine (Eds.), *Diet and subsistence: Current archaeological perspectives* (pp. 243–259). *Proceedings of the 19th Annual Chacmool Conference*. Calgary: Archaeological Association of the University of Calgary.

Buikstra, J. E., & Douglas, H. U. (1994). *Standards for data collection from human skeletal remains*. Fayetteville: Arkansas Archaeological Survey, Research Series, No. 44. A copy of standards is required in order to fill out these forms accurately. It may be obtained from the Arkansas Archeological Survey, 2475 N. Hatch Ave., Fayetteville, AR 72704, http://www. uark.edu/campus-resources/archinfo/.

Buikstra, J. E., King, J., & Nystrom, K. C. (2003). Forensic anthropology and bioarchaeology in the American anthropologist: Rare but exquisite gems. *American Anthropologist, 105*(1), 38–52.

Buikstra, J. E., & Ubelaker, D. H. (Eds.). (1994). *Standards for Data Collection from Human Skeletal Remains.* Fayetteville: Arkansas Archaeological Survey, Research Series, No. 44. A copy of Standards is required in order to fill out these forms accurately. It may be obtained from the Arkansas Archeological Survey, 2475 N. Hatch Ave., Fayetteville, AR 72704, http://www.uark. edu/campus-resources/archinfo/.

Cameron, C. M. (2011). Captives and culture change. *Current Anthropology, 52*(2), 169–209.

Capasso, L., Kennedy, K. A. R., & Wilczak, C. A. (1999). *Atlas of Occupational Markers on Human Remains.* Teramo: Edigrafital S.P.A.

Choi, J. K., & Bowles, S. (2007). The coevolution of parochial altruism and war. *Science, 318,* 636–640.

Clarke, N. G., Carey, S. E., Srikandi, W., Hirsch, R. S., & Leppard, P. I. (1986). Periodontal Disease in Ancient Populations. *American Journal of Physical Anthropology, 71*(2), 173–183.

Cohen, M. N., & Armelagos, G. J. (1984). *Paleopathology at the origins of agriculture.* Orlando: Academic.

Cohen, M. N., & Crane-Kramer, G. M. M. (2007). Ancient health: Skeletal indicators of agricultural and economic intensification. In C. S. Larsen (Ed.), *Bioarchaeological interpretations of the human past: Local, regional, and global perspectives.* Gainesville: University Press of Florida.

Cole, T. M., III. (1994). Size and Shape of the Femur and Tibia in Northern Plains. In R. L. Jantz & D. W. Owsley (Eds.), *Skeletal Biology in the Great Plains: Migration, warfare, health, and subsistence* (pp. 219–233). Washington, D.C.: Smithsonian Institution Press.

Colton, H. S. (1936). The rise and fall of the prehistoric population of northern Arizona. *Science, 84*(2181), 337–343.

Conkey, M. W., & Spector, J. (1984). Archaeology and the study of gender. In M. B. Schiffer (Ed.), *Advances in archaeological method and theory* (9th ed., pp. 1–38). New York: Academic.

Cronk, L. (1991). Human behavioral ecology. *Annual Review of Anthropology, 20,* 25–53.

Crowder, C., & Austin, D. (2005). Age Ranges of Epiphyseal Fusion in the Distal Tibia and Fibula of Contemporary Males and Females. *Journal of Forensic Sciences, 50*(5), 1001–1007.

Crumley, C. L. (2006). Archaeology in the new world order: What we can offer the planet. In E. C. Robertson, J. D. Seibert, D. C. Fernandez, & M. U. Zender (Eds.), *Space and spatial analysis in archaeology* (pp. 383–396). Calgary: University of Calgary Press.

Dahlberg, F. (1981). *Woman the gatherer.* New Haven: Yale University.

Deleuze, G. (1990). *The logic of sense.* New York: Columbia University Press.

Deleuze, G., & Guattari, F. (1987). *A thousand plateaus: Capitalism and schizophrenia* (B. Massumi, Trans.) Minneapolis: University of Minnesota Press.

deVries, M. (1996). Trauma in cultural perspective. In B. van der Kolk, A. McFarlane, & L. Weisaeth (Eds.), *Traumatic stress: The effects of overwhelming experience on mind, body, and society* (pp. 398–413). New York: The Guilford Press.

DeWitte, S. N., & Bekvalac, J. (2011). The Association Between Periodontal Disease and Periosteal Lesions in the St. Mary Graces Cemetery, London, England A.D. 1350–1538. *American Journal of Physical Anthropology, 146*(4), 609–618.

DiGangi, E. A. (2012). Library research, presenting, and publishing. In E. A. DiGangi & M. K. Moore (Eds.), *Research methods in human skeletal biology* (pp. 483–506). Oxford: Academic.

DiGangi, E. A., & Moore, M. K. (2012). Introduction to skeletal biology. In E. A. DiGangi & M. K. Moore (Eds.), *Research methods in human skeletal biology* (pp. 3–28). Oxford: Academic.

Dongoske, K. E. (1996). The Native American Graves Protection and Repatriation Act: A new beginning, not the end, for osteological analysis—A Hopi perspective. *American Indian Quarterly, 20*(2), 287–297.

Douglas, M. (1973). *Natural symbols: Exploration in cosmology.* New York: Vintage Books.

Douglas, M. (1992). *Purity and danger: An analysis of the concepts of pollution and taboo*. New York: Routledge. (Original work published 1966)

Dupras, T. L., Schultz, J. J., Wheeler, S. M., & Williams, L. J. (2011). *Forensic recovery of human remains: Archaeological approaches* (2nd ed.). London: CRC.

Durrant, R. (2011). Collective violence: An evolutionary Perspective. *Aggression and Violent Behavior, 16*, 428–436.

Ember, C. R., & Ember, M. (1997). Violence in the ethnographic record: Results of cross-cultural research on war and aggression. In D. L. Martin & D. W. Frayer (Eds.), *Troubled times: Violence and warfare in the past* (pp. 1–20). Amsterdam: Gordon and Breach.

Erdal, Ö. D. (2012). A possible massacre at Early Bronze Age Titriş Höyük, Anatolia. *International Journal of Osteoarchaeology, 22*(1), 1–21.

Etkins, N. L. (2003). The co-evolution of people, plants, and parasites: Biological and cultural adaptations to malaria. *The Proceedings of the Nutrition Society, 62*, 311–317.

Farmer, P. (1999). *Infections and Inequalities: The modern plagues*. Berkeley: University of California Press.

Fausto-Sterling, A. (1993). The five sexes: Why male and female are not enough. *The Sciences, March/April*, 20–24.

Ferguson, R. B. (1997). Violence and war in prehistory. In D. L. Martin & D. W. Frayer (Eds.), *Troubled times: Violence and warfare in the past* (pp. 321–355). Amsterdam: Gordon and Breach.

Ferguson, T. J., Dongoske, K. E., & Kuwanwisiwma, L. J. (2001). Hopi perspectives on southwestern mortuary studies. In D. R. Mitchell & J. L. Brunson-Hadley (Eds.), *Ancient burial practices in the American southwest* (pp. 9–26). Albuquerque: University of New Mexico Press.

Ferllini, R. (2007). *Forensic archaeology and human rights violations*. Springfield: Charles C Thomas.

Finegan, O. (2008). Case Study: The interpretation of skeletal trauma resulting from injuries sustained prior to, and as a direct result of freefall. In E. H. Kimmerle & J. P. Baraybar (Eds.), *Skeletal Trauma: Identification of injuries resulting from human remains abuse and armed conflict* (pp. 181–195). Boca Raton: CRC Press.

Finzsch, N. (2008). Extirpate or remove that vermine: Genocide, biological warfare, and settler imperialism in the eighteenth and early nineteenth century. *Journal of Genocide Research, 10*(2), 215–232.

Frayer, D. W. (1997). Ofnet: Evidence for a Mesolithic massacre. In D. L. Martin & D. W. Frayer (Eds.), *Troubled times: Violence and warfare in the past* (pp. 181–216). Amsterdam: Gordon and Breach.

Galloway, A., Birkby, W. H., Jones, A. M., Henry, T. E., & Parks, B. O. (1989). Decay rates of human remains in an arid environment. *Journal of Forensic Sciences, 34*, 607–616.

Galtung, J. (1990). Cultural violence. *Journal of Peace Research, 27*(3), 291–305.

Gampel, Y. (1996). The interminable uncanny. In L. Rangell & R. Moses-Hrushovski (Eds.), *Psychoanalysis at the political border*. Madison: International Universities Press.

Gampel, Y. (2000). Reflections on the prevalence of the uncanny in social violence. In A. C. G. M. Robben & M. M. Suárez-Orozco (Eds.), *Cultures under Siege: Collective violence and trauma* (pp. 48–69). Cambridge: Cambridge University Press.

Gargett, R. H. (1999). Middle Paleolithic burial is not a dead issue: The view from Qafzeh, Saint-Césaire, Keba Amud, and Dederiyeh. *Journal of Human Evolution, 37*, 27–90.

Geller, P. L. (2005). Skeletal analysis and theoretical complications. *World Archaeology, 37*(4), 597–609.

Geller, P. L. (2008). Conceiving sex: Fomenting a feminist bioarchaeology. *Journal of Social Archaeology, 8*(1), 113–138.

Geller, P. L., & Stockett, M. K. (2006). *Feminist anthropology: Past, present, and future*. Philadelphia: University of Pennsylvania Press.

Genovés, S. (1967). Proportionality of the Long Bones and their Relation to Stature among Mesoamericans. *American Journal of Physical Anthropology, 26*, 67–78.

Gero, J. M., & Conkey, M. W. (1991). *Engendering archaeology: Women and prehistory*. Oxford: Blackwell.

Giddens, A. (1979). *Central problems in social theory: Action, structure, and contradiction in social analysis*. London: Macmillan.

Gill, G. W. (1998). The beauty of race and races. *Anthropology News, 39*(3), 1–5.

Gnecco, C. (2003). Against ecological determinism: Late Pleistocene hunter-gatherers in the tropical forests of northern South America. *Quaternary International, 109–110*, 13–21.

Goodman, A. H. (1998). The biological consequences of inequality in antiquity. In A. H. Goodman & T. L. Leatherman (Eds.), *Building a new biocultural synthesis: Political-economic perspectives on human biology* (pp. 147–169). Ann Arbor: University of Michigan Press.

Goodman, A. H., & Rose, J. C. (1991). Dental Enamel Hypoplasias as Indicators of Nutritional Status. In M. A. Kelley & C. S. Larsen (Eds.), *Advances in Dental Anthropology* (pp. 279–293). New York: Wiley-Liss.

Gould, R. A. (2005). Archaeology prepares for a possible mass-fatality disaster. *The SAA Archaeological Record, 5*, 10–12.

Gould, R. A. (2007). *Disaster archaeology*. Salt Lake City: The University of Utah Press.

Grauer, A. L., & Stuart-Macadam, P. (1998). *Exploring the differences: Sex and gender in paleopathological perspective*. Cambridge: Cambridge University Press.

Green, L. (1999). *Fear as a way of life: Mayan widows in rural Guatemala*. New York: Columbia University Press.

Guyomarc'h, P., Campagna-Vaillancourt, M., Kremer, C., & Sauvageau, A. (2010). Discrimination of Falls and Blows in Blunt Head Trauma: A multi-criteria approach. *Journal of Forensic Sciences, 55*(2), 423–427.

Harrod, R. P. (2011). Phylogeny of the southern Plateau–An osteometric evaluation of inter-tribal relations. *HOMO-Journal of Comparative Human Biology, 62*(3), 182–201. doi:10.1016/j.jchb.2011.01.005.

Harrod, R. P. (2012). Centers of control: Revealing elites among the Ancestral Pueblo during the "Chaco Phenomenon". *International Journal of Paleopathology* http://dx.doi.org/10.1016/j.ijpp.2012.09.013.

Harrod, R. P., Liénard, P., & Martin, D. L. (2012). Deciphering violence: The potential of modern ethnography to aid in the interpretation of archaeological populations. In D. L. Martin, R. P. Harrod, & V. R. Pérez (Eds.), *The bioarchaeology of violence* (pp. 63–80). Gainesville: University of Florida Press.

Hillson, S. W. (2008). Dental Pathology. In M. A. Katzenberg & S. R. Saunders (Eds.), *Biological Anthropology of the Human Skeleton, second edition* (pp. 301–340). Hoboken: John Wiley & Sons, Inc.

Hollimon, S. E. (2011). Sex and gender in bioarchaeological research: Theory, method, and interpretation. In S. C. Agarwal & B. A. Glencross (Eds.), *Social bioarchaeology* (pp. 149–182). Malden: Wiley-Blackwell.

Hunter, J. (2002). Foreword: A pilgrim in archaeology—a personal view. In W. D. Haglund & M. H. Sorg (Eds.), *Advances in forensic taphonomy: Method, theory, and archaeological perspectives* (pp. 25–32). Boca Raton: CRC.

Hunter, J., & Cox, M. (2005). *Forensic archaeology: Advances in theory and practice*. London: Routledge.

Irish, J. D., & Nelson, G. C. (Eds.). (2008). *Technique and Application in Dental Anthropology*. Cambridge: Cambridge University Press.

Johnson, J. L., & Repta, R. (2012). Sex and gender: Beyond the binaries. In J. L. Oliffe & L. Greaves (Eds.), *Designing and conducting gender, sex, and health research* (pp. 17–37). Thousand Oaks: Sage.

Kelly, R. C. (2005). The evolution of lethal intergroup violence. *Proceedings of the National Academy of Sciences of the United States of America, 102*(43), 15294–15298.

Klaus, H. D., & Tam, M. E. (2010). Oral Health and the Postcontact Adaptive Transition: A contextual reconstruction of diet in Mórrope, Peru. *American Journal of Physical Anthropology, 141*, 594–609.

Knauft, B. M. (1991). Violence and sociality in human evolution. *Current Anthropology, 32*, 391–428.

Knudson, K. J., & Stojanowski, C. M. (2009). *Bioarchaeology and identity in the Americas*. Gainesville: University Press of Florida.

Kremer, C., Racette, S., Dionne, C.-A., & Sauvageau, A. (2008). Discrimination of Falls and Blows in Blunt Head Trauma: Systematic study of the Hat Brim Line Rule in relation to skull fractures. *Journal of Forensic Sciences, 53*(3), 716–719.

Krieger, N. (1994). Epidemiology and the web of causation: Has anyone seen the spider? *Social Science & Medicine, 39*(7), 887–903.

Krystal, H. (1968). Studies of concentration camp survivors. In H. Krystal (Ed.), *Massive psychic trauma* (pp. 23–46). New York: International Universities Press.

Kuckelman, K. A., Lightfoot, R. R., & Martin, D. L. (2002). The bioarchaeology and taphonomy of violence at Castle Rock and Sand Canyon Pueblos, Southwestern Colorado. *American Antiquity, 67*, 486–513.

Larsen, C. S. (1997). *Bioarchaeology: Interpreting behavior from the human skeleton*. Cambridge: Cambridge University Press.

Larsen, C. S. (2001). *Bioarchaeology of Spanish Florida: The impact of colonialism*. Gainesville: University Press of Florida.

Larsen, C. S., Shavit, R., & Griffin, M. C. (1991). Dental Caries Evidence for Dietary Change: An archaeological context. In M. A. Kelley & C. S. Larsen (Eds.), *Advances in Dental Anthropology* (pp. 179–202). New York: Wiley-Liss.

LeBlanc, S. A. (1999). *Prehistoric warfare in the American southwest*. Salt Lake City: The University of Utah Press.

Lee, R. B., & Devore, I. (1968). *Man the hunter*. Chicago: Aldine.

Litwack, L. (2000). Hellhounds. In J. Allen (Ed.), *Without sanctuary: Lynching photography in America* (pp. 8–37). Santa Fe: Twin Palms.

Maples, W. R., & Browning, M. (1994). *Dead men do tell tales: The strange and fascinating cases of a forensic anthropologist*. New York: Doubleday.

Marchi, D., & Shaw, C. N. (2011). Variation in Fibular Robusticity Reflects Variation in Mobility Patterns. *Journal of Human Evolution, 61*(5), 609–616.

Martin, D. L. (1998). Owning the sins of the past: Historical trends in the study of southwest human remains. In A. H. Goodman & T. L. Leatherman (Eds.), *Building a new biocultural synthesis: Political-economic perspectives on human biology* (pp. 171–190). Ann Arbor: University of Michigan Press.

Martin, D. L., Akins, N. J., Goodman, A. H., Toll, W., & Swedlund, A. C. (2001). *Harmony and discord: Bioarchaeology of the La Plata valley*. Santa Fe: Museum of New Mexico Press.

Martin, D. L., Goodman, A. H., Armelagos, G. J., & Magennis, A. L. (1991). *Black Mesa Anasazi health: Reconstructing life from patterns of death and disease*. Carbondale: Southern Illinois University Press.

Martin, D. L., Harrod, R. P., & Pérez, V. R. (2012). Introduction: Bioarchaeology and the study of violence. In D. L. Martin, R. P. Harrod, & V. R. Pérez (Eds.), *The bioarchaeology of violence* (pp. 1–10). Gainesville: University of Florida Press.

Martin, D. L., & Pérez, V. R. (2001). Dead bodies and violent acts. Commentary. *Anthropology News, 42*(7), 8–9.

Martin, D. L., Harrod, R. P., & Fields, M. (2010). Beaten Down and Worked to the Bone: Bioarchaeological investigations of women and violence in the ancient Southwest. *Landscapes of Violence, 1*(1), 3.

Martin, D. L., Ryan P. H., & Misty Fields. (2010). Beaten down and worked to the bone: Bioarchaeological investigations of women and violence in the ancient Southwest. *Landscapes of Violence, 1*(1):Article 3.

Mariotti, V., Facchini, F., & Belcastro, M. G. (2007). The Study of Entheses: Proposal of a standardised scoring method for twenty-three entheses of the postcranial skeleton. *Collegium Antropologicum, 31*(1), 291–313.

McCaffrey, R. J., William, A. D., Fisher, J. M., & Laing, L. C. (1997). *The practice of forensic neuropsychology: Meeting challenges in the courtroom*. New York: Plenum.

McGee, R. J., & Warms, R. L. (1996). *Anthropological theory: An introductory history* (3rd ed.). McGraw Hill: Boston.

Micozzi, M. (1997). Frozen environments and soft tissue preservation. In W. D. Haglund & M. H. Sorg (Eds.), *Forensic taphonomy: The postmortem fate of human remains* (pp. 171–180). Boca Raton: CRC Press.

Moore, J. H. (1994). Putting anthropology back together again: The ethnogenetic critique of cladistic theory. *American Anthropologist, 96*(4), 925–948.

Mummert, A., Esche, E., Robinson, J., & Armelagos, G. J. (2011). Stature and Robusticity During the Agricultural Transition: Evidence from the bioarchaeological record. *Economics and Human Biology, 9*(3), 284–301.

Neitzel, J. E. (2000). Gender hierarchies: A comparative analysis of mortuary data. In P. L. Crown (Ed.), *Women and men in the prehispanic southwest: Labor, power, and prestige* (pp. 137–168).

Nicholas, G. P. (2004). Editor's notes. *Canadian Journal of Archaeology, 28*(1), 3–4.

Nordstrom, C. (2009). Fault lines. In B. Rylko-Bauer, L. Whiteford, & P. Farmer (Eds.), *Global health in times of violence* (pp. 63–87). Santa Fe: School for Advanced Research Press.

O'Rourke, D. H., Geoffrey Hayes, M., & Carlyle, S. W. (2005). The Consent process and a DNA research: Contrasting approaches in North America. In T. R. Turner (Ed.), *Biological anthropology and ethics: From repatriation to genetic identity* (pp. 231–240). Albany: State University of New York Press.

Ogilvie, M. D., & Hilton, C. E. (2000). Ritualized violence in the prehistoric American southwest. *International Journal of Osteoarchaeology, 10*, 27–48.

Ortner, D. J. (2003). *Identification of Pathological Conditions in Human Skeletal Remains*. New York: Academic Press.

Osterholtz, A. J. (2012). The social role of hobbling and torture: Violence in the prehistoric southwest. *International Journal of Paleopathology*. http://dx.doi.org/10.1016/j.ijpp.2012.09.011.

Ousley, S. D., & Jantz, R. L. (2005). FORDISC 3.0: Personal computer forensic discriminant functions. Knoxville: University of Tennessee.

Ousley, S. D., Jantz, R. L., & Freid, D. (2009). Understanding race and human variation: Why forensic anthropologists are good at identifying race. *American Journal of Physical Anthropology, 139*(1), 68–76.

Palkovich, A. M. (1980). *The Arroyo Hondo skeletal and mortuary remains*. Santa Fe: School of American Research Press.

Parsons, T. (1938). The role of theory in social research. *American Sociological Review, 3*(1), 13–20.

Pérez, V. R. (2012). The politicization of the dead: Violence as performance, politics as usual. In D. L. Martin, R. P. Harrod, & V. R. Pérez (Eds.), *The bioarchaeology of violence*. Gainesville: University of Florida Press.

Perry, E. M. (2008). Gender, labor, and inequality at Grasshopper Pueblo. In A. L. W. Stodder (Ed.), *Reanalysis and reinterpretation in southwestern bioarchaeology* (pp. 151–166). Tempe: Arizona State University, Anthropological Research Papers.

Pinhasi, R., & Stock, J. T. (2011). *Human bioarchaeology of the transition to agriculture*. Chichester: Wiley-Blackwell.

Potter, W. E. (2010). *Evidence for a Change in the Rate of Aging of Osteological Indicators in American Documented Skeletal Samples*. Unpublished Ph.D. dissertation, University of New Mexico, Albuquerque.

Powell, M. L. (1991). Ranked status and health in the Mississippian chiefdom at Moundville. In M. L. Powell, P. S. Bridges, & A. M. W. Mires (Eds.), *What mean these bones? Studies in southeastern bioarchaeology* (pp. 22–51). Tuscaloosa: The University of Alabama Press.

Quesada, J. (1998). Suffering children: An embodiment of war and its aftermath in post-Sandinista Nicaragua. *Medical Anthropology Quarterly, 12*, 51–73.

Rhine, S. (1998). *Bone voyage: A journey in forensic anthropology*. Albuquerque: University of New Mexico Press.

Riches, D. (1986). The phenomenon of violence. In D. Riches (Ed.), *The anthropology of violence* (pp. 1–27). New York: Blackwell.

Robb, J. E. (1998). The interpretation of skeletal muscle sites: A statistical approach. *International Journal of Osteoarchaeology, 8*, 363–377.

Robb, J. E., Bigazzi, R., Lazzarini, L., Scarsini, C., & Sonego, F. (2001). Social "status" and biological "status": A comparison of grave goods and skeletal indicators from Pontecagnano. *American Journal of Physical Anthropology, 115*(3), 213–222.

Robbins Schug, G. (2011). Bioarchaeology and climate change: A view from south Asian prehistory. In C. S. Larsen (Ed.), *Bioarchaeological interpretations of the human past: Local, regional, and global perspectives*. Gainesville: University Press of Florida.

Roseberry, W. (1998). Political economy and social fields. In A. H. Goodman & T. L. Leatherman (Eds.), *Building a new biocultural synthesis: Political-economic perspectives on human biology* (pp. 75–92). Ann Arbor: University of Michigan Press.

Ruff, C. B. (1987). Sexual dimorphism in human lower limb structure: Relationship to subsistence strategy and sexual division of labor. *Journal of Human Evolution, 16*, 391–416.

Ruff, C. B. (2008). Biomechanical Analyses of Archaeological Human Skeletons. In M. A. Katzenberg & S. R. Saunders (Eds.), *Biological Anthropology of the Human Skeleton, 2nd edition* (pp. 183–206). Hoboken: John Wiley & Sons, Inc.

Sahlins, M. (1981). *Historical metaphors and mystical realities: Structure in the early history of the Sandwich Islands Kingdom*. Ann Arbour: University of Michigan Press.

Sauer, N. J. (1992). Forensic anthropology and the concept of race: If races don't exist, why are forensic anthropologists so good at identifying them? *Social Science & Medicine, 34*(2), 107–111.

Sauer, N. J. (1993). Applied anthropology and the concept of race: A legacy of Linneaus. *National Association for the Practice of Anthropology, Bulletin, 13*, 79–84.

Saul, J. M., & Saul, F. P. (2002). Forensics, archaeology, and taphonomy: The symbiotic relationship. In W. D. Haglund & M. H. Sorg (Eds.), *Advances in forensic taphonomy: Method, theory, and archaeological perspectives* (pp. 71–98). Boca Raton: CRC.

Scheper-Hughes, N., & Bourgois, P. (2004). *Violence in war and peace, an anthology*. Malden: Blackwell.

Scheper-Hughes, N., & Lock, M. M. (1987). The mindful body: A prolegomenon to future work in medical anthropology. *Medical Anthropology Quarterly, 1*(1), 6–41.

Scheuer, L., & Black, S. M. (2000). *Developmental Juvenile Osteology*. San Diego: Elsevier Academic Press.

Schiffer, M. B. (2000). Preface and acknowledgements. In M. B. Schiffer (Ed.), *Social theory in archaeology* (pp. 7–8). Salt Lake City: The University of Utah Press.

Scott, G. R., & Turner, C. G., II. (1988). Dental Anthropology. *Annual Review of Anthropology, 17*, 99–126.

Shuler, K. A. (2011). Life and death on a Barbadian sugar plantation: Historic and bioarchaeological views of infection and mortality at Newton plantation. *International Journal of Osteoarchaeology, 21*(1), 66–81.

Sigler-Eisenberg, B. (1985). Forensic research: Expanding the concept of applied archaeology. *American Antiquity, 50*, 650–655.

Skinner, M., Alempijevic, D., & Djurić-Srejic, M. (2003). Guidelines for international forensic bio-archaeology monitors of mass grave exhumations. *Forensic Science International, 134*, 81–92.

Sluka, J. A. (1992). The anthropology of conflict. In C. Nordstrom & J. Martin (Eds.), *The paths to domination, resistance, and terror* (pp. 18–36). Berkeley: University of California Press.

Smith, E. A. (2000). Three styles in the evolutionary analysis of human behavior. In L. Cronk, N. A. Chagnon, & W. Irons (Eds.), *Adaptation and human behavior: An anthropological perspective* (pp. 27–46). New York: Aldine.

Smith, E. A., & Winterhalder, B. (1992). *Evolutionary ecology and human behavior*. New York: Aldine de Gruyter.

Sofaer, J. R. (2006). *The body as material culture: A theoretical osteoarchaeology*. Cambridge: Cambridge University Press.

Sorg, M. H., & Haglund, W. D. (2002). Advancing forensic taphonomy: Purpose, theory, and practice. In W. D. Haglund & M. H. Sorg (Eds.), *Advances in forensic taphonomy: Method, theory, and archaeological perspectives* (pp. 3–30). Boca Raton: CRC.

Spurr, K. (1993). *NAGPRA and archaeology on Black Mesa Arizona.* Window Rock: Navajo Nation Papers in Anthropology, No. 30, Navajo Nation Archaeological Department.

Steadman, D. W. (2009). *Hard evidence: Case studies in forensic anthropology* (2nd ed.). Upper Saddle River: Pearson Education.

Steckel, R. H. (1995). Stature and the Standard of Living. *Journal of Economic Literature, 33,* 1903–1940.

Steckel, R. H., & Rose, J. C. (2002). *The backbone of history: Health and nutrition in the Western hemisphere.* Cambridge: Cambridge University Press.

Steele, C. (2008). Archaeology and the forensic investigation of recent mass graves: Ethical issues for a new practice of archaeology. *Journal of the World Archaeological Congress, 4*(3), 414–428.

Stefanović, S., & Porčić, M. (2011). Between-group differences in the patterning of musculo-skeletal stress markers: Avoiding confounding factors by focusing on qualitative aspects of physical activity. *International Journal of Osteoarchaeology, 21*(2), 187–196.

Stefanović, S., & Porčić, M. (2013). Between-Group Differences in the Patterning of Musculo-skeletal Stress Markers: Avoiding confounding factors by focusing on qualitative aspects of physical activity. *International Journal of Osteoarchaeology, 23*(1), 94–105.

Steward, J. (1955). The concept and method of cultural ecology. In J. Steward (Ed.), *Theory of cultural change.* Urbana: University of Illinois Press.

Stock, J. T., & Shaw, C. N. (2007). Which Measures of Diaphyseal Robusticity are Robust? A comparison of external methods of quantifying the strength of long bone diaphyses to cross-sectional geometric properties. *American Journal of Physical Anthropology, 134*(3), 412–423.

Stodder, A. L. W. (2012). Data and data analysis issues in paleopathology. In A. L. Grauer (Ed.), *A companion to paleopathology* (pp. 339–356). Malden: Blackwell.

Stodder, A. L. W., & Palkovich, A. M. (2012). The bioarchaeology of individuals. In C. S. Larsen (Ed.), *Bioarchaeological interpretations of the human past: Local, regional, and global perspectives.* Gainesville: University Press of Florida.

Suárez-Orozco, M. M., & Robben, A. C. G. M. (2000). Interdisciplinary perspectives on violence and trauma. In A. C. G. M. Robben & M. M. Suárez-Orozco (Eds.), *Cultures under Siege: Collective violence and trauma* (pp. 194–226). Cambridge: Cambridge University Press.

Sutton, M. Q., & Anderson, E. N. (2010). *Introduction to cultural ecology* (2nd ed.). Lanham: Altamira.

Taussig, M. (1984). Culture of terror-face of death: Roger Casement's Putumayo report and the explanation of torture. *Comparative Studies in Society and History, 26,* 467–497.

Thompson, D. D. (1982). Forensic anthropology. In F. Spencer (Ed.), *A history of American physical anthropology, 1930–1980* (pp. 357–365). New York: Academic.

Titiev, M. (1972). *The Hopi Indians of old Oraibi.* Ann Arbor: University of Michigan Press.

Trotter, M. (1970). Estimation of Stature from Intact Long Bones. In T. D. Stewart (Ed.), *Personal Identification in Mass Disasters* (pp. 71–83). Washington, D.C.: Smithsonian Institution Press.

Trotter, M., & Gleser, G. (1952). Estimation of Stature from Long Bones of American Whites and Negroes. *American Journal of Physical Anthropology, 10,* 463–514.

Tung, T. A. (2007). Trauma and violence in the Wari empire of the Peruvian Andes: Warfare, raids, and ritual fights. *American Journal of Physical Anthropology, 133,* 941–956.

Turpin, J., & Kurtz, L. R. (1997). *The web of violence: From interpersonal to global.* Urbana: University of Illinois Press.

Ubelaker, D. H. (1999). Aleš Hrdlicka's role in the history of forensic anthropology. *Journal of Forensic Sciences, 44*(4), 724–730.

Van Gerven, D. P., Carlson, D. S., & Armelagos, G. J. (1973). Racial history and bio-cultural adaptation of Nubian archaeological populations. *The Journal of African History, 14*(4), 555–564.

Vercellotti, G., Caramella, D., Formicola, V., Fornaciari, G., & Larsen, C. S. (2010). Porotic Hyperostosis in a Late Upper Paleolithic Skeleton (Villabruna 1, Italy). *International Journal of Osteoarchaeology, 20*, 358–368.

Verdery, K. (1999). *The political lives of dead bodies: Reburial and postsocialist change.* New York: Columbia University Press.

Wakely, J. (1997). Identification and Analysis of Violent and Non-Violent Head Injuries in Osteo-Archaeological Material. In J. Carman (Ed.), *Material Harm: Archaeological studies of war and violence* (pp. 24–46). Glasgow: Criuthne Press.

Walker-Bone, K., & Palmer, K. T. (2002). Musculoskeletal Disorders in Farmers and Farm Workers. *Occupational Medicine, 52*(8), 441–450.

Walker, P. L. (1989). Cranial injuries as evidence of violence in prehistoric southern California. *American Journal of Physical Anthropology, 80*(3), 313–323.

Walker, P. L. (2001). A bioarchaeological perspective on the history of violence. *Annual Review of Anthropology, 30*, 573–596.

Walker, P. L., Bathurst, R. R., Richman, R., Gjerdrum, T., & Andrushko, V. A. (2009). The Causes of Porotic Hyperstosis and Cribra Orbitalia: A reappraisal of the iron-deficiency-anemia hypothesis. *American Journal of Physical Anthropology, 139*, 109–125.

Walker, P. L., & Cook, D. C. (1998). Brief communication: Gender and sex: Vive la difference. *American Journal of Physical Anthropology, 106*(2), 255–259.

Wapler, U., Crubezy, E., & Schultz, M. (2004). Is Cribra Orbitalia Synonymous with Anemia? Analysis and Interpretation of Cranial Pathology in Sudan. *American Journal of Physical Anthropology, 123*, 333–339.

Watkins, R. (2012). Variations in health and socioeconomic status within the W. Montague Cobb skeletal collection: Degenerative joint disease, trauma and cause of death. *International Journal of Osteoarchaeology, 22*(1), 22–44.

Weston, D. A. (2008). Investigating the Specificity of Periosteal Reactions in Pathology Museum Specimens. *American Journal of Physical Anthropology, 137*(1), 48–59.

White, L. A. (1943). Energy and the evolution of culture. *American Anthropologist, 45*(3), 335–356.

White, T. D., Folkens, P. A., & Black, M. T. (2012). *Human osteology* (3rd ed.). Burlington: Academic.

Whitehead, N. L. (2004a). Introduction: Cultures, conflicts, and the poetics of violent practice. In N. L. Whitehead (Ed.), *Violence* (pp. 3–24). Santa Fe: School of American Research Press.

Whitehead, N. L. (2004b). *Violence.* Santa Fe: School of American Research Press.

Whitehead, N. L. (2005). War and violence as cultural expression. *Anthropology News, 46*, 23–26.

Wood, J. J., & Powell, S. (1993). An ethos for archaeological practice. *Human Organization, 52*(4), 405–413.

Woolhouse, M. E. J., Webster, J. P., Domingo, E., Charlesworh, B., & Levin, B. R. (2002). Biological and biomedical implications of the co-evolution of pathogens and their hosts. *Nature, 32*(4), 569–577.

Zollikofer, C. P. E., de León, M. S. P., Vandermeersch, B., & Lévêque, F. (2002). Evidence for interpersonal violence in the St. Césaire Neanderthal. *Proceedings of the National Academy of Sciences of the United States of America, 99*(9), 6444–6448.

Zuckerman, M. K., & Armelagos, G. J. (2011). The origins of biocultural dimensions in bioarchaeology. In S. C. Agarwal & B. A. Glencross (Eds.), *Social bioarchaeology* (pp. 15–43). Malden: Wiley-Blackwell.

Zuckerman, M. K., Turner, B. L., & Armelagos, G. J. (2012). Evolutionary thought in paleopathology and the rise of the biocultural approach. In A. L. Grauer (Ed.), *A companion to paleopathology* (pp. 34–57). Malden: Blackwell.

Chapter 4
Best Practices: Excavation Guidelines and Taphonomic Considerations

No single text can prepare bioarchaeologists for what they may encounter in the field while working at an archaeological site where there are human remains—there are simply too many factors affecting the situation. There will be the ethical and legal considerations which will guide the contours of how the human remains will be approached (see Chap. 2). And, it will depend on the context of the field site, whether it is a field school, a research site, or a CRM project. There may be a research question that is under investigation by the individuals running the field site, or there may not be. The idea of "best practices" as covered here is to paint a broad brush stroke of things to consider prior to being in the field. It is driven by the belief that a better understanding of where and how human remains are excavated from archaeological sites and the important roles that bioarchaeologists (and forensic anthropologists) play in the recovery and preservation of these remains is key to improving bioarchaeology as an integrative approach. We include forensic anthropology with bioarchaeology because some of the goals are similar especially in the careful recovery and recording of human remains (see Chap. 3).

4.1 Understanding the Origin and Condition of the Human Remains

In bioarchaeology and forensic anthropology the importance of developing an appreciation of where the human remains are situated, both temporally and spatially, is critical to enhancing what can be learned from them. While the bones are important, it is imperative that researchers do everything they can to understand the larger context. The term context is a rather vague concept. According to Yarrow (2008:126), context is a particular moment in time that is defined by certain actions that we as archaeologists believe gave rise to a particular event. Context is revealed by looking at as many lines of evidence as possible in order to try to recreate the "potential" actions that resulted in the deposition and creation of the site.

D.L. Martin et al., *Bioarchaeology: An Integrated Approach to Working with Human Remains*, Manuals in Archaeological Method, Theory and Technique, DOI 10.1007/978-1-4614-6378-8_4, © Springer Science+Business Media New York 2013

The result is that numerous factors can contribute to context so what is important is identifying and recording as many of these factors as possible. This includes understanding as much as possible the environmental, biological, and cultural factors that affect the remains from around the time of death through to recovery and analysis. Ideally, after this long and complex process, the remains are properly curated in a repository or are reinterred.

4.2   Levels of Participation

Bioarchaeologists play any number of roles in the excavation process. However, whether as a collaborator or supervisor, the mere presence of a bioarchaeologist on site where human remains are being recovered is important.

4.2.1   The Role of Bioarchaeologists

Participation from the onset of an excavation allows not only for the proper retrieval of the remains but helps in the formation of the analysis and in the interpretation of the burial. By having bioarchaeologists present at the initial stages of a project, bioarchaeological principles and methods for the study of ethnicity, identity, and biological affinity along with a variety of skeletal health indices can be built into the project. These kinds of data can be used to test hypotheses of a broader scope concerning biocultural responses to interregional and intra-regional interactions. If the remains are fragmentary and poorly preserved, some information may need to be obtained during the excavation process. Without having a research agenda prior to excavation, important data may be lost forever.

Human skeletal remains provide an invaluable primary source of information on past populations given the developmental plasticity of bone tissue and the sensitivity of bone and teeth to environmental and physiological stress experienced during an individuals' lifetime. Given their basis in physiological processes of growth, development, and acclimatization to environmental change, human hard tissues are a record of the body's response to mechanical stress and conditions resulting in metabolic insults (e.g., dietary deficiencies/nutritional deprivation, trauma, disease).

In order to collect skeletal health indices that integrate data on nutrition, diet, growth, pathology, biomechanics, and trauma, bioarchaeologists must have at the ready their data collection sheets. There is no guarantee that all of the skeletal indicators of the above processes will survive the excavation, retrieval, and movement to a laboratory setting. Also, at this early stage, it is crucial to document and understand the kinds of taphonomic changes that have affected the human remains. Understanding the multiple variables that can impact skeletal material from excavation techniques to taphonomic processes is crucial in teasing apart the life history of the individuals being studied. The best case scenario for the recovery of human remains is to have a trained bioarchaeologist present from consultation and design of the project through excavation and analysis.

There are some excellent examples of this emerging model for bioarchaeologists. Tung's (2012) work in Peru provided an extraordinary record of integration of the human remains within the broader archaeological context. Her analysis of the Wari empire (AD 600–1000) is a model for how to integrate research questions about power and influence with the ways that bodies in the Wari culture were used to control communities. Her analysis showed how Wari elites controlled the region through violent actions such as raiding for men, women, and children from groups who they considered as outsiders. Some of these individuals were often sacrificed, dismembered, and body parts were turned into trophies used for displays. Tung integrated a number of different levels of analysis including the mortuary and archaeological context of the human remains; the analysis of the remains for age, sex, and other identifying features; and DNA extraction to distinguish local from nonlocal individuals.

4.2.1.1 Bioarchaeologists as Consultants

Many bioarchaeologists work as consultants on projects, in that they come in once remains are found and identified. This role is arguably the most commonly held position of bioarchaeologists and the most traditional. There are limitations of this approach however as it is only one step up from the involvement of a trained osteologist who would receive the remains in a laboratory site far removed from the original excavation site. In these situations that were much more common prior to the 1990s, much of the context was lost. Despite the limitations of being brought on to the project after the finding of human remains, this does represent the majority of projects that bioarchaeologists (and forensic anthropologists) work on. The reason for this is that bioarchaeologists are typically highly trained and specialized. They may be more costly to include from the ground level on up. Unlike adding additional crew members who may have specialization in geomorphology, soil, faunal, or floral analysis, ceramic analysis, or lithic identification, many bioarchaeologists are typically not seen as critically useful to the overall excavation process. And in some cases, bioarchaeologists may not be viewed as "real" archaeologists, but instead they are seen as osteologists whose expertise does not go beyond analysis of human remains. While this is not the case, it is a historical remnant from the early days when physical anthropologist and osteologists were not trained in archaeological field techniques.

There are many examples of projects where bioarchaeologists were employed primarily in a consultant capacity. There is also a large number of unpublished works that have been done in the context of cultural resource management (CRM) projects carried out under the auspices of a government or private contracting firm. What is extremely common is that the analyses carried out by bioarchaeologists contracted to do the skeletal analysis as consultants are often filed away as "gray literature," and the results are not disseminated to the public. These documents are often difficult to locate, or they are brief appendices attached to larger reports, or they languish as unpublished papers on file often authored by individuals no longer within this field. Separating the analyses of the human remains from other aspects of archaeological interpretation and synthesis is highly problematic.

For a number of reasons some projects are often not disclosed to the public. This can be due to the work being conducted on land where classified research is or was carried out (e.g., Los Alamos National Laboratory, the Idaho National Laboratory, and the Nevada Test Site). Two examples of projects where the information was not made public are the bioarchaeological excavation of City Hall Park in New York City (Anderson 2000) and the multitude of forensic excavations that have been carried out by the Central Identification Laboratory, Hawaii (CILHI) for the Joint Prisoners of War, Missing in Action Accounting Command in Hawaii (JPAC). Both of these projects involve the removal and analysis of hundreds of burials that are of interest to the larger community, but because of the nature of the work the information is not made public.

In the working life of active bioarchaeologists (and forensic anthropologists), it is likely that they will do a variety of kinds of work, some of which will be purely as consultants brought on to a project in the latter stages. But there is no reason to believe that this work is not extremely valuable even though it may present overwhelming challenges. Each situation will present different ways that consultant-bioarchaeologists can still attempt to integrate the data within a larger context and to use the data to engage with broader issues.

4.2.1.2 Bioarchaeologist as Collaborators

Bioarchaeologists working as collaborators in the form of codirectors or as co-principle investigators (PIs) are in very good positions to influence the ways that human remains are handled. The truth is that most archaeologists can and often do excavate and retrieve human remains. But most archaeologists are not trained in the analysis of human remains and therein lies the difference. The value of having bioarchaeologists involved in both the design and implementation of a project is that the more they know from the beginning about context and taphonomy, the better their analyses will be. While less common than a bioarchaeologist being brought on to the project as a consultant, there are many more projects that include bioarchaeologists as collaborators.

This type of work is typified by the presence of bioarchaeologists on large-scale, long-term archaeological projects, such as Larsen at Çatalhöyük (Pilloud and Larsen 2011), Martin at Black Mesa (Martin et al. 1991), and Palkovich at Arroyo Hondo (Palkovich 1980). In Mexico, Tiesler and Cucina have created and promoted the Joint Agendas in Maya Bioarchaeology (Tiesler and Cucina 2008). They realized that there was an underutilization of bioarchaeological expertise on many of the archaeological projects conducted on Mayan sites in Mexico. They have focused on promoting a relationship of collaboration with the archaeologists.

4.2.1.3 Bioarchaeologists as Supervisors

The least common role that bioarchaeologists hold is that of the sole director or PI. However, there are a number of researchers that are designing and supervising archaeological projects. This approach is especially useful if the project is primarily designed around and focused on human remains, but even this is not always the case.

The most prominent and perhaps earliest example of a bioarchaeologist directing a project is the long-term and ongoing field school Buikstra directs at Kampsville (Buikstra 1981). The field school is a collaborative program of scholars from Arizona State University, the Center for American Archaeology, the Illinois State Museum, and the Center for Advanced Spatial Technology at the University of Arkansas (http://shesc.asu.edu/kampsville). It has trained hundreds of students since the 1980s in archaeological techniques and is one of the best training programs for bioarchaeologists in the United States.

Other bioarchaeologists are leading the way in directing bioarchaeology projects. The African American Burial Ground excavation and analysis led by bioarchaeologist Blakely is another example of how bioarchaeologists can and do take a supervisory role on archaeological projects (Blakey 1998). Klaus directs the Lambayeque Valley Biohistory Project (Klaus and Tam 2009), and Tung directs the Beringa Bioarchaeology and Archaeology Project (Tung 2007), both of which are in Peru.

4.2.2 The Different Roles and their Varying Outcome: Case Studies

The examples provided here only represent a portion of the work being conducted by bioarchaeologists who are engaged in work as consultants, collaborators, and supervisors. A characteristic that all these roles share is that researchers are increasingly recognizing the importance of being present and assisting in the excavation of both archaeological and bioarchaeological material in order to better understand the context of the burials. Future bioarchaeologists have to be more than experts in human anatomy, they also need to be well-trained, highly skilled archaeologists. Having a strong foundation in archaeological field methods is increasingly being seen as a requirement that is essential for bioarchaeology. They will need to be trained in how to recover human remains from archaeological and forensic contexts (excavation) and with the challenges associated with analyzing the remains once they are recovered (identifying taphonomic changes).

4.2.2.1 Case Study: CILHI

The Central Identification Laboratory, Hawaii (CILHI) was created to help determine the fate of all Americans lost in combat and to provide closure for their families. During the final stages of the Vietnam War, US authorities made considerable efforts to locate missing Americans (prisoners of war and the missing in action) in Vietnam, Laos, and Cambodia (Davis 2000). The Central Identification Laboratory, Thailand was created in 1973 to focus on Americans missing in Southeast Asia. Three years later, the U.S. Army Central Identification Laboratory, Hawaii was established with the explicit goal of locating, retrieving, and identifying the human remains of Americans lost in all previous military operations.

The Joint Task Force—Full Accounting (JTF-FA) was formed in 1992 and was responsible for establishing the "fullest possible accounting" for US MIAs from the Vietnam War (Davis 2000:547). The overlapping goals of JTF-FA and CILHI led the Department of Defense to combine their efforts under the joint POW/MIA Accounting Command (Ainsworth 2003). At the time of their merging there were more than 88,000 Americans still missing, 1 from the Gulf War, 1,800+ from the Vietnam War, 120 from the Cold War, 8,100+ from the Korean War, and 78,000+ from World War II.

The procedures that CILHI follows resemble those of other state and federal forensic anthropologists. Once the JPAC investigative teams identify a site for recovery, the CILHI recovery team is sent in to excavate and bring the remains back to the lab for identification. The team includes a team leader, forensic anthropologist, team sergeant, linguist, medic, life support technician, forensic photographer, explosive ordnance disposal technician, and sometimes specialists with diverse training such as mountaineering. Deployments are typically 35–60 days and take place all over the world (Webster 1998).

The excavations that CILHI conduct are very different from standard archaeological excavations. In most CILHI excavations the identity of the individual is already known. This is significantly different from most forensic anthropology cases where little or no case-specific information is known. CILHI anthropologists are not constructing a crime scene nor are they attempting to reconstruct past behavior through material culture and site formation within an archaeological context. As such, it is not necessary for the team to create a three-dimensional map of site in order to establish the relationship and distribution of artifacts, with the notable exception of mass graves that require more traditional archaeological techniques (Hoshower 1998).

Once the American remains have been excavated by CILHI team, they are brought back to the lab in Hawaii for identification. Despite the team's best effort it is not always possible to make a positive identification. The many bioarchaeologists that have trained and worked at CILHI speak to the power of bioarchaeologists as consultants. Over the past two decades, CILHI has continued to work to identify US service men and women while reaching out to the scientific community through its "Visiting Scientist Program" and by encouraging its staff to attend conferences and to publish in peer-reviewed journals. The academic isolation that initially plagued CILHI has all but vanished, and the transparency in their research has earned greater trust from the people it hopes to best serve.

4.2.2.2   Case Study: Bioarchaeologists as Engaged Researchers

Bioarchaeologists often find themselves engaged with potentially politically charged research. For many bioarchaeologists it has become clear that there is a need to integrate their work at various universities and other organizations (e.g., teaching, research, and service), engage with those outside the academy, and synthesize and use what they learn as a catalyst for change. Boyer (1996:11) poses the challenge this way when he states that "… the academy must become a more vigorous partner in the search for answers to our most pressing social, civic, economic, and moral problems, and must reaffirm its historic commitment to what I call the scholarship of engagement." Academics and academic departments have often struggled with the tensions that can come from trying to balance research teaching and community outreach. This is particularly true when the scholar moves from the relative safety of the academy into the complex and often convoluted realm of activist research.

The criticism that follows many of those who dare to engage in these often politically charged research agendas is that activist research lacks objectivity and that it is often simplistic, under-problematized, and under-theorized. If bioarchaeologists are to take seriously the mandate to be engaged researchers as articulated by Boyer and others, they cannot turn away from the challenges of taking their science out of academia and into the world.

The example of the Yaqui repatriation in Chap. 2 highlights the challenges, importance, and level of gratification that can come from an engaged bioarchaeological project. Among the Yaqui Indians of Mexico, identity and ethnicity are intertwined and connected to space, place, and history. Understanding of space and place play a critical role in the ethnic self-identification of the Yaqui people. These ideas are central to the production of the concept of homeland and the nearly 500 years of ongoing struggle to maintain it. Stories are still shared between elders and children about the war years, and the mothers who had their infant children taken from their breast only to watch as their heads were smashed against the very trees the mothers were to be hanged from. The life-giving milk that drained from the breasts of these women as they slowly died serves as a poignant metaphor for the Mexican nation's forced hegemonic discourse regarding its ingenious population.

The Yaqui project underscores the potential role bioarchaeology can play in embracing an activist research agenda within the framework of scholarly engagement. The analysis of the human remains and the massacre site along with the transnational repatriation of the fallen heroes of Sierra Mazatan all helped address what Martin (1998) has referred to as owning the sins of the past. This in turn begins to restore a level of autonomy to the Yaqui people. It can be argued that the Yaqui struggle for sovereignty is a struggle against the racism of Mexican society that is part of the structure promoting and maintaining violence and needs to be understood in the context of intellectual and political power. For the Yaqui and for us, the "real value" of this project lies in the scholarly engagement that recognizes and acknowledges the structural violence embedded in the political discourse raised by this research.

There are many examples of bioarchaeology and/or forensic anthropology taking the forefront as consultants and engaging multiple stakeholders in the practice of excavation and analysis. The current exhumations of mass graves from the Spanish Civil War are another excellent example (Ferrándiz 2006). During the Spanish Civil War, thousands fell under the category of desaparecidos (forced disappearance). The public memory of the Republican dead was silenced, and as such, the investigation and exhumations are making private and suppressed memories public. The stories of these resurfaced bodies are being connected to communities. The multiple narratives surrounding the event constitute an intangible heritage which becomes intertwined with the tangible heritage of the human remains. These stories are being told not only by the family member's narratives but also from the process that identifies the previously unidentified, through the transcript of the body. Thus the body has the potential to create order and disorder depending on the specific social context within which these narratives are expressed.

Osteological analyses are being transformed into heritage, where the skeletal remains are serving as a bridge to link multiple generations. These skeletons are sites of remembrance and contestation as Spain decides how to address the many unanswered questions from the past 70 years. The dead are encoded with meanings and are part of the cultural, social, and political logic of present-day Spain. This project demonstrates how bioarchaeologists who are engaged with multiple stakeholders at the onset of project can improve research, teaching, and integration thus incorporating reciprocal practices of civic engagement into the production of knowledge. This provides for more inclusiveness and truly collaborative projects that benefit all parties.

4.3 Excavation of Human Remains

Excavation of human remains is one of the most crucial aspects of the process of analysis because as mentioned above this is where context is preserved or lost. Yarrow (2008:125) describes excavation as the "... process of taking the site apart, whilst producing a record that preserves not only the artifacts or finds, but also the various kinds of relationships or 'contexts' through which they are related." Excavation is similar in many ways to research that relies on destructive analysis because both are moments when what information can be gathered about the remains cannot be collected ever again. Once the body is removed from where it was interred there is no way to put it back exactly the way it was. This is why documenting everything with detailed notes, numerous photos, and precise measurements are crucial.

4.3.1 Locating the Body

It is not often that bioarchaeologists locate burials or human remains, but instead, the remains are found during the process of excavating the material and architectural components of the site. In other words, human remains are discovered as a result of research projects unrelated to the burials themselves. This is also true in forensic cases where the body is discovered by accident. Despite how the remains are found and by whom they are discovered, researchers analyzing human remains in the field must still understand how to identify and survey a site. This information is crucial for both reconstructing the mortuary context and in many instances for assisting in the location of other burials at the site.

The key to understanding how to locate a body is comprehending how the search was performed and for what purpose. For example, a quick survey can reveal changes in disturbances in things like terrain and vegetation. However, there is a high likelihood that once the bioarchaeologist arrives to recover the body, an additional survey will need to be conducted to look for miscellaneous isolated bones on the surface that may have been missed. Surface finds are important because they can indicate the location of other burials. It is critical that the area be resurveyed with special attention being paid toward looking at taphonomic factors (discussed in detail below) that could have transported the remains from their original location.

4.3.2   Procedures for Removing the Body

The recovery of human remains, whether from a bioarchaeological or forensic context, requires attention to detail and sound methodology. Several researchers provide detailed descriptions of how to excavate human remains (Dupras et al. 2011; Connor 2007; Reichs 1998), and these should be consulted prior to any excavation. The two most important rules are to record and to obtain accurate measurements of as much mortuary information as possible. From the moment human remains are discovered, no one who has not been properly trained in the removal of skeletal material should be involved in the excavation. It is crucial that all archaeologists/recovery personnel involved in the excavation, photographing, mapping, and recoding of the burial be fully trained in human skeletal identification. Effective planning is key to facilitating the proper removal of the remains regardless if it is a forensic case or a historic or ancient burial. Indeed this is one of the many areas were the methodologically driven science of forensic anthropology and the more holistic science of bioarchaeology have helped shaped the best practices of excavation (Fig. 4.1).

4.3.2.1   Maintain Detailed Records and Photograph Everything

It is critical to take detailed notes and to photograph extensively the excavation of human remains from the time of discovery to the point they are removed from the site. "The rule of thumb is to be generous in making a photographic record as the chance for the same photographs will never happen twice" (Connor 2007:205). Beyond the human remains, photographs and notes of the entire mortuary site are important for understanding the mortuary context. Recent research has found that with detailed photography, it is possible to generate a three-dimensional representation of the site that will permit future researchers the ability to produce a more detailed reconstruction (Koistinen 2000).

Fig. 4.1 Students excavating at the Field and Laboratory Methods in Bioarchaeology and Forensic Anthropology field school at the University of Massachusetts, Amherst

4.3.2.2   Measure and then Measure Some More

From helping to set up, map, and lay out the grid used to control the excavation of the remains to documenting the location of each element of the skeletal anatomy as it is removed from the ground, the ability to accurately and precisely measure is critical to bioarchaeologist. It is crucial to be able to locate the body within the larger spatial context of the site (Charles and Buikstra 2002), as well as the orientation the body was in before it is removed from the ground (Binford 1971).

4.3.2.3   Case Study: Peñasco Blanco

Although an extreme example, the "excavation" of Peñasco Blanco illustrates the need to document everything. Peñasco Blanco ("White Cliff Point" or "White Rock Point") was named by Carravahal, a guide from San Juan Pueblo who was employed by Lt. Simpson, a United States Calvary officer who visited the site in 1849 (Lekson 1984:94). Lt. Simpson published the first report of the massive ruins located within Chaco Canyon in 1850. For a detailed early overview of Chaco Canyon, see Lister and Lister (1981). The site is located 100 m above the convergence of the Escavada and the Chaco wash and dates between AD 900 and 1125. This Chacoan Great House consists of a 180 m arc of rooms that was up to three stories in height and was five rows deep (Lekson 1984:94). Lister and Lister (1981:235) indicate that Peñasco Blanco contained 150 ground-level rooms and nine or more kivas. Peñasco Blanco was one of three great houses built at the beginning of the Chaco phenomenon circa AD 900. It, along with Pueblo Bonito and Una Vida, was situated where agricultural conditions were exceptional for the region (Lekson 1999:51). By AD 1050 the population within Chaco had grown, and the social organization was complex enough to warrant the inclusion of allies beyond the canyon's borders (Lekson 1999:62). This created a series of redistribution networks within the larger basin, which ultimately would lead to a shift from a subsistence economy to a political prestige economy (Lekson 1999:63).

Two distinct sets of burials were excavated from the site of Peñasco Blanco, a collection of disarticulated and culturally modified remains, and a collection of crania. The disarticulated and culturally modified human remains from Peñasco Blanco were "excavated" by Old Wello (also spelled "Waylo" and "Wylo"), and some other Navajo workmen employed by Richard Wetherill during the 1898 Hyde expedition field season. George Pepper and Richard Wetherill were excavating at Pueblo Bonito that summer, but Wetherill never claimed responsibility for authorizing excavation of Peñasco Blanco (Lekson 1984:104). Pepper (1920:378) makes reference to the disarticulated and culturally modified materials: "During the period of our work in Pueblo Bonito some of our Navajo workman cleaned out a number of rooms in Peñasco Blanco and in one of these a great number of human bones were found. Some of these, including portions of a skull, were charred, and the majority of long bones were cracked open...."

The exact room from which the human remains were excavated is not known, and the collection is now housed at the American Museum of Natural History in New York City. There is very little published on this material. Even less is known about the cranial remains removed from the site. Catalogued as Peñasco Blanco at the American Museum of Natural History, the only record of the origin of these remains is from an appendix of a manuscript on a different site (Brand et al. 1937) that suggests that they were recovered outside of the great house.

Turner and Turner (1999) include the disarticulate remains from the site in their review of violence and cannibalism in the Southwest and offered the first published examination of the material. White (1992:337–338) makes reference to Pepper's suggestion of cannibalism, but he does not include the site in his survey of cannibalized sites within the Southwest because at the time there were no published details on the human remains.

Turner and Turner (1999) treated their analysis of the remains as if it were from a single depositional unit based on Pepper's (1920:378) account of Waylo's description of his "excavation" of a single room at Peñasco Blanco. However, there is no way to know if that was the case. In fact, it is impossible to know if all of these remains are from the same temporal period, let alone their spatial distribution within the site. This is made clear in the weathering variability of some of the remains.

Fig. 4.2 Innominates from
Peñasco Blanco showing
extreme weathering
variability

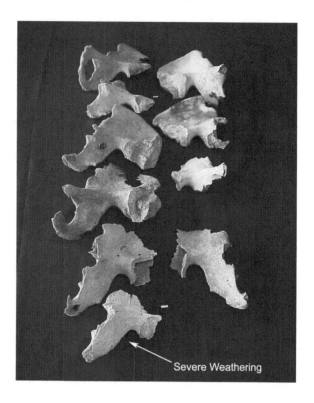

Figure 4.2 shows all the innominates represented in the collection. Note the extreme difference in the weathering of these elements. This variation compounded with the low frequency of carnivore damage suggests that the material at the very least came from several different strata and should not be analyzed as a single depositional event.

It is important to acknowledge the quality of the data set being analyzed. The "integrity" of the assemblage can be seen as a function of its deposition and diagenesis along with the excavation methods utilized during the retrieval of the material while *in situ* (O'Connor 1996:6). The Peñasco Blanco collection lacks site contextualization and cannot be placed into categories such as trash deposits, house floors, or hearth contents. Such information allows for specific analysis of mortuary behavior through the examination of the placement and preparation of the corpse, which is often related to politics, gender, power, and ritual.

Bioarchaeologists, taphonomists, and zooarchaeologists have long touted the importance of understanding the cultural and natural factors that influence the formation of skeletal assemblages and the distribution of cutmarks on skeletal material (Lyman 2010). Poor contextual information impedes the development of strong working hypotheses, scientific methodology, and strong inference. The strengths of bioarchaeology are severely limited when multiple lines of evidence are unavailable in forming interpretations. Cultural factors such as the number of people involved in processing, their mortuary methods, the number of people processed in any given period of time, the tools used in processing, and corpse taboos are all lost. Natural variables, including processing site and habitation area, time of year, ambient temperature, precipitation, and amount of natural light available, are also lost, because we have no *in situ* information (Yellen 1977, 1991; Binford 1981; Lyman 1987).

Without detailed field notes, maps, and photographs, it is not possible to reconstruct the specific and often unique cultural meanings associated with the burial assemblages. Models are only as good as the explanatory data from which they are created (Hockett 2002). If bioarchaeology is to accurately reconstruct past lifeways and infer behaviors that can have far-reaching consequences for the descendent populations, it is critical that the discipline be cautious and meticulous in evaluating taphonomic data. In cases such as Peñasco Blanco, where we have almost no context, future researchers were deprived the opportunity to infer behavioral intent of the processing of the disarticulated human remains.

Archaeology and specifically bioarchaeology have moved light-years beyond the methodologies employed 100 years ago at Peñasco Blanco. However, crucial mistakes are made every day in the field that severely impact the analysis of human remains being excavated. It is not enough that one is able to reconstruct the burial but future researchers must be able to understand the spatial reality of the data that is generated.

4.4 Taphonomy: The History of an Individual and their Remains After Death

It is important to have a comprehensive understanding of the variables that can alter the skeletal remains, and care must be given to accurately identify all of the taphonomic variables responsible for bone modification. While excavation is the means for recovering remains, taphonomy explains the state the remains are recovered in and why some remains are recovered and others are not. Haynes (1990) and Walker (2000) point out that skeletal data are as susceptible to interpretative error as the archaeological and historical sources of information researchers draw upon to contextualize them. It is for this reason that taphonomy is one of the principles to understand when working with human skeletal material recovered from either an archaeological or forensic setting.

Taphonomy involves understanding environmental conditions (abiotic), animal activity (biotic), and human activities (cultural) that effect the remains from the time of death to the day they become of interest to researchers. The body is unique from other material culture that is analyzed to understand people's lives in the past because it represents the person. Sofaer (2006) argues that the body though unique is still simply a representation of a person's lived experience and as such is the product of their and other people's actions and ideology. Understanding that the body is a record of the individual's lived experience, including biological, cultural, and social factors (Scheper-Hughes and Lock 1987) is key to understanding the importance of identifying human-related taphonomic factors.

4.4.1 A Brief History of Taphonomy

The field of taphonomy begins with Efremov (1940) who coined the words *taphos* (burial) and *nomos* (laws). In its broadest usage, taphonomy refers to everything that impinges on the physical characteristics of bone from the time of the animal or human's death up to the point of its analysis. For a more thorough explanation on the evolution of the definition of taphonomy, see Lyman (2010) and Bonnichsen (1989).

The objective of the science of taphonomy is to recognize the variables that can affect bone in order to reconstruct the environment that the animal or human occupied during its life. This includes human and animal manipulation along with soil acidity levels, erosion, soil compaction, and fluvial action. Taphonomy has been instrumental to paleontologists, archaeologists, bioarchaeologists, and forensic archaeologists. One reason for this is that taphonomy research provides the ability to distinguish between human activities and natural influences on bone. The contextualization of this information within the physical and cultural setting of the site (the spatial, temporal, and cultural context) gives a more complete understanding of the behaviors of past populations.

A "taphonomic agent" refers to the "immediate physical cause" of modifications to animal remains and skeletal tissues (Gifford-Gonzalez 1991:228). It is also vital to be aware that multiple agents can leave similar and/or overlapping signatures on a single bone, creating complex patterning. Research on archaeological faunal remains has been conducted since the nineteenth century in both Europe and America. In fact, the concept of taphonomy introduced by Efremov (1940) is actually the merger of two older disciplines, actuopaleontology and biostratinomy. Actuopaleontology (known today as the field of neotaphonomy) was based on the idea that studying the changes that affect living and recently dead life-forms as they enter the lithosphere provides a model for understanding and interpreting similar events in the fossil and/or archaeological record. Biostratinomy was concerned with the spatial relationships of faunal assemblages and how the environment interacts and alters the skeletal material from the time of death to deposition. What started as an obscure archaeological subdiscipline concerned primarily with the identification of animal remains has evolved into a complex science concerned with multiple issues. This is particularly true when one examines the influence taphonomy has had on forensic anthropology. Dirkmaat et al. (2008) argue that forensic anthropology shifted from a laboratory science primarily concerned with generating a positive ID to a field-based science that has fully embraced the principles of taphonomy and archaeology. The combination of these three subdisciplines helped to create a significant shift in the core theoretical paradigm of forensic anthropology to one that mirrors bioarchaeology.

4.4.2 Taphonomic Influences to the Bone

Inanimate forces of nature such as shifts in the tectonic plates, rock falls, sediment loading, soil pH, flooding, rainfall, ambient temperature, and wind all lead to the deterioration of organic material and are known as abiotic or nonliving forces that affect the preservation of bone. These can impact and speed the disintegration of both soft and hard tissues and leave marks and patterning on bone. Biotic forces that affect bone are living or biological organisms, such as plants and animals (including humans). Insects, scavengers, and human antemortem, perimortem, and postmortem behaviors relating to corpse processing all can impact the survival of skeletal elements and leave identifiable marks on bone. What follows is a brief description of some of the key abiotic and biotic factors that impact taphonomic interpretations of animal and human remains. Specifically, it is most appropriate here to focus on those factors that are useful in teasing apart cultural from natural taphonomic marks. This is helpful when trying to interpret behavioral practices of past and present populations.

The body of a dead animal undergoes a series of biological changes that facilitate the decomposition process. Micro- and macroorganisms begin to break down the soft tissue, and these actions denote the beginning of the taphonomic process. The survival of any or all of the skeletal tissue is going to be contingent upon a series of variables that include but are not limited to bone structural densities, weathering, animal activity, diagenesis, and transport (including abiotic disturbances such as fluvial action and sediment loading and/or biotic disturbances such as trampling). Finally, preservation is also impacted by the anatomy and physiology of the organism or in the case of humans, the individual. This includes body size, amount of soft tissue present, and pathological conditions (González et al. 2012; Ubelaker 1974).

4.4.2.1 Abiotic Factors

Bones exposed to the atmosphere begin to lose protein collagen and this leads to fracturing and destruction. As the organic and inorganic matrix of bone are destroyed by chemical and physical forces, they are reduced to soil nutrients. Bone weathering follows a general pattern, and the effects are readily identifiable. Over the last several decades, researchers have developed a very useful series of stages to identify and describe bone weathering (Behrensmeyer 1978; Madgwick and Mulville 2012). The scale goes from greasy intact bone with soft tissue still attached (weathering stage 0) to cracking, splintering, and disintegration of the bone (weathering stage 5). The rates of weathering depend on a series of variables that include taxon, body size, exposure time, ambient temperature, seasonality, and soil pH. Additionally, an understanding of the microenvironments for understanding bone weathering is important because even a difference within only a few meters can dramatically alter bone preservation. The result is that estimating time since death based on bone weathering is a geographically specific enterprise.

Once the remains are buried, they are affected by chemical interactions with the soil that leads to physical changes in the structure of the bone. This interaction between bone and soil chemistry and the resulting transformation to the bone matrix is known as diagenesis. The resulting chemical corrosion of the bone is influenced by the size and porosity of the bone, length of time in the ground, soil acidity levels, bacteria, water, drainage, and ambient temperature. Understanding how diagenesis operates can provide information about burial practices and about dietary reconstruction based on trace element and isotopic analysis (Hollund et al. 2012).

Earthquakes, rockslides, sediment loading, and erosion can all deform, break, or crush bone (Lyman 1994). Movement of soil through sequences of freezing and thawing can also lead to the dissociation and loss of elements. Bone density along with shape and size determines the amount and type of damage that will occur from the weight and movement of sediments. Dense bones such as femurs and humeri tend to survive this process intact whereas the cranium and os coxae are more likely to be crushed or fractured. Recognition of postmortem fractures caused by sediment loading can be accomplished by looking for the presence of partial fractures and fragmentation of a single element within close proximity (Villa and Mahieu 1991). Excavation can also produce postmortem fracturing of elements. These are easily recognizable due to their clean and white fracture surfaces that contrast with the darker bone surface (Ubelaker and Adams 1995; Villa and Mahieu 1991).

Hydrological processes form a significant number of archaeological assemblages (Schiffer 1987:243–256). Understanding how to recognize which bone deposits were affected by these processes and which were not is essential in determining the sequence of disarticulation as well as inferring agency. This does not just include the movement of the water but also the composition of that fluid. Recent research has found that pH level of the water or other fluid that bone is submerged in has a significant impact on bone (Christensen and Myers 2011).

The movement of fluid often referred to as fluvial action is the most common hydrological process analyzed as it can affect the bone in numerous ways, such as burying, transporting, braking, and/or abrading the remains. Researchers have identified three phases to fluvial transportation (1) movement of the body prior to disarticulation, (2) movement of disarticulated body parts, and (3) movement of isolated bones (Nawrocki et al. 1997; Voorhies 1969). The majority of research has focused on the third-phase fluvial transportation. Experimental studies have focused on single skeletal elements as they pertain to accumulation and movement in water environments. These studies have found that particular elements are more likely to be carried by water. What determines hydrodynamic transport is the size, shape, and density of the bone (Behrensmeyer 1984; Lyman 1994). Bone density appears to be the most significant factor in determining the distance an element will travel.

4.4.2.2   Biotic Factors

Both plants and animals impact the taphonomic process of all dead bodies. From the very beginning of the decomposition process, insects, soil acidity levels, plant growth, and rodent and carnivore activity all have the potential to disturb or destroy carcasses.

Animal bioturbation or disturbance can be found to affect uncovered burials. Typically research has focused on birds, rodents, and carnivores (Lyman 1994) as they have all had their bone and bone fragment collecting documented. However, research suggests that other animals should also be considered as even amphibians have been found to impact bone preservation (Stoetzel et al. 2012). Since many of these animals make their homes in caves and rock shelters, they have the potential of causing confusion with regard to human activity patterns as to who may have also occupied these locations before, during, or after the animal habitation.

Plant bioturbation is also of concern as root expansion can have a considerable impact on both the structural integrity of the bone as well as the positioning of the individual skeletal elements of the burial. Furthermore, plant decay can also change the acidity of the surrounding soil and this can affect the bone.

4.4.3 Human-Caused Taphonomy: Understanding that there are Cultural Factors that Affect Bone After Death

While bone is affected after death by a range of biotic factors, some of the most important are the changes caused by humans. These activities can include secondary burial along with other mortuary practices (e.g., ancestor veneration), and activities related to violence (e.g., warfare, cannibalism, and other direct physical violence). While mortuary practices and violence are things to consider when looking at human-caused taphonomic changes that occurred in the past, one factor to consider that happens frequently is the damage that archaeological excavation, laboratory analysis, and curation can have on the remains. This final human-caused category is known as field or laboratory taphonomy.

One of the best ways to identify human assemblages that have been modified as a result of mortuary practice is to differentiate between human and animal bone assemblages by identifying the elements present, types of modification, and taxonomic diversity. The patterns differ because animals tend to select specific portions of the body for transport that are related to consumption, while humans are much more variable in their selection of body parts. What body part is selected for transportation and display by humans is culturally specific, and these variables are often archaeologically invisible.

4.4.3.1 Violence

Trauma provides evidence of accidental injury or violent encounters. Fractures can result from accidental falls (e.g., Colles fracture) in hazardous terrain and/or high-risk activities. Fractures, such as parry fractures in the forearm and weapon wounds, serve as records of conflict ranging from domestic violence to warfare (Martin et al. 2012; Martin and Frayer 1997). By examining traumatic lesions and weapon wounds for size, location (side), extent of healing, weapon type (sharp/blunt force trauma), type of fracture (e.g., simple, comminuted), and approximate size of the affected area (maximum diameter), it is often possible to establish pattern recognition and differentiate between violent injury and accidental trauma.

From our earliest human ancestry, disarticulated bone assemblages evidencing perimortem cultural processing have been part of the history of our species (Holbrook 1982; Chacon and Dye 2007). Although it is difficult to establish specific reasons for such behavior, plausible explanations include mortuary practices, ritual destruction, mutilation, cannibalism, and violence.

The destruction of body symbolizes political dismemberment. Complete annihilation of corpses equals complete success (e.g., transition from war to victory) and power of the victor. Also, the mutilation of the vanquished emphasizes total subjugation and dominance of the victors. This type of violence is fairly prevalent in the past. Looking at just North America, there are numerous cases of dismemberment, mutilation, torture, and destruction of past populations, for example, in the American Southwest (Stodder et al. 2010; White 1992; Billman et al. 2000; Kuckelman et al. 2002), the Great Plains (Willey 1990), and the Arctic (Melbye and Fairgrieve 1994). However, this is not just a phenomenon of the past as there are cases of dismemberment and destruction in the present as well. These include Bosnia's "ethnic cleansing" that promoted the use of rape and death by mallets and hammers, the taking of hands and feet of prisoners during the Sierra Leone civil war, the knee-capping in Northern Ireland, and the dismembering and anal impalements of the Rwandan genocide are all forms of cultural performances embedded in local socio-cultural relationships (Whitehead 2004:74).

Cutmarks and chop marks related to dismemberment and destruction during massacre or "cannibalism" events can be identified using criteria established in bio-archaeology, zooarchaeology, and forensic and taphonomic sciences. In the past, cutmarks and chop marks on human bones have almost universally been identified as evidence of violence. The identification of violence is not straightforward however as what may appear to be butchery or "cannibalism" can in fact be the result of mortuary practice (Fig. 4.3).

4.4.3.2   Mortuary Practice

Recent research has shown that cutmarks can be caused by secondary processing of burials for defleshing, transporting, and displaying bodies as a form of ancestor veneration (Pickering 1989; Pérez et al. 2008; Pérez 2006). At the site of La Quemada there is evidence that the likelihood that ancestor veneration along with the ritualized destruction of enemy remains accounts for the multiple mortuary behaviors present (Nelson et al. 1992; Pérez et al. 2000; Pérez 2002). Many of the remains, particularly skulls and long bones, appear to have been placed on or suspended from racks located in several residential and ceremonial centers throughout the site (Nelson et al. 1992). Thus, although the assemblages as a whole reflect an abundance of evidence that individuals were dismembered and defleshed, analysis of the patterning of the types of bone with cuts revealed differences in both the frequency and morphology of the cutmarks at each of the deposits along with the importance placed on specific elements.

In cultures that practice corpse dismemberment, parts of bodies come to represent whole bodies. Displayed or otherwise memorialized parts of bodies are often

Fig. 4.3 Examples of cutmark and chop mark variability from two ancient human mandibles (*top*) and a primate femur (*bottom*)

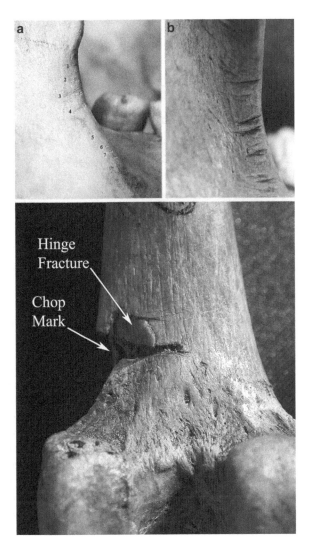

considered comforting, and have powerful symbolic messages about how to remember and obtain power from the dead.

4.4.3.3   Field or Laboratory Taphonomy

A well-placed trowel or scalpel mark can often appear similar to a cutmark or chop mark if careful consideration of taphonomy is not a part of the analysis. To the untrained eye these marks can appear to be cutmarks as opposed to tool-induced alterations and sedimentary scratches. However, trowel marks have their set of diagnostic criteria that allow for their identification. These characteristics include the general shape of the mark, differences in how they modify the bone, and the logic

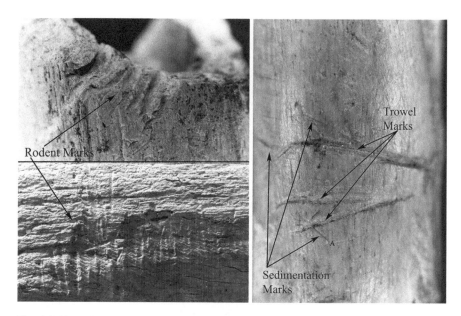

Fig. 4.4 Examples of nonhuman (*top* and *bottom left*—rodent gnawing) and human (*right*—trowel damage) taphonomic agents affecting bone

behind why they were made. In terms of shape, trowel marks are generally very straight, and the trough of the cut is u-shaped as trowels are not true blades. A trowel mark made during an excavation or a scalpel cut during analysis affects the bone differently than marks made at the time of death or shortly thereafter. The more recent marks are made as the bone is being removed from the soil, which over time has stained (Ubelaker and Adams 1995) and abraded (Thompson et al. 2011) the surface of the bone. Thus, any mark made recently is going to result in color differences between the surface of the bone (which is often darker) and the interior surface of the cut. Cutmarks made during excavation or analysis are often accidental, so they tend to have a lack of logic in regards to their patterning (Fig. 4.4).

4.5 Summary

There are many factors that act to modify the body after death, and all of these must be considered—from the excavation through to the analysis. The lack of detailed taphonomical processes can lead to erroneous conclusions. Not knowing the mortuary context can forever prohibit a full understanding of human remains. Not being able to distinguish culturally modified human remains from field or laboratory taphonomy will inhibit the ability to make accurate interpretations.

Many of these best practices can be learned by attending an archaeological or bioarchaeological field school. Participating in archaeological excavations as often as is possible is strongly encouraged for bioarchaeologists in order to obtain expertise in many facets of the excavation process.

References

Ainsworth, B. L. (2003, 6 October). CILHI, JTF-AF combines to form JPAC. www.GlobalSecurity. org, Electronic News article. Retrieved October 20, 2012, from http://www.globalsecurity.org/ military/library/news/2003/10/mil-031006-mcn01.html.

Anderson, M. (2000). Under City Hall Park, Online Features. Archaeology. Accessed October 15, 2012. http://archive.archaeology.org/online/features/cityhall/.

Behrensmeyer, A. K. (1978). Taphonomic and ecological information from bone weathering. *Paleobiology, 4*, 150–162.

Behrensmeyer, A. K. (1984). Taphonomy and the fossil record. *American Scientist, 72*, 558–566.

Billman, B. R., Lambert, P. M., & Leonard, B. L. (2000). Cannibalism, warfare, and drought in the Mesa Verde Region in the twelfth century AD. *American Antiquity, 65*, 1–34.

Binford, L. R. (1981). *Bones: Ancient men and modern myths*. New York: Academic.

Binford, L. R. (1971). Mortuary practices: Their study and their potential. In J. A. Brown (Ed.), *Approaches to the social dimensions of mortuary practices* (p. 25). Washington, DC: Memoirs of the Society for American Archaeology.

Blakey, M. L. (1998). The New York African burial ground project: An examination of enslaved lives, a construction of ancestral ties. *Transforming Anthropology, 7*, 53–58.

Bonnichsen, R. (1989). Constructing taphonomic models: Theory, assumptions, and procedures. In Robson B. & Marcella H. S. (Eds.), *Bone modification* (pp. 515–526). Orono: Institute for Quarternary Studies, University of Maine, Center for the Study of the First Americans.

Boyer, E. L. (1996). The scholarship of engagement. *Journal of Public Service & Outreach, 1*(1), 11–20.

Brand, D. D., Hawley, F. M., Hibben, F. C. et al. (1937). *Tseh so, a small house ruin Chaco Canyon, New Mexico (preliminary report)*. Albuquerque: Anthropological Series, (Vol. 2, No. 2), University of New Mexico.

Buikstra, J. E. (1981). Mortuary practices, paleodemography and paleopathology: A case study from the Koster Site (Illinois). In R. Chapman, I. Kinnes, & K. Randsborg (Eds.), *The archaeology of death* (pp. 123–132). Cambridge: Cambridge University Press.

Chacon, R. J., & Dye, D. H. (2007). *The taking and displaying of human body parts as trophies by Amerindians*. New York: Springer Science and Business Media.

Charles, D. K., & Jane E. B. (2002). Siting, sighting, and citing the dead. In *Special issue: The place and space of death* (pp. 13–25). Archeological Papers of the American Anthropological Association.

Christensen, A. M., & Myers, S. W. (2011). Macroscopic observations of the effects of varying fresh water pH on bone. *Journal of Forensic Sciences, 56*(2), 475–479.

Connor, M. A. (2007). *Forensic methods: Excavation for the archaeologist and investigator*. Plymouth: AltaMira.

Davis, V. (2000). *The long road home: US prisoner of war policy and planning in Southeast Asia*. Washington, DC: Office of the Secretary of Defense.

Dirkmaat, D. C., Cabo, L. L., Ousley, S. D., & Symes, A. (2008). New perspectives in forensic anthropology. *Yearbook of Physical Anthropology, 51*, 22–52.

Dupras, T. L., Schultz, J. J., Wheeler, S. M., & Williams, L. J. (2011). *Forensic recovery of human remains: Archaeological approaches* (2nd ed.). London: CRC.

Efremov, I. A. (1940). Taphonomy: A new branch of paleontology. *Pan American Geologist, 74*, 81–93.

Ferrándiz, F. (2006). The return of civil war ghosts: The ethnography of exhumations in contemporary Spain. *Anthropology Today, 22*(3), 7–12.

Gifford-Gonzalez, D. (1991). Bones are not enough: Analogues, knowledge, and interpretive strategies in zooarchaeology. *Journal of Anthropological Archaeology, 10*, 215–254.

González, M. E., Álvarez, M. C., Massigoge, A., Gutiérrez, M. A., & Kaufmann, C. A. (2012). Differential bone survivorship and ontogenetic development in Guanaco (Lama guanicoe). *International Journal of Osteoarchaeology, 22*, 523–536.

Haynes, G. (1990). Taphonomy: Science and folklore. *Tempus, 2*, 7–16.

Hockett, B. (2002). Advances in paleolithic zooarchaeology: An introduction. *Journal of Archaeological Method and Theory, 92*(2), 97–100.

Holbrook, S. J. (1982). *Skeletal evidence of stress in subadults: Trying to come of age at Grasshopper Pueblo*. Unpublished PhD dissertation, University of Arizona, Tucson.

Hollund, H. I., Jans, M. M. E., Collins, M. J., Kars, H., Joosten, I., & Kars, S. M. (2012). What happened here? Bone histology as a tool in decoding the postmortem histories of archaeological bone from Castricum, The Netherlands. *International Journal of Osteoarchaeology, 22*, 537–548.

Hoshower, L. M. (1998). Forensic archaeology and the need for flexible excavation strategies: A case study. *Journal of Forensic Sciences, 43*, 53–56.

Klaus, H. D., & Tam, M. E. (2009). Contact in the Andes: Bioarchaeology of systemic stress in colonial Mórrope, Peru. *American Journal of Physical Anthropology, 138*, 356–368.

Koistinen, K. (2000). 3D Documentation for archaeology during the Finnish Jabal Haroun project. *The International Archives of the Photogrammetry, Remote Sensing and Spatial Information Sciences, 38*(5), 440–445.

Kuckelman, K. A., Lightfoot, R. R., & Martin, D. L. (2002). The bioarchaeology and taphonomy of violence at Castle Rock and Sand Canyon Pueblos, Southwestern Colorado. *American Antiquity, 67*, 486–513.

Lekson, S. H. (1984). *Great Pueblo architecture of Chaco Canyon, New Mexico*. Albuquerque: University of New Mexico Press.

Lekson, S. H. (1999). *The Chaco Meridian: Centers of political power in the ancient Southwest*. Walnut Creek: AltaMira.

Lister, R. H., & Lister, F. C. (1981). *Chaco Canyon*. Albuquerque: University of New Mexico Press.

Lyman, R. L. (1987). Archaeofaunas and butchery studies: A taphonomic perspective. *Advances in Archaeological Method and Theory, 10*, 249–337.

Lyman, R. L. (1994). *Vertebrate taphonomy*. Cambridge: Cambridge University Press.

Lyman, R. L. (2010). What taphonomy is, what is isn't, and why taphonomists should care about the difference. *Journal of Taphonomy, 8*(1), 1–16.

Madgwick, R., & Mulville, J. (2012). Investigating variation in the prevalence of weathering in faunal assemblages in the UK: A multivariate statistical approach. *International Journal of Osteoarchaeology, 22*, 509–522.

Martin, D. L. (1998). Owning the sins of the past: Historical trends in the study of Southwest human remains. In A. H. Goodman & T. L. Leatherman (Eds.), *Building a new biocultural synthesis: Political-economic perspectives on human biology* (pp. 171–190). Ann Arbor: University of Michigan Press.

Martin, D. L., & Frayer, D. W. (1997). *Troubled times: Violence and warfare in the past*. Amsterdam: Gordon and Breach.

Martin, D. L., Goodman, A. H., Armelagos, G. J., & Magennis, A. L. (1991). *Black Mesa Anasazi health: Reconstructing life from patterns of death and disease*. Carbondale: Southern Illinois University Press.

Martin, D. L., Harrod, R. P., & Pérez, V. R. (Eds.) (2012). The bioarchaeology of violence. Gainesville: University Press of Florida.

Melbye, J., & Fairgrieve, S. I. (1994). A Massacre and possible cannibalism in the Canadian Arctic: New evidence from the Saunaktuk Site (NgTn-1). *Arctic Anthropology, 31*(2), 57–77.

Nawrocki, S. P., Pless, J. E., Hawkley, D. A., & Wagner, S. A. (1997). Fluvial transport of human crania. In W. D. Haglund & M. H. Sorg (Eds.), *Forensic Taphonomy: The postmortem fate of human remains* (pp. 529–552). Boca Raton: CRC.

Nelson, B. A., Andrew Darling, J., & Kice, D. A. (1992). Mortuary practices and the social order at La Quemada, Zacatecas, Mexico. *Latin American Antiquity, 3*(4), 298–315.

O'Connor, T. P. (1996). A critical overview of archaeological animal bone studies. *World Archaeology, 28*, 5–19.

Palkovich, A. M. (1980). *The Arroyo Hondo skeletal and mortuary remains*. Santa Fe: School of American Research Press.

Pepper, G. H. (1920). *Pueblo Bonito*. New York: Anthropological Papers, No. 27, American Museum of Natural History.

Pérez, V. R. (2002). La Quemada tool induced bone alterations: Cutmark differences between human and animal bone. *Archaeology Southwest, 16*(1), 10.

Pérez, V. R. (2006). *The politicization of the dead: An analysis of cutmark morphology and culturally modified human remains from La Plata and Peñasco Blanco (AD 900–1300)*. Unpublished PhD dissertation, University of Massachusetts Amherst, Amherst.

Pérez, V. R., Martin, D. L., & Nelson, B A. (2000). Variations in Patterns of Bone Modification at La Quemada. *American Journal of Physical Anthropology, 111* (Suppl 30), 248–249.

Pérez, V. R., Nelson, B. A., & Martin, D. L. (2008). Veneration of violence? A study of variations in patterns of human bone modification at La Quemada. In D. L. Nichols & P. L. Crown (Eds.), *Social violence in the Prehispanic American Southwest* (pp. 123–142). Tucson: University of Arizona Press.

Pickering, M. P. (1989). Food for thought: An alternative to "cannibalism in the Neolithic". *Australian Archaeology, 28*, 35–39.

Pilloud, M. A., & Larsen, C. S. (2011). "Official" and "Practical" Kin: Inferring social and community structure from dental phenotype at Neolithic Çatalhöyük, Turkey. *American Journal of Physical Anthropology, 145*(4), 519–530.

Reichs, K. J. (1998). Postmortem dismemberment: Recovery, analysis and interpretation. In K. J. Reichs (Ed.), *Forensic osteology: Advances in the identification of human remains* (2nd ed., pp. 218–228). Springfield: Charles C. Thomas.

Scheper-Hughes, N., & Lock, M. M. (1987). The mindful body: A prolegomenon to future work in medical anthropology. *Medical Anthropology Quarterly, New Series, 1*(1), 6–41.

Schiffer, M. B. (1987). *Formation processes of the archaeological record*. Albuquerque: University of New Mexico Press.

Sofaer, J. R. (2006). *The body as material culture: A theoretical osteoarchaeology*. Cambridge: Cambridge University Press.

Stodder, A. L. W., Osterholtz, A. J., & Mowrer, K. (2010). The bioarchaeology of genocide: The mass grave at Sacred Ridge, site LP0. *American Journal of Physical Anthropology, 141*(S50), 224.

Stoetzel, E., Denys, C., Bailon, S., El Hajraoui, M. A., & Nespoulet, R. (2012). Taphonomic analysis of amphibian and squamate remains from El Harhoura 2 (Rabat-Témara, Morocco): Contributions to palaeoecological and archaeological interpretations. *International Journal of Osteoarchaeology, 22*, 616–635.

Thompson, C. E. L., Ball, S., Thompson, T. J. U., & Gowland, R. (2011). The abrasion of modern and archaeological bones by mobile sediments: The importance of transport modes. *Journal of Archaeological Science, 38*(4), 784–793.

Tiesler, V., & Cucina, A. (2008). Joint Agendas in Maya bioarchaeology: Conducting collaborative research at the autonomous University of Yucatan, Mérida, Mexico. *The SAA Archaeological Record, 8*(2), 12–14.

Tung, T. A. (2012). Violence against women: Differential treatment of local and foreign females in the heartland of the Wari Empire, Peru. In D. L. Martin, R. P. Harrod, & V. R. Pérez (Eds.), *The bioarchaeology of violence* (pp. 180–198). Gainesville: University of Florida Press.

Tung, T. A. (2007). Trauma and violence in the Wari Empire of the Peruvian Andes: Warfare, raids, and ritual fights. *American Journal of Physical Anthropology, 133*, 941–956.

Turner, C. G., & Turner, J. A. (1999). *Man Corn: Cannibalism and violence in the prehistoric American Southwest*. Salt Lake City: The University of Utah Press.

Ubelaker, D. H. (1974). Reconstruction of demographic profiles from Ossuary skeletal samples: A case study from the Tidewater Potomac. In *Smithsonian contributions to anthropology, No. 18*. Washington, DC, Smithsonian Institution Press.

Ubelaker, D. H., & Adams, B. J. (1995). Differentiation of perimortem and postmortem trauma using taphonomic indicators. *Journal of Forensic Sciences, 40*(3), 509–512.

Villa, P., & Mahieu, E. (1991). Breakage patterns on human long bones. *Journal of Human Evolution, 21*, 27–48.

Voorhies, M. R. (1969). *Taphonomy and population dynamics of early pliocene vertebrae fauna, knox county nebraska*. Laramie: Contributions to Geology, Special Papers No.1, University of Wyoming.

Walker, P. L. (2000). Bioarchaeological ethics: A historical perspective on the value of human remains. In M. Anne Katzenberg & S. R. Saunders (Eds.), *Biological anthropology of the human skeleton* (pp. 3–39). Hoboken: Wiley.

Webster, A. D. (1998). Excavation of a Vietnam-era aircraft crash site: Use of cross-cultural understanding and dual forensic recovery method. *Journal of Forensic Sciences, 43*(2), 277–283.

White, T. D. (1992). *Prehistoric cannibalism at Mancos 5MTUMR-2346*. Princeton: Princeton University Press.

Whitehead, N. L. (2004). On the poetics of violence. In N. L. Whitehead (Ed.), *Violence* (pp. 55–77). Santa Fe: School of American Research Press.

Willey, P. S. (1990). *Prehistoric warfare on the Great Plains: Skeletal analysis of the Crow Creek massacre victims*. New York: Garland.

Yarrow, T. (2008). In context: Meaning, materiality and agency in the process of archaeological recording. In L. Malafouris & C. Knappett (Eds.), *Material agency: Towards a non-anthropocentric approach* (pp. 121–138). New York: Springer.

Yellen, J. E. (1977). Cultural patterning in faunal remains: Evidence from the !Kung Bushman. In D. Ingersoll, J. E. Yellen, & W. MacDonald (Eds.), *Experiential archaeology* (pp. 271–331). New York: Columbia University Press.

Yellen, J. E. (1991). Small mammals: !Kung San utilization and the production of faunal assemblages. *Journal of Anthropological Archaeology, 10*, 1–26.

Chapter 5
The Mortuary Component and Human Remains

Although archaeologists almost always document the context from which human remains are recovered, that information is almost always available to researchers once analysis of the remains is undertaken. Historically this was almost always the case. Often there was a disconnection between the skeletal and mummified remains and the mortuary and cultural context. This set the stage for a particular intellectual trajectory in biological anthropology and osteology that focused solely on the remains as biological entities. Little attention was paid to the culture, identity, and lived experience of the individuals themselves. The analyses were more focused on the body and not on reconstructing social identity or interpreting behavior (Martin 1998:174–176). Also, there was no emphasis on population-level analyses. Rather, single specimens with unusual pathologies or anomalies were more often the focus. These earlier studies reported on individual bone elements with detailed descriptions of their morphological features and any pathological or unusual characteristics.

This medicalized approach to ancient human remains was divorced from the archaeological context, which resulted in an inability to place individuals within a larger context or to use the data to answer questions about population-level dynamics. More importantly with this approach, it was difficult to make the case that ancient human remains had relevance for problems faced by people today, particularly the living descendants of those ancient "specimens." As bioarchaeology began to replace these more descriptive osteological studies, the mortuary context became a crucial part of the analyses. For example, Rakita and coeditor's volume entitled *Interacting with the Dead: Perspectives on Mortuary Archaeology for the New Millennium* (2005) provides an abundance of case studies that demonstrate the importance of systematic consideration of the mortuary context.

D.L. Martin et al., *Bioarchaeology: An Integrated Approach to Working with Human Remains*, Manuals in Archaeological Method, Theory and Technique, DOI 10.1007/978-1-4614-6378-8_5, © Springer Science+Business Media New York 2013

5.1   Mortuary Archaeology

It was primarily archaeologists who were interested in the architectural, structural, physical, and material aspect of graves and burial sites. Archaeologists have incorporated burial contexts into their analyses, but they have not incorporated the human remains themselves. There are many studies of mortuary contexts from an archaeological perspective that detail mortuary analyses and scientific studies of everything within and surrounding the cultural space created for dead bodies (Binford 1971; Saxe 1970; O'Shea 1981; Tainter 1975; Pepper 1909; Morris 1924). Yet, in almost all of these studies on mortuary archaeology or the archaeology of burial treatments, the human remains are not integrated into the studies or even discussed.

As this literature on the archaeology of mortuary spaces attests, there is a large body of important data that can be gleaned from the mortuary context. An excellent example of this is in the provocative and theoretically sophisticated work *Mummies and Mortuary Monuments*, Isbell's (2010) study on rethinking the role of ancient Andean kinship groups using a post-processual approach to how individuals were buried. Grave goods and offerings, preparation of the bodies, and analyses of grave structure, function, and contents inform the analysis, but the actual mummified human remains are not integrated into the interpretation. This is not a criticism. Isbell's expertise and interest is in theorizing about human behavior through the analysis of material culture. It would have taken a collaborator who was trained in the analysis of mummified and skeletonized remains to incorporate demographic and forensic/medical information into the study.

In a similarly engaging and important volume, Malone et al. (2009) present a detailed case study of the *Mortuary Customs in Prehistoric Malta* where they focus on a subterranean burial temple and cave cemeteries. Using a well-formulated set of ideas about the relationship between subsistence, status, and social organization, the authors are able to make many well-supported claims. Information about the complex mortuary behaviors of these early nonurban farmers helps to explain the complexity that emerged on the island of Malta during the Neolithic. Again, there is no data from the human remains (and likely these were very poorly preserved), but the study provides an example of the kinds of ways mortuary features can be used, in conjunction with other archaeological data, to reconstruct time- and space-specific cultural patterns.

An edited volume by Fitzsimmons and Shimada (2011) entitled *Living with the Dead: Mortuary Ritual in Mesoamerica* examines from a number of different perspectives the ways that the dead were highly politicized and how they remained an essential part of everyday reciprocal relations. The studies in this collection provide detailed and important data that permits the reconstruction of social relations by examining the articulation of the dead with their still-living descendants. Documented are the ways that tombs were continually used by the living where they would routinely carry out periodic feasts and sacrifices. And, revealed is the practice of the living sometimes retrieving grave offerings that were originally offered to the dead at the time that they died. This is an important caveat for the interpretation of the presence or absence of grave offerings. In the Mesoamerican context, the dead are seen to have important functions and roles to play, from defining territorial boundaries to witnessing ceremonial events. This collection of studies in particular highlights the enormous value of studying the mortuary context from every possible angle. Yet, it is interesting to note that there are no chapters of comparably complex analyses of the human remains from these sites. Again, this is not a criticism. It simply shows the historically created and maintained separation of the human remains from their archaeological context by archaeologists and biological anthropologists/osteologists.

Parker Pearson (2000) presents one of the more comprehensive approaches to conceptualize mortuary contexts in *The Archaeology of Death and Burial*. In this thorough account of variables that need to be considered in putting the mortuary context thoughtfully into larger studies about human behavior, he covers everything from funerary rituals to cultural views of the afterlife by providing numerous cross-cultural examples. Parker Pearson states the obvious (as have others before him) when he says "The dead do not bury themselves but are treated and disposed of by the living" (2000:3). Archaeologists have been on the vanguard of developing this into a body of method and theory that incorporates everything about the disposition of the dead to its meaning for the living. This concept was expanded in the edited volume entitled *The Archaeology of Death* (Chapman et al. 2009). The editors provide ten very specific and detailed case studies on aspects of funerary and mortuary archaeology across temporally and spatially diverse cultures. Again, although there is no mention of human remains, it is a very useful collection of studies necessary for understanding theoretical and methodological ways of approaching mortuary context.

Mytum (2004) provides a detailed methodological approach to *Mortuary Monuments and Burials Grounds of the Historic Period*. This collection of observations and practical guidelines for analysis of historic monuments is particularly useful because it covers all possible aspects within a framework that helps archaeologists and others working directly with these kinds of data sets to envision how they can carry out a study. Particularly in the historic period, it may often be the case that the human remains are legally or ethically *unavailable* for excavation and study. Mytum demonstrates how social changes are correlated with places reserved for the dead and that the study of the aboveground structures can reveal information on demography, social status, social conflict, and ethnic identity. These are some of the same areas of study that bioarchaeologists attempt to reconstruct from the analysis of the remains (see especially Chaps. 6 and 7 for the methods of analysis having to do with age, sex, health status, social class, and identity). Mytum's volume underscores the value of mortuary analyses even when there are no available human remains. This is a very important consideration when working with historic or living groups as there are increasingly more ethical and legal considerations regarding the excavation of human remains in the United States but also in many other parts of the world (discussed in Chap. 2).

In summary, archaeological approaches to mortuary context are both valuable and necessary, and they often are relied upon by bioarchaeologists for not only the data they bear but because they offer insights into what kinds of mortuary data to collect to answer particular kinds of questions. Furthermore, bioarchaeologists who can obtain access to the human remains for study can use the mortuary archaeology to broaden their interpretations regarding the human remains. The practice of bioarchaeology remedies the decoupling of biological remains from their archaeological (and cultural) context, but it also means that bioarchaeologists need to have a great deal of archaeological training or be in close collaboration with archaeologists. While ideal, this is not always possible. Yet, the more bioarchaeologists know about mortuary archaeology, the better their studies will be.

5.2 Linking Mortuary Context with Human Remains

The mortuary component of human remains provides the most immediate cultural information regarding the person who died, and thus it reveals a wealth of crucial information that can help expand the understanding not only of the dead but of the living. Although death is the end result of an accumulated set of biological, behavioral, and cultural responses to challenges in the social and physical environment, its inevitability does not mean that humans deal with the dead in similar ways. In fact, there is so much variability across time and space that it is difficult to find too many universals in the way that the dead are handled.

Because bioarchaeologists generally utilize some kind of framework that is biocultural in nature (see Chap. 1 for a review), integrating these two realms (the biological information from the human remains and the cultural information from the mortuary context) is sometimes challenging, but it is always productive. It is the mortuary data that provides information about the *interplay* between the living and the dead. It provides information about the cultural realm of the living and the ideas behind the preparation and disposition of the dead. And, it provides information on the much larger context regarding ideology and social structures as discussed in the preceding section.

Numerous studies are available on mortuary analyses from a bioarchaeological perspective. Two edited volumes considered classics in the field provide a number of case studies as examples for how to methodologically and theoretically approach the mortuary context (Beck 1995; Rakita et al. 2005). Through a case study approach, these both provide a wealth of information on the history and current uses of theory in bioarchaeology and mortuary studies.

In *Bioarchaeology, the Contextual Analysis of Human Remains*, Buikstra and Beck (2006) provide one view of the intellectual history of bioarchaeology in the United States through a series of detailed explorations of different aspects of the field of bioarchaeology as it emerged over the last 25 years. Beginning in the late 1970s and throughout the 1980s, the first generation of bioarchaeologists began integrating mortuary context into their analyses. These pioneers worked on the challenges of directing excavations of the human remains themselves, and fully documenting the archaeological context. These researchers set the standards for how excavation should proceed and for the importance of meticulously documenting the mortuary context. Many of the works covered by experts in the field in this volume are considered the standards in the field, and should be consulted early and often in the designing of a new study.

There is a larger literature on bioarchaeology and mortuary context, and this chapter aims to present a broad brushstroke of how to begin to incorporate mortuary variables in a systematic fashion. Classic handbooks and manuals that provide methodologies for incorporating mortuary context into bioarchaeological approaches are crucial to use in approaching any bioarchaeological study. The most important of these include *Burial Terminology: A Guide for Researchers* (Sprague 2006); *Human Osteology* (Bass 2005); *Human Bone Manual* (White and Folkens 2005); *Human Osteology* (White et al. 2012); *Human Skeletal Remains, Excavation, Analysis and Interpretation* (Ubelaker 1999); *Standards for Data Collection from Human Skeletal Remains* (Buikstra and Ubelaker 1994); and *Human Remains in Archaeology: A Handbook* (Roberts 2009). While focused largely on the analysis of human remains, these also provide important aspects of excavation, mortuary analysis, and curation.

Having bioarchaeologists in the field at the time of excavation is ideal and is becoming more prevalent (see Chap. 4). But in many cases, bioarchaeologists work with human remains for which the mortuary component must be reconstructed from archaeological field notes and reports (if available) or from repository documentation. Often the human remains of interest were excavated many years ago, and in some cases, there is little or no information about mortuary context. This is especially true when working with the vast collections housed within museums and other international, national, or state repositories. The lack of mortuary context is problematic not only in the sense that information is lost about the burial and grave goods, but there is often no information on provenience beyond the region and time period. The lack of provenience makes it extremely difficult to affiliate the remains with living people. What to do in this case is to attempt to locate records, publications (both in the gray and main publishing venues), and any other anthropological or other information about the context of the remains.

5.2.1 Best to Worst Case Scenarios

The best case scenario is that the bioarchaeologist is in the field, excavates the human remains, has full access to all archaeological information and reconstructions from those remains, and does the analysis and interpretation in consultation with other archaeologists and specialists. There are many case studies available in the literature that demonstrate the full methodology for doing these kinds of integrative studies, and they represent the "best practices" within bioarchaeology. Buikstra (1977:71–82) was among the first to propose in great detail a biocultural model for integrating burials into the larger archaeological context, thereby providing a means for getting at larger issues of regional cultural adaptation. Her legacy is still visible in the more recent bioarchaeological studies where mortuary archaeology is included as an integral part of the bioarchaeological study.

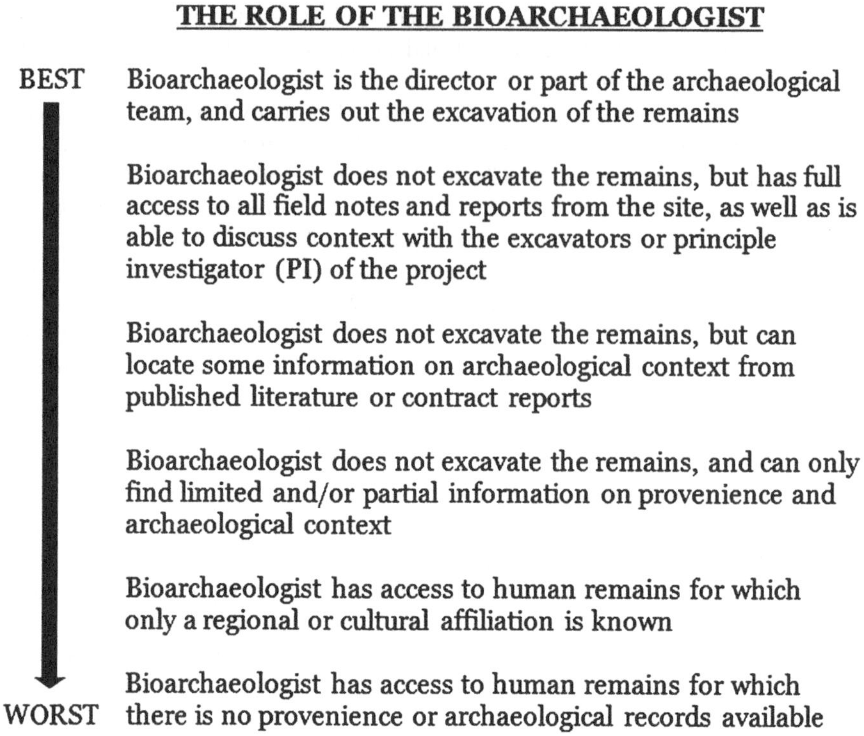

THE ROLE OF THE BIOARCHAEOLOGIST

BEST   Bioarchaeologist is the director or part of the archaeological team, and carries out the excavation of the remains

Bioarchaeologist does not excavate the remains, but has full access to all field notes and reports from the site, as well as is able to discuss context with the excavators or principle investigator (PI) of the project

Bioarchaeologist does not excavate the remains, but can locate some information on archaeological context from published literature or contract reports

Bioarchaeologist does not excavate the remains, and can only find limited and/or partial information on provenience and archaeological context

Bioarchaeologist has access to human remains for which only a regional or cultural affiliation is known

WORST   Bioarchaeologist has access to human remains for which there is no provenience or archaeological records available

Fig. 5.1 A summary of best to worst case scenarios regarding bioarchaeological integration of human remains with the cultural context

In everyday practice, bioarchaeologists work in any number of scenarios with differing access to additional information on the archaeology of the site the human remains are from (Fig. 5.1). Even in the worst case situation, where there is nothing much known about the archaeological context, there should at least be some attempt to incorporate that which is known through an aggressive and extensive literature search. Studies that focus *only* on the human remains without any contextual information are technically not bioarchaeological projects; rather, they are descriptive osteology or paleopathology reports. Bioarchaeology as an intellectual activity demands some integration across the biological-cultural divide. While this is not always possible, at the very minimum, it should be attempted.

Deciding on the most important aspects of mortuary context can be difficult because of the variability in the specifics of *when* the bodies were interred (or not interred), *how* the bodies were prepared and dealt with, *where* they were placed, and *what* kinds of rituals or practices were carried out by the living. Dealing with variability in recording of information is important because there is much value in comparing and contrasting death across different cultures and also across different time periods. To do this usually entails a data capture system that uses fairly broad and general categories of information. The following provides a very general overview of these major categories of variables and the ways that they have proven useful in bioarchaeology studies. This is not meant to be a complete listing of every possible detail regarding mortuary context, but rather it is an overview of some of the more important features. For an exhaustive and detailed set of criteria and variables to consider, Sprague (2006) is one of the best sources to consult.

5.3 Mortuary Features

Careful documentation of all possible features of the mortuary context is crucial if there is a desire to answer questions regarding how and under what circumstances a person who died is treated and interred by the living of that group. Collecting information on these aspects often aids in answering questions about why a dead person was dealt with in that particular way. Human remains come in many different arrangements. Fully articulated burials are only one of many possibilities. Bodies can be interred in one place and then moved later creating a secondary burial location. Bodies can be manually altered with extensive cultural modification such as defleshing and dismembering. Remains may be concentrated or scattered across an area. Individuals can be placed in shallow pits within habitation zones (such as designated midden or garbage areas), or they may be placed in elaborately excavated pits that are lined with stone and other materials. Burials may be within living areas (intramural) or outside of the habitation spaces (extramural). Though these are the typical ways of describing burials, it is important to remember that the variation in burial pattern is *almost limitless* in its expression. For example, Martin (1996) describes Tibetan "sky burials" as the intentional exposure of recently dead individuals to vultures and other birds of prey. The body is left in specific areas reserved for this such as mountain tops. After the soft tissue is gone, the remaining bones may be partially reduced by pounding with mallets into smaller pieces and dispersed across the area. In this example, locating and reconstructing the burial context is much more difficult, but not impossible (see Heller 2003).

Thus, every archaeological excavation that yields human remains will have its own character and specific features that define it. The following mortuary features form the broadest, and typically the initial, approach to collecting data. Different contexts will require a more nuanced approach. Photos and description of every aspect not covered by these standardized categories should complement retrieval of human remains. There are web postings of many different kinds of data collection sheets for use in the field or lab, and many of the texts cited in this chapter provide versions of data collection (see Roberts 2009:39–54; Ubelaker 1999:3–38). If the burials have already been excavated and are now in a repository, these kinds of data should be located in excavation reports or other documents. Depending on the type of archaeological site and the general patterning of the burials, there may be more or less detailed inclusions in the follow categories.

Most configurations of human remains fall into overlapping categories so that the delineation below of mortuary features is somewhat artificial. Skeletons are often found in multiple and overlapping categories. The following descriptions only begin to frame the important distinctions among types of human bone assemblages, and do not contain full literature reviews of each type. Judicious use of examples from the literature is provided as a way to show temporal and spatial diversity in mortuary configurations.

5.3.1 Interment Type

Taylor (2002) has noted that the act of burial creates the idea of the soul and its importance to humans particularly when it is monumentalized and preserved. The act of interring an individual relays information to all future descendants about where the dead are physically located in this world and how they should be treated. Ideology and notions about the cosmos provide information about where the dead are located in other worlds. The importance of burial for humans is illustrated by the fact that intentional and often elaborate burials are found from as early as the middle Upper Paleolithic.

White et al. (2012:323) say it best: "In an archaeological context it is important to recognize that there is a very large culturally determined and ethnographically observed range of variation in human mortuary practices." The variety in interment type from ancient to modern times is staggering. There are norms for each different culture, and while some patterns are more regularized than others, there is much diversity in cross-cultural comparisons. There can also be diversity within sites, suggesting the complexity in ideology that forms the behaviors of the living with respect to the dead. There are also "deviant burials" that can only be understood against the backdrop of "normal burials" (Murphy 2008).

Iserson (1994) provides a broad overview of historical, ethnic, and geographic variation in practices associated with dead bodies. These include practices such as mortuary cannibalism, grave offerings made out of scalps for Scythian warriors, disposing of the dead of the Parsees by leaving the bodies exposed to vultures, and burials at sea.

The point here is that bioarchaeologists are likely to have access to only a small proportion of the total number of humans that have died. This is because of issues of natural preservation (as discussed in the Chap. 4 on taphonomy) as well as the fact that the great majority of the death rituals as practiced over the last 150,000 years can also make recovery and interpretation of the remains challenging. However, some general guidelines are routinely used by archaeologists and bioarchaeologists to collect quantitative and qualitative data on the mortuary context.

As discussed with more specificity in Chap. 4, the goals of excavation include identifying and analyzing every human bone, no matter how fragmentary, burned, or poorly preserved they are. This point needs to be emphasized that fragmentary, or commingled, or disarticulated human remains cannot be disregarded for analysis. While these fall outside the category of discrete, articulated burials, fragmentary and incomplete burials must always be fully considered. A wealth of data is still possible to retrieve, and bioarchaeologists must view these more challenging deposits with the same systematic and scientific study (Blau 2001). The more thick the description during the excavation phase, the better the reconstruction of what happened around the time of death to these individuals.

5.3.1.1 Inhumation: Primary

This category is used for burials that are recovered in a mostly articulated fashion. Although they may be disturbed or otherwise missing bone elements due to poor preservation, primary interments suggest the placement of a whole body in the ground. Observations that should be noted include the orientation of the burial (discussed later in more detail) as well as all of the ways that the burial has been disturbed by natural taphonomic processes or by cultural interference. The more disturbed the primary burial, the more difficult it is to reconstruct all the possibilities for how the disturbance occurred. That is why when recording observations about the primary burial, meticulous attention to the details of natural weathering and animal agents is called for.

Primary burials are usually in highly variable states of preservation both between burials at the same site and within individual burials. To capture as much information as possible on how the remains came to be where they are and under what circumstances, each bone element should be recorded as to its location (point provenience) and state of preservation (weathering, root etching, carnivore gnawing, water damage, and peri- and postmortem breakage). Without very precise information on the range of variability in these kinds of factors, it will be difficult to reconstruct depositional and postdepositional features of the mortuary context.

5.3.1.2 Inhumation: Secondary

This category is used for human remains that are disarticulated and that do not show any features of being a primary burial. Secondary burials are often human remains that have been moved (possibly from an original primary context) and placed in a different location. For example in northern Australia, some cultures like Gidjingal conduct elaborate secondary burial rituals which involve the carrying around, painting, and smashing certain parts of the skeleton before reinterring the elements in a log or cave (Pickering 1989). Once human remains lose all of their soft tissues, the skeletonized remains are not easily scooped up. Remember that there are 206 skeletal elements in adults and many more elements in children whose bones have not yet fused. Once the tendons (that connect bones together) and the ligaments (that anchor muscles to bones) are gone, individual bone elements will separate. Small hand and foot bones are those that are usually displaced or lost during a secondary burial.

There are many cultural practices that include rituals at the time of death and placement of the body in a location where it will naturally desiccate. Tree and scaffold burials were common among several indigenous Native American groups in the United States such as the Crow, Dakota, and Blackfoot (Ubelaker 1999:7–10). Tree burials were also used in the Old World as well. The Colchiens, Tartars, and Scythians put their dead in leather bags and hung them from trees (Bendann 1930/2003). Long after the bodies were reduced to bones, they were bundled together and reburied in a second ceremonial function.

5.3.1.3 Inhumation: Multiple

There are cases where burials are found in different kinds of arrangements, with several bodies in one location. These multiple burials can be all primary interments, or they may be mixed, with some secondary deposits mixed in with primary. Whatever the arrangement, more than one person in a context makes documentation more difficult and time consuming. But, in these cases, it is crucial for being able to interpret how several individuals came to be placed in a single mortuary features. Multiple burials are controversial in the sense that it is usually under special or unusual circumstances that several people die at the same time and are interred together. Epidemics, warfare, and massacres are situations that may account for multiple burials, but often the context is less clear. For example, Formicola and Buzhilova (2004) demonstrate the complexities in analysis for Upper Paleolithic burials from the site of Sungir (Russia). Two children were buried together with spectacular grave offerings. Combined with the skeletal analysis, it appears that one of the children had an unusual pathological condition of the femur bones, and the authors find links to other burials in this time period where there appears to be "...a patterned relationship between physical abnormality and extraordinary Upper Paleolithic funerary behavior" (2004:189). In the case of multiple burials, the data from the biological and the cultural reconstruction needs to be integrated in order to specify the circumstances under which more than one individual may be interred.

5.3.1.4 Cremation

Ubelaker (1999:35–38) presents a very thorough overview of all of the relevant aspects of excavation, recording, and analysis of burned bone that is most useful for a bioarchaeologist. Intentional burning of the body will result in what has come to be called cremains. If the heat applied is sufficiently high, the boney remains will be fragmentary, small, and likely to have changed in chemical composition. Based on the number of identification of cremains, it may be possible to establish some basic information on the identity of the individual or individuals who were cremated. The position of the cremains may reveal something about the type of pyre or other aspect of how the body was prepared and burned. Fire also produces many changes to the size and surface structure of bones that remain. The overall color of the cremains shows great variability, and this information can be used to reconstruct how long the bodies were burned, the heat of the fire, and the overall condition of the individual (e.g., whole when burned or dismembered and then burnt). Finally fracture patterns of the bones may help reconstruct if the body was burned shortly after death or if the body was cremated long after death. Often burials may be partially burned as well, so it is important to do a bone-by-bone assessment of surface changes that might be related to natural or cultural factors. There are many new techniques being applied to the study of cremated human remains (e.g., Harvig et al. 2012), yet they are so challenging and problematic that many cremated remains are never thoroughly analyzed.

5.3.1.5 Cultural Modification

When the bodies of individuals who have died are processed or in any way acted upon with implements in order to modify the body, it makes it extremely important for the contextual information to be scrupulously recorded. One of the more detailed analyses of culturally modified human remains is presented in White's (1992) case study from a site in Mancos, Colorado (circa AD 1100). In terms of methodology and rigor, this study is important because it is the first time that a bioarchaeological study was presented using very systematic data collection similar to the ways that faunal analysts analyze animal bones that have been culturally modified (i.e., cut, chopped, broken, reduced).

Other studies have continued to improve upon this original method for data collection from highly processed human remains, most notably the Animas-La Plata case study by Perry et al. (2010) who have excavated, documented, analyzed, and interpreted a very large deposit of human remains representing at least 33 individuals who were tortured, defleshed, dismembered, chopped, cut, smashed, and in some cases burned. White provides a very useful methodology (1992:116) for coding bone elements that are cut and fractured. Others have continued to fine-tune this basic methodology for different cultural contexts (see especially Perry et al. 2010, Chaps. 12 and 13).

In a very complicated burial assemblage from the Middle and Late Iron Age in England, Redfern (2008) integrates a great deal of taphonomic, biological, archaeological, and cultural data to interpret secondary burial practices where perimortem bone modification was also practiced. Reduction of the dead bodies included excarnation (defleshing), burial, retrieval, secondary burial, and selection of some bones for special treatment. Blunt force trauma on some of the male crania provided additional information about the nature of the bone assemblage. This study is a very good example of the ways that a systematic and detailed study can reveal an extraordinary amount of information about the identities of those buried, the sustained treatment and ritual around burial and reburial, and the cultural context of those behaviors.

5.3.1.6 Isolated Bones

Every bioarchaeologist who has worked at an archaeological site in the United States or abroad is aware that seemingly random and largely isolated human bones and bone fragments are common. Tracing the perimortem and postmortem history of isolated bones is often difficult because they usually lack a defined mortuary context. It gets worse because often isolated human remains will get bagged with faunal remains by mistake, or they may show up in ethnobotanical material being screened, or they may simply be found in what archaeologists refer to as "fill." Archaeological fill is usually artifact-free soil that has settled into cultural features and is presumed to have come from someplace else such as through wind. There are few methods or standards available for properly documenting the context of isolated human bones. Patterns in these anomalous specimens can contribute to a better understanding of burials across the site. Deciphering the process by which the bone became isolated from individual skeletons is very difficult.

Often isolated bones are not identified or catalogued in the field. They may be bagged with a provenience, or they may be added to other bags from the same square that hold artifacts or faunal remains. However, it would be easy enough to establish a systematic method for bioarchaeological purposes. What most needs to be documented is the cultural and noncultural formation processes. Thus, relevant attributes of the isolated bone need to be recorded. This includes the precise location of the bone; the relationship of the bone to other artifacts and features; the soil types around the bone; the completeness of the bone; fracturing and breakage; assessment of premortem/antemortem, perimortem, and postmortem changes to the bone; weathering; burning; tool marks; and tooth marks.

If, as example, there was a low frequency of rodent gnawing it would suggest that the isolated bones were not on a living surface for any length of time. If there was weathering and sun bleaching, then it would be clear that the isolated bones had laid on the surface for some length of time. Thus, observations made in the field of isolated human bone remains could greatly extend bioarchaeological interpretation. These observations might also make possible the linking of isolated bones with the rest of the skeleton. Isolated human bones retrieved from archaeological sites have the potential to aid in the overall understanding of mortuary behavior included with the analysis of primary to secondary burials, disarticulated and fragmentary bone, and culturally modified assemblages. Anomalous specimens, patterns in location, and other characteristics of isolated bones from archaeological sites can be invaluable.

Margolis (2007) conducted a systematic and rigorous study on isolated bone that was collected throughout many years of excavation at Grasshopper Pueblo (AD ~1400). He noted that isolated human bones were often bagged along with faunal bone, or were found in the context of articulated burials. His study included over 1,800 isolated bones that were not designated as a burial or included in with other burials. One conclusion was that site formation processes were at play. One of these processes was the ancient cultural disturbance of burials by the activity of digging by the original inhabitants of Grasshopper Pueblo. Digging may have occurred for many reasons, but the most prominent at Grasshopper would have been for the placement of subsequent burials and to a lesser degree for features (e.g., pits and pit structures). Similar to rodent gnawing, the disturbance of burials by later burials is well documented by the excavation field notes. Thus this study of isolated bones provided important cultural information about the community and about the ways that bones become displaced. This study strongly argued for the analysis of all isolated human bones in conjunction with burials.

Isolated bones also have taken on an increased importance since the passage of NAGPRA. Although the legislation concerns itself with burials, isolated bones do need to be addressed within the framework of NAGPRA. Salvage archaeology, limited testing, and other cultural resource management (CRM)-type excavations increase the possibility of finding only isolated human bone. The inadvertent discovery of an isolated bone usually brings work to halt. Having a sense of how human remains become separated, displaced, and isolated across a site could help in establishing the larger context of the site.

5.3.2 Bone Concentrations

Concentrated collections of human bone are found in many different kinds of context, and they present some unusual excavation challenges. They demand specialized data collection, both for the mortuary context as well as for the information collected from the bones. The standard observations made for bones from articulated burials (such as in Buikstra and Ubelaker 1994) are applicable to fully or partially disarticulated bone concentrations and should be used for comparability to burial data, as well as to the isolated bone data. These include the observations of element, side, segment, completeness, age class, numerical age, paleopathology, and sex when possible.

5.3.2.1 Ossuaries

An ossuary is a special form of secondary burial that contains human remains of more than one individual. Sometimes the bones are collected from an original location and put into a pot or urn. Ossuaries also could be large numbers of individuals all placed in a single pit feature. Over time, the bones become commingled in these kinds of contexts, and they present a different kind of problem in interpretation. The term can be applied to situations where there are disarticulated or secondary burials, and so it is not a precise category in that it may overlap substantially with other designations such as multiple interments or secondary burials.

For example, the dead in ancient Arabia during the Bronze Age were normally buried in collective circular tombs. At Tell Abraq, such a tomb (Potts 2000) was constructed of beach rock, and it had a single, internal wall running south from the northern side of the tomb that divided the internal space into two chambers with an entry door placed at the juncture. Such an entry would have been required for the repeated interment of the dead over time. Upwards of 400 individuals have been identified in the tomb, and these bones were entirely commingled creating a dense bone bed. In this case, this very large ossuary was likely a secondary burial for individuals who died although this interpretation has not yet been verified.

5.3.2.2 Intentionally Disarticulated Assemblages

Bioarchaeologists share many intellectual interests with forensic anthropologists, and one of the biggest areas of overlap is in interpreting the nature of violent encounters that include disarticulation of the body or reduction of the body into smaller pieces. Both in antiquity as well as today, these kinds of activities create human bone concentrations. Acts of anthropophagy (or cannibalism more popularly) have resulted in human bone assemblages that are fully commingled and disarticulated (White 1992; Turner and Turner 1999). However, violence is not the only reason why remains are intentionally disarticulated, as certain burial practices often require the disarticulation of remains (e.g., Tibetan sky burial discussed earlier).

In terms of the mortuary context of intentionally disarticulated bodies, it is crucial to collect as much detailed data about the placement and deposition of the bones. The end result will be to say something about how those bones came to be in the deposit, the various activities of the living around and after the time of death, and the order and sequence of processing the bodies. If the bones are splintered and highly fragmentary, it will be important to try and refit or conjoin pieces that may come from one individual. While this activity may be done later in a laboratory setting, it may be much easier to do in the field with bones that are in close proximity to each other within excavation units. Point provenience of each and every fragment will provide a body map for the whole site. This is important for understanding the possible relationship between human remains in different contexts across the site.

5.3.2.3 Massacres

In *Broken Bones, Anthropological Analysis of Blunt Force Trauma*, Galloway (1999) presents a full range of modern cultural activities that produce broken bones, and she also provides a bone-by-bone description of the characteristic ways that bone fractures and breaks under particular kinds of forces. Using a wide range of clinical and forensic case studies, this book provides an important starting point for analysis of assemblages that exhibit perimortem trauma and breakage. Excavating ancient sites of warfare and massacres is challenging because although context is important for understanding the event, the end result of blunt force trauma and other perimortem fractures and marks on bones will not reveal who the victims are from who the perpetrators are. Only extremely meticulous documentation of the mortuary site as well as related areas of the site will help in distinguishing victims from aggressors, executions from hand-to-hand combat, and torture from captivity (see Martin and Frayer 1997 for a wide range of examples of massacre sites and bone analyses; Kimmerle and Baraybar 2008).

Excavation of mass graves presents major challenges in recording. Fiorato et al. (2000) provide one of the more complete bioarchaeological analyses of a massacre site at Towton, North Yorkshire, during the fifteenth century. Recording methods used during the retrieval of the human remains were extremely systematic and utilized several innovative techniques including the use of rectified vertical photography and the use of the Harris matrix to illustrate the order of interment of the skeletons from the mass grave.

5.3.3   Position, Orientation, and Dimensions

Every burial recording form prompts the excavator to detail the position (flexed, semi-flexed, or extended) and orientation of the body and which direction it is facing (using standard compass readings based on true north). These are most important to document for primary burials, since these are largely complete individuals that were intentionally placed within mortuary features. The intentionality behind how individuals are placed is important because it often can reveal ideology about both the living and the dead. If the burials all face a certain cardinal direction, that is likely a meaningful piece of a larger cosmology. Burials placed facedown may reflect "bad deaths" or burial of criminals or other societal outcasts (see Roberts 2009).

5.3.4   Grave Goods

Grave goods have always been an important part of the archaeological artifact assemblage from sites being excavated. In situations where there are single individuals within well-defined pits or structures, it is usually possible to associate cultural artifacts such as stone tools, pots, or organic matter with the individual burial. The proximity of these items and their shared stratigraphic location within the burial feature provides solid evidence for the placement of these items around the time of the burial. These grave offerings or grave goods are important features of the mortuary context because they reveal much about the cultural belief system of the group.

There has been a tendency to equate rich grave offerings with high or higher status than burials without similar offerings. Many studies have been done that compare various aspects of the mortuary context (including the quality and quantity of grave goods) with social identity, social status, or social organization and level of complexity. However, these kinds of correlations are never straightforward (Binford 1971; Tainter 1978; Braun 1981). Robb et al. (2001) compared social status (as determined by quality and quantity of grave goods) with biological status (as determined by the presence of skeletal indicators of disease and poor health) for a burial population from Pontecagnano, Italy (circa fifth century AD). They found a statistically significant correlation between sex, grave goods, and the level of occupationally related physiological stress. They interpreted these data to suggest that the grave goods were related to the division of labor and possibly gender more than simple social status. This study shows the importance of combining skeletal and archaeological data because ". . . the result is a deeper picture of the social and economic life of the community than can be obtained from either source" (Robb et al. 2001:213).

Much care must be taken in the interpretation of grave goods because they could be the possessions of the dead, gifts from the living, symbolic representations of the dead person's life, or items to help transport the dead into another world (Parker Pearson 1999:7–11). Also, different burial contexts such as multiple burials or secondary burials create challenges for associating artifacts with the human remains. As example, burials placed in middens may have many pieces of pottery sherds, broken stone tools, and other kinds of material culture near to the human remains, but it is difficult because of the context to know if they are grave goods or just debitage from daily life. It is also difficult to distinguish between deliberately placed items and those that may end up in the burial through chance or random placement. Even more important is to understand when graves are revisited or robbed in antiquity. In both of these cases, grave goods may be removed.

Uerpmann and Uerpmann (2006) meticulously reconstructed the burial of two individuals buried with iron arrowheads (and one with a camel) that may represent warriors at Jebel al-Buhais (circa AD 200). The interpretation goes on to suggest the possibility that these warriors may have taken part in battles related to the spread of Islam in the region. These kinds of hypotheses laid out to account for the observed skeletal findings and mortuary configurations provide a lively way to engage the reader in thinking about the implications of these post-Neolithic burials for understanding larger patterns at the population level.

5.3.5 Representativeness of the Mortuary Sample: Biases by Age, Sex, and Status

Researchers have attempted to address the age and sex biases that may result in mortuary contexts. Age-associated biases appear to one of the major areas where there may be clear discrepancies. Walker et al. (1988) demonstrated that discrepancies occurred when comparing skeletal remains from a historic Indian cemetery in California that had documentation on the number of individuals placed in the cemetery and age at death. Conducting a demographic study based on the recovered skeletal remains, they found that infants and elderly adults were markedly underrepresented when compared with the historic records.

Pinhasi and Bourbou (2008) present a very thorough overview of the assessment of the representativeness of skeletal collections. They suggest that variables such as bone survival and recovery, the underrepresentation of subgroups such as infants and children, burial practices, and the size and composition of the living population all can have profound effects on recovery and analysis of the human remains. Biases in the skeletal assemblages are not always simple due to a wide range of taphonomic and cultural processes that have profound effects on where bones are preserved and whether bones are recovered. One of the biggest problems is assessing if some individuals are buried elsewhere, beyond the confines of the archaeological site.

An example of the complexity introduced into burial patterns by the living is provided in modern-day Mexico. In a cemetery in Culiacán, Sinaloa, the center of one of the larger drug cartels in Mexico, there are lavish mausoleums built upon the burials of Mexican drug traffickers who have died. Meanwhile, the victims of these drug lords (both random and calculated retributions for perceived wrongs) are dumped in the desert or are dismembered and scattered throughout the city, or are retrieved by family members and placed in modest graves. This modern-day mortuary behavior provides a caveat for making quick judgments about the status of the buried person via analysis of the associated mortuary monument and grave accoutrements. In this context, the murderers and aggressors have the high status burials while the victims have plain or unmarked burials (Pachico 2011).

Another further complicating twist to mortuary behavior and social processes comes from modern-day China. The New York Times presented a story that provided another caveat for associated grave types with status and class (LaFraniere 2011). Many large and very ostentatious mausoleums and tombstones in China are regularly attended to by state workers who are paid to clean and maintain the mortuary monuments and grounds. This show of wealth with ornate tombs is a relatively recent phenomenon in China, starting only in the 1970s. But these are becoming rare as conservation of land becomes more of an issue and as the gap increases between the rich and the poor. Given a recent state decree, low-cost small burials have become the national policy, and plots for ashes measuring 4 by 4 ft are now the norm. Tombstones can be no higher than about 40 in. In this case, a political move by those in power is creating a shift in the relationship between the type of grave and wealth.

It is important to remember that bioarchaeological studies are based on incomplete and biased sources of information. It is incumbent upon those carrying out bioarchaeological studies to be aware of the ways that the mortuary sample may be biased or nonrepresentative of the living populations. Partial help in remedying this situation can come from having ethnographic and ethnohistoric data to complement the study and fill in any gaps. But there also needs to be some creativity in how the data are approached. The example of the drug lord's high status burials and the political decree in China that all burials must now be exactly the same are two cultural innovations that go against how mortuary data might be interpreted without thoroughly understanding the cultural context.

5.3.6 *Ritual*

Ritual behavior, ritual items, ritualized performances, and ritual spaces are all important aspects of death and mortuary, yet it is often difficult to see ritual in the archaeological record. There are many definitions of ritual behavior and its causal relationship to various outcomes, but there is actually no encompassing definition of ritual behavior (see Liénard and Boyer 2006; Smith and Stewart

2011 for more ideas about what ritual behavior is and is not). There is some agreement that rituals organize society in particular ways, and when performed or engaged with, rituals are bound by rules of use and specific (usually rigid and well defined) material objects and behaviors. Walker (2002) suggested that evidence for ritual behavior is present but often obscured in the archaeological record because it is difficult to know what to look for. Howey and O'Shea (2006) go so far as to say that whenever something cannot be attributed to an everyday activity such as diet and trade, it is labeled as a ritual object or ritual activity. Yet there can be rituals that are performed within the mundane realms of diet and trade, and so distinguishing ritual from normal routine is often very difficult.

Ritual is profoundly associated with notions of religion and the supernatural, and these realms come together in dealing with death. Charles and Buikstra (2002) demonstrate even more complexity in the notion of burial rituals by trying to tease apart mortuary ritual from ancestor cults in their innovative study on Mississippian funerary activity. By looking at changing mortuary behavior over time (circa 7000 BC–AD 1200), they were able to show that mortuary ritual mirrors ideas about current economic and political contexts, and any changes in these will constitute a change in mortuary ritual as well. This study also discussed the ways that the living "... actively manipulate the rituals and symbols surrounding death as they negotiate their lives through ever-changing social and political contexts" (2002:22).

The bioarchaeology of ritual may be challenging to interpret from the evidence of the human remains and the contexts they are found in, but it is a necessary component to build into the study. Ethnographic and ethnohistoric literature as well as folklore and oral tradition may all be productively utilized in aiding in the interpretation of ritual objects and ritual behaviors.

5.3.6.1 Incorporating the Dead into Rituals and Ceremonies

In the Moche culture, burial is often delayed for prolonged periods of time while ceremonies and ritualized activities take place. Once the Moche are buried, they are often retrieved by the living, and more offerings and ceremonial tributes are made (see Millaire 2004 for a very detailed bioarchaeological study of Moche mortuary ritual). In this context, mortuary spaces were not static places where the dead resided; rather, they were dynamic sites that were often revisited and where the human remains were handled and manipulated in complex way. The practice of grave reopening and the taking of some skeletal elements are seen as a way to control, recognize, and acknowledge the continued *active* relationship between the living and the dead. Here there is a complete lack of an ontological divide between the categories living and dead as it relates to personhood and social influence. This kind of complexity in mortuary ideology was only figured out and understood because of very careful excavation and observation of the location and orientation of various human remains across the site.

5.3.6.2 Rituals Involving Manipulation of the Dead

Bioarchaeologists must incorporate the collection of systematic and standard data as well as take a more nuanced approach tailored to the complexities of the culture at hand. As discussed in other sections, it is important to know if a body found with a smashed in skull was the victim of a violent action, or was the aggressor who met her or his match. Duncan (2005) provides a good set of methodologies for teasing violence from veneration. In addition to looking for ethnographic and ethnohistoric information that may help clarify the context for these different behaviors, he suggests the use of a model that helps to conceptualize behaviors before, during, and after the death of people that are loved vs. people that are hated.

There are many different kinds of culturally constructed notions about the ability of the dead and their ability to continue to exert power over the living. Mortuary cannibalism as practiced by the Wari in Brazil was practiced up until the 1960s, and Conklin (1995) provides a cautionary note about the motivations and cultural significance of it. She demonstrates how eating the dead fits in with cultural behaviors associated with grief and mourning. The social significance of the body plays a role in how the living organized their thoughts about the dead. Consuming of some of the flesh of the dead provides a mechanism for binding the living to the dead in specific and important ways. Specifically, the dead body literally becomes the ground on which emotions are reconciled and renegotiated. Thus, bioarchaeologists need to factor in the possibilities of positive and life-affirming aspects of behaviors such as anthropophagy in addition to the more common notions of it being something done as part of the theater of war and violence.

Other rituals and ceremonies involve the constant movement of the dead from one place to another particularly in cultures where the ideology is in a belief that the dead remain a vital part of the ongoing daily life of the community. In case of the movement of remains from one resting place to another over a period of years, as is seen in Madagascar and Chinchorro cultures of South America, mortuary context may appear very confused. Detailed documentation while the human remains are in situ is the only way that the bioarchaeologists will be able to reconstruct the various uses of the dead in ceremonies, dances, governance, or routine seasonal rituals (Arriaza 1995).

Jacobi (2003) presents a very thorough overview of the belief in and evidence for such things as zombies, vampires, ghosts, and reanimated corpses. By using a cross-cultural approach, Jacobi carefully outlines the salient features within a variety of cultures that show a strong and enduring belief in the continued presence of those who are known to be dead. From the Maya to the Netsilik and to the Navajo, there are strong notions about the ways that the dead can still have tangible effects on the living. Understanding these belief systems can help the bioarchaeologist differentiate between mortuary behavior that is meant to honor and venerate the dead, and those that are meant to keep the dead from reentering the realm of the living.

5.3.6.3 Animal Graves, Animal Offerings, and Animal Companions

Inclusion of animal remains in mortuary contexts is common, and it is important to document the location and proximity during excavation. Oftentimes animal bones associated with human burials are bagged separately and sent to the faunal analyst. When this happens, the information about the relatedness of the animals can be lost.

Animals may be included as sacrifices and grave offerings, or they may be included as guides or companions. The animal bones may be inclusions as part of ritual and ceremonial feasting done at the funeral. Finally, animals included in human graves may be symbolic of warlike attributes or fierceness. To distinguish among these kinds of culturally motivated inclusions in human mortuary contexts, detailed information during excavation needs to be recorded that may capture the ethnic identity or clan membership of the individuals.

Dog and wolf remains deposited with human remains is a worldwide phenomenon (see Morey 2006 for a cross-cultural review). Ethnographic and ethnohistoric accounts of the uses of dogs and wolves in rituals and ceremonies suggest that there is a great deal of symbolic and cultural meaning attached to these animals. Certainly bears played an important role in burials during the Paleolithic, and again, understanding the significance is important, and therefore the precise orientation and location of animal remains must be on par with that of human remains. During and following the Neolithic Revolution, there was an emphasis placed on cats, evidenced by the discovery of several cat remains with or near human burials (Pennisi 2004; Linseele et al. 2007; Vigne et al. 2004; Wang et al. 2010). Owls, birds, and foxes are also found in human burials.

Animal burials and animals in their own mortuary contexts are also of interest because they reveal information about cultural ideology regarding human-animal relations and may provide insight into human burials. For example, pigs were an important part of highly ritualized ceremonies in the Mycenaean culture (Hamilakis and Konsolaki 2004). Camel and horse skeletons were found in their own graves in the protohistoric era in the United Arab Emirates (Uerpmann 1999). Dogs found in their own necropolises in ancient Turkey suggested that hunting and working dogs were highly valued, and when they died they were given their own burial chambers (Onar et al. 2002).

5.3.7 Spatial and Locational Information

Graves and burials need to be understood collectively, as if looking at their placement from high above. The routine use of certain spaces or landscapes plays an active role in the cultural meaning of these spaces and that can influence decisions and behaviors. Furthermore, these contextualized places become part of the ideological heritage of a group. Once this happens, the landscape is no longer just the physical landscape but a conceptualized one (also known as a "perceived" environment) (Butzer 1982:253–257; Knapp and Ashmore 1999).

Understanding burials across these landscapes can help understand both the ideology of special or symbolic places in the culture and ideas in regard to the person's status. This same idea can also be applied when human-made structures are built such as walls, wells, and monuments, and these are often referred to as "built" environments (Ceruti 2004). Burials on the landscape often leave important built environments such as formal cemeteries and monuments (Mytum 1989).

Constructed landscapes are important for bioarchaeological analyses because they often leave easily discernable remains which show important distinctions of space that usually reflect political or social organization. Buikstra and Charles (1999) document how a particular ideology about the natural topography and landscape (the perceived environment) was routinely used and transformed into a constructed environment over time. They provide evidence that higher political and socioeconomic status was associated with burials on bluffs, hills, and ridges in the earlier archaic groups of the Mississippian. In later time periods, high status individuals were associated with burials in the constructed burial mounds, showing that these ideologies of landscape and status were transferred to the new constructed environments.

Spatial perspectives can be conceptualized on a variety of different scales. Many of the concepts regarding landscapes can be applied to the smaller scales of space such as intra-site and intra-household spatial organization. Routine occupancy of these areas shapes the meaning of different spaces according to cultural traditions. These traditions have accompanying behaviors which leave resulting archaeological assemblages that can clue us in to things like subsistence activities and social relations at a village site and within a household. Sayer and Williams (2009) present a very useful compendium of chapters; this is very helpful for understanding the spatial component of burials.

As just one example of the importance of space, residential burials as a spatial category regarding the placement of burials represent the interment of the dead in living spaces. These may include burial rooms, subfloor graves, and burials within walls or in ventilator shafts (Adams and King 2010). While this designation falls within the broader categories, intramural and extramural, Adams and King take the idea of the use of residential space and residential burials to have special meaning, and they present many examples of how this is so. They conclude that "As an extension to social memory, burial and mortuary practice can foster the integration of individuals and related households that form social groups and the continuity of these groups through time" (2010:5).

5.3.8 Ancestor–Descendant Relationships and Perspectives

One of the places bioarchaeology has the potential to make huge inroads into being a more inclusive and collaborative enterprise is to always consider the historical relationship between living descendants and their ancestors who are being studied through their mortuary patterns and skeletal remains. There is often much that can be learned by doing research into the ethnohistory through archival documents and through contemporary ethnography with the living descendants.

For example, in an edited volume on *Ancient Burial Practices in the American Southwest* (Mitchell and Brunson-Hadley 2001), one of the chapters provides a descendant group perspective on mortuary studies of their people (see the chapter on the Hopi by Ferguson et al. 2001) and is the only chapter to do so. It is a compelling read because the indigenous perspectives on death, dying, and the afterlife are quite at odds with the usual bioarchaeological interpretations. For example, Hopi people believe that the dead continue to play important roles in everyday life and that the connection between the living and dead is dynamic and ever changing. It is an obligation of Hopi people to care for these ancestors, and excavation and analysis prohibit the continuation of this decade's long process. While archaeologists assume certain things about the ancient Hopi people regarding migration and abandonment of sites, Hopi understanding of their ancestors rarely maps on to those interpretations. In multiple ways, Hopi concepts about death and dying conflict with concepts of archaeological research and analysis of biological remains. While NAGPRA has been hailed as an innovative and progressive move towards inclusion, it is also a huge financial and emotional burden to place on tribes. Thus, understanding mortuary behavior through contemporary and ethnohistoric research is part of bringing bioarchaeology into a more ethical domain.

5.4 Ideology and Death

Mortuary studies are often found within archaeology, and there are a number of volumes that define it within the context of the archaeology of death and burial. Rakita and Buikstra (2005) provide one of the better overviews of the theoretical paradigms that the study of mortuary architecture and context have been influenced by. Because mortuary studies, and here we are referring primarily to ancient, precolonial/precontact populations, have been situated within archaeology and not biological anthropology, it was greatly influence by processual and post-processual perspectives. Mytum provides a very thorough overview of mortuary studies from the historic period, and he presents an overview of different theoretical approaches that can be utilized, from culture-history, functional and structural approaches, to Marxist and symbolic perspectives (2004:5–10). Chapter 3 also covers a number of theoretical approaches one can use to approach ideological and cultural factors affecting human behavior around death and dead bodies.

5.5 Case Study: La Plata Skeletal Analysis—How to Integrate the Different Levels of Analysis

This abbreviated case study provides a cautionary tale about the importance of linking information gleaned from the human remains with information regarding the archaeological and mortuary context. As part of a large CRM project operated out of the Office of Archaeological Services in Santa Fe, New Mexico, the limited excavation of several habitation areas yielded 65 burials. These were excavated by

Table 5.1 Frequency of healed trauma for La Plata

	Children	Males	Females
Cranial	1/16 [6.2%]	3/13 [23.1%]	6/10 [60%]
Postcranial	0/16 [0.0%]	3/15 [20%]	6/12 [50%]

the archaeological team and transported to a repository in Santa Fe, New Mexico. A bioarchaeology team led by Martin et al. (2001:13–32) analyzed the human remains about a year after the excavations had ended. The remains are from a group of Ancestral Pueblo people (formerly referred to as the Anasazi).

Located near the borders of New Mexico and Colorado, the La Plata River Valley in northwestern New Mexico (near the Colorado border) was a permanently watered, productive agricultural area in which more than 900 sites have been reported. The valley was continuously occupied from AD 200 until about AD 1300. Large communities were maintained throughout the occupation. This area was lush by local and regional standards, and density of available resources was high. Agricultural potential was likewise very good; there is also ample evidence of hunted and domesticated game in the diet. This area is located in the middle of a large and interactive political sphere of influence with Mesa Verde to the north and Chaco Canyon to the south. Trade items and nonutilitarian goods are present. Some have suggested that this region was a "breadbasket" compared with other contemporaneous sites, in part because of the well-watered location and accessibility to major population centers to the north and south (Martin et al. 2001:196–197).

5.5.1   The Osteological Findings

Data from individuals included all of the standard osteological conditions of age, sex, metrics, and pathology. One of the more dramatic findings was that there was a great deal of evidence for cranial and postcranial trauma from the La Plata burial series that included healed fractures and traumatic injuries that were in the state of healing or fully healed (therefore, injuries that were nonlethal in nature but still quite serious). The cranial wounds at La Plata fit the description of depression fractures caused by blows to the head (Merbs 1989; Walker 1989).

A summary of individuals from this population with trauma and pathology is provided in Table 5.1. Individuals with cranial and/or postcranial pathology related to trauma suggest different patterns between adult males and females.

Eight females (out of 10) have healed cranial trauma (largely in the form of depression fractures), and the ages of these women range from 22 to 38 (Table 5.2). The inventory of healed nonlethal cranial wounds for the females is longer and more extensive than that of the males, with 3 of the 6 cases involving multiple head wounds. The youngest female (age 20) has a healed broken nose (65030 B8). Another young female (age 22) with a cranial trauma demonstrates two depression fractures, one on the forehead and one on the back of the head (65030 B15). A 25-year-old has

Table 5.2 A listing of the La Plata individuals with trauma and related pathologies

Site, age, sex	Trauma and pathology
LA 37592, female age unknown	Healed right lower leg bone (fibula)
LA 65030 B8, female age 20	Healed broken nose, healed fractures on first and second vertebrae at base of the back of the head
LA 37601, female age 25	Healed depression fracture on left side of head and 6 rounded depression fractures center of the forehead above the eyes. Healed fractures on two right and one left rib, depression fracture on right shoulder blade, 3 neck vertebrae with trauma-induced reaction boney growths, healed lower back fracture in lumbar vertebrae
LA 65030, female age 28	Healed cranial depression fracture on the right forehead and healed cranial depression fracture on the back of the head
LA 37603 B2.1, female age 30	Healed Colle's fracture near right wrist (due to breaking a fall, distal end of radius)
LA 65030 B9, female age 33	Severe healed cranial depression fracture at the top of the head, with involvement of several bones, uneven healing due to un-reunited sutures. Healed fracture on left hip
LA 37601 B10, female age 38	Healed cranial depression fracture on forehead above right eye
LA 65030 B6, female age 38	Healed cranial depression fracture back of head. Trauma to left hip leaving the joint arthritis and asymmetrical

multiple depression fractures on the front and side of her head (37601 B4). A 33-year-old has a large un-reunited but healed series of fractures at the top of her head (65030 B9). Of the two 38-year-old females, one has a healed fracture above her right eye (37601 B10), and one has a depression fracture at the back of the head (65030 B6).

These nonlethal head fractures are quite different from that found in age-matched males. Wounds are not located on the sides of the head (as is the male pattern) but are on the top, back, and front of the head. These injuries are very similar to those documented in forensic cases of wife-beating in historic and contemporary samples (Walker 1997:161). It is primarily the face that gets struck or the back of the head as the victim attempts to flee. Also, the wounds sustained by the females are largely not the typical circular depression left by use of blunt force. These wounds are highly variable in size, location, and depth, suggesting the use of more common and expedient implements (sticks, pottery, stones) than clubs and weapons.

Other characteristics of the females with cranial trauma are that these women as a group generally have more frequent involvement with anemia and systemic infection. Several of the women with cranial trauma exhibit more left/right asymmetry of 2–6 mm in long bone proportions (3 individuals in particular are asymmetrical, LA 65030 B6, B8, and B9) and more pronounced cases of postcranial ossified ligaments, osteophytes at joint surfaces (unrelated to general osteoarthritis or degenerative joint disease), and localized periosteal reactions (enthesopathies). Whether these observations are the result of occupational stress (Capasso et al. 1999) or the sequelae of injuries that caused abnormal biomechanical problems is not clear. They do, however, indicate muscle strain and biomechanical stress. In fact, the one physical characteristic that most distinguishes several of the women with trauma is the pattern of musculoskeletal markers associated with muscle stress or habitual use of select muscle groups. For example, both females in pit structure 1 from site LA 65030

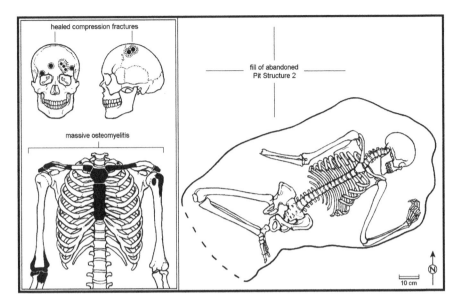

Fig. 5.2 Female, age 25, cranial and postcranial trauma, pit structure 2, middle fill. LA 37601, B4 (Courtesy Robert Turner, Office of Archaeological Studies, Department of Cultural Affairs, Santa Fe, NM)

demonstrate asymmetrical measurements for many of the width proportions of the long bones. Particularly, the humerus, radius, and ulna are most affected. Trinkaus et al. (1994) examined modern, extant, and extinct groups and found that humeral bilateral asymmetry related most often to activity-related functional changes.

5.5.2   The Mortuary Context

After concluding the osteological analyses, it was noted that there seemed to be a pattern of interpersonal violence involving reproductive-aged women. Had there not been access to the archaeological forms from the excavation, the interpretation may have been simply that this is a case of domestic abuse or small-scale conflict. However, included in the analysis was the recording of burial location, strata, position, grave type, grave goods, and completeness and preservation.

The majority of the burials from La Plata were intentional burials with individuals in a flexed or semi-flexed position in shallow pits or within abandoned structures or storage pits. Almost always, there were associated objects, usually ceramic vessels or ground stone, placed in close proximity of the individual. This pattern is somewhat the norm for this region and time period.

An association emerged when the mortuary contexts of the individuals with cranial trauma and the osteological and mortuary data were integrated. Every female at La Plata *with cranial trauma* was buried in a different way from the rest of the population (Figs. 5.2, 5.3, 5.4, 5.5, and 5.6). All were found in positions that were loosely

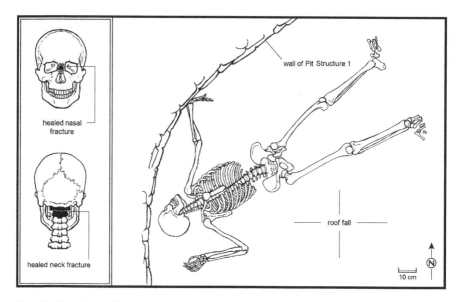

Fig. 5.3 Female, age 20, cranial and postcranial trauma, pit structure 1, lower fill. LA 65030, B8 (Courtesy Robert Turner, Office of Archaeological Studies, Department of Cultural Affairs, Santa Fe, NM)

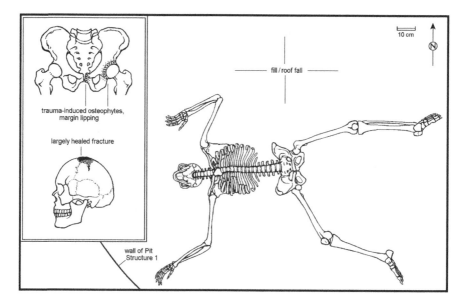

Fig. 5.4 Female, age 33, cranial and postcranial trauma, pit structure 1, lower fill, LA 65030, B9 (Courtesy Robert Turner, Office of Archaeological Studies, Department of Cultural Affairs, Santa Fe, NM)

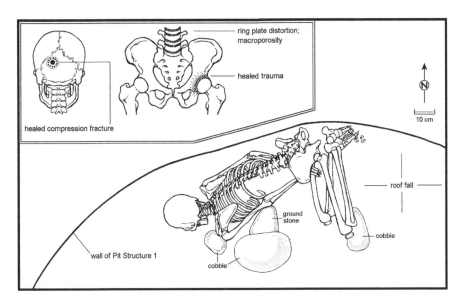

Fig. 5.5 Female, age 38, cranial and postcranial trauma, pit structure 1, lower fill. LA 65030, B6 (Courtesy Robert Turner, Office of Archaeological Studies, Department of Cultural Affairs, Santa Fe, NM)

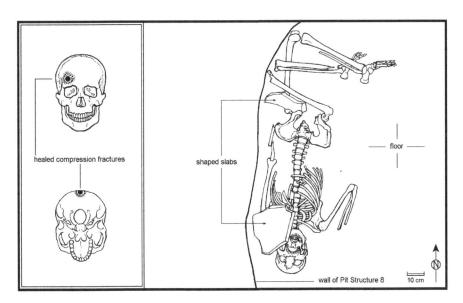

Fig. 5.6 Female, age 28, cranial trauma, pit structure 8, lower fill. LA 65030, B16 (Courtesy Robert Turner, Office of Archaeological Studies, Department of Cultural Affairs, Santa Fe, NM)

flexed, prostrate, or sprawled. The mortuary context of females with cranial trauma reveals that, unlike their counterparts without signs of trauma, they were generally haphazardly placed in abandoned pit structures, and in each case there were no associated grave goods.

At site LA 37601, a 25-year-old female with cranial and postcranial trauma was located in a similar position with no grave goods (Fig. 5.2). At the nearby site LA 65030, four adult females were interred (Figs. 5.3, 5.4, 5.5, and 5.6). All are in haphazard positions, some as if thrown from a higher elevation. The cause of death could not be ascertained for any of these individuals. As a group, these females are distinguished not only by their unintentional burial but also by having no associated grave goods or offerings, as the great majority of the other burials at these sites demonstrated. Also, these females also had other health problems such as active infections and traumatic bone anomalies.

To summarize the association of healed cranial trauma and mortuary context, out of a total sample size of 10 adult females with crania, 6 show trauma and were buried without grave goods and in either sprawled or semi-flexed positions. Six females had no trauma and were in a flexed or semi-flexed positions with associated grave goods; one female had no cranial trauma, a semi-flexed burial but no grave goods; and one female with no cranial trauma had an unknown mortuary context (see Martin and Akins 2001 for complete discussion of the mortuary treatment of the La Plata individuals).

In contrast, for the 13 males that could be assessed for cranial trauma, 6 had no cranial trauma and were in flexed burials with grave goods, while 4 had no cranial trauma or grave goods and a variety of positions ranging from extended to flexed. Of the 3 males with cranial trauma, 1 had grave goods, and 2 did not although all were in a semi-flexed position. There is more variability in the relationship between cranial trauma and burial treatment for the males, with most males in general having no grave goods, but males are placed in formal burial contexts with flexed or semi-flexed positions.

5.5.3 Integrating Osteology and Mortuary Analyses, Ideology, and Culture

This subgroup of women collectively demonstrates what may be the bioarchaeological signature of forced captivity and slavery. An examination of other attributes suggests that these women were local or regional and not from a distinctively different cultural group. Isotopic analysis was conducted on rib samples from the women with cranial fractures and compared with the other adult males and females. For the sample of 13 adults, there were no statistically significant differences between males and females, or between individuals with and without evidence of traumatic injury. This suggests that the women with trauma were local, or at least they came from an environment similar to that at La Plata. Cranial metrics and cranial and postcranial discrete traits cannot be used to characterize the subgroups because of small sample size.

The picture that emerges is one of an agricultural population that was doing relatively well given the circumstances of crowded living and subsistence farming. Anemia and infectious disease are expected outcomes of group living and agrarian lifeways. These disease conditions are not pronounced at La Plata. In comparison to nearby groups in the Mesa Verde and Chaco Canyon regions, La Plata individuals seem to be faring quite well, if not better than expected. However, all of this is shadowed by the high frequencies of trauma found in the female subpopulation. The high frequencies of cranial trauma suggest signs of strife and troubled times for some living at La Plata even with abundant resources and low disease stress.

Why would this underclass be primarily women who were routinely battered? Spouse abuse or domestic violence can be ruled out because there are other adult females who demonstrate no trauma and who were placed in prepared graves in a flexed position with grave goods. Also, the association of women with injuries and a young child in close association suggests that their children were also part of the underclass, and when they died, the child was slayed as well. Witch executions detailed by Walker (1998) and Darling (1999) likewise do not fit the pattern seen at La Plata.

The most parsimonious explanation is that these women represent captive or enslaved women. A more secure economy may have led to increased population density, decreased mobility, increased political centralization, and the need for more routes of exchange and prestige-garnering with competing population centers. Without putting together the mortuary context, the position and orientation of the bodies, the lack of grave goods, and the forensic reconstruction of traumatic wounds, there would have been no opportunity to see the multiple lines of evidence supporting that these women were outcasts of some sort.

This example highlights the importance of integrating data across different domains—in this case, excavation, taphonomy, mortuary, burial analyses, and cultural context. None of data from any one of these could have produced the complex picture of inequality and human suffering that emerged when the combined data sets were interpreted together. Engaging with theories about the nature of captivity and slavery, the subordinate role of women in some cultures, and the distribution of goods and resources provided a window to piecing together what life was like for a segment of ancestral Pueblo society. In this sense, the whole was much greater than a sum of the parts.

5.6 Summary

Important information is often lost when human remains are not fully documented while they are still in situ. It may seem ironic that the analysis of the dead provides so much insight into the living. Humans have not always buried their dead, and so the appearance of this practice approximately 10,000 years ago signaled a shift in how people were thinking about the significance of dying.

Analysis of burial ritual and mortuary sites has revealed a great deal about social ranking, variability in location and positioning, social organization, and treatment of elites. On the one hand, these and other studies have revealed a great deal about how different cultures have dealt with their dead. On the other hand, mortuary studies are far fewer than many other kinds of archaeological or bioarchaeological studies.

In many ways, NAGPRA legislation in the United States and other kinds of repatriation efforts in international settings serve to underscore the importance of the connections between the dead and the living. Descendant communities care deeply about their ancestors, and being able to retain a relationship demands that bioarchaeologists honor the desires of the living. By facilitating a relationship with descendant populations, bioarchaeologists can be part of the process of furthering these important connections between the dead and the living. And, bioarchaeologists need to acknowledge that in some cases, it would be inappropriate for them to study certain mortuary contexts and the human remains if specifically asked not to by tribal or indigenous representatives.

References

Adams, R. L., & King, S. M. (2010). Residential burial in global perspective. In *Residential Burial: A multiregional exploration*, eds. Ron L. Adams, and Stacie M. King, 1–16. Washington, D. C.: Archeological Papers of the American Anthropological Association.

Arriaza, B. T. (1995). *Beyond death, the Chinchorro mummies of ancient Chile*. Washington, DC: Smithsonian Institution Press.

Bass, W. M. (2005). *Human osteology: A laboratory and field manual* (5th ed.). Columbia: Missouri Archaeological Society.

Beck, L. A. (1995). *Regional approaches to mortuary analysis*. New York: Plenum Press.

Bendann, E. (2003). *Death customs: An analytical study of burial rites*. Kila: Kessinger Publishing Company. (Original work published 1930).

Binford, L. R. (1971). Mortuary practices: Their study and their potential. In J. A. Brown (Ed.), *Approaches to the social dimensions of mortuary practices*. Washington, DC: Memoirs of the Society for American Archaeology 25.

Blau, S. (2001). Limited yet informative: Pathological alterations observed on human skeletal remains from third and second millennia bc collective burials in the United Arab Emirates. *International Journal of Osteoarchaeology, 11*(3), 173–205.

Braun, D. P. (1981). A critique of some recent North American mortuary studies. *American Antiquity, 46*(2), 398–416.

Buikstra, J. E. (1977). Biocultural dimensions of archaeological study: A regional perspective. In R. L. Blakely (Ed.), *Biocultural adaptation in prehistoric America* (pp. 67–84). Athens: Southern Anthropological Society Proceedings, No. 11, University of Georgia Press.

Buikstra, J. E., & Beck, L. A. (2006). *Bioarchaeology: The contextual analysis of human remains*. Burlington: Academic.

Buikstra, J. E., & Charles, D. K. (1999). Centering the ancestors: Cemeteries, mounds, and sacred landscapes of the ancient North American midcontinent. In W. Ashmore & A. B. Knapp (Eds.), *Archaeologies of landscape: Contemporary perspective* (pp. 210–228). Malden: Blackwell Publishers Ltd.

Buikstra, J. E., & Ubelaker, D. H. (1994). *Standards for data collection from human skeletal remains*. Fayetteville: Arkansas Archaeological Survey, Research Series, No. 44. A copy of Standards is required in order to fill out these forms accurately. It may be obtained from the Arkansas Archeological Survey, 2475 N. Hatch Ave., Fayetteville, AR 72704, http://www.uark.edu/campus-resources/archinfo/.

Butzer, K. W. (1982). *Archaeology as human ecology: Method and theory for a contextual approach.* Cambridge: Cambridge University Press.

Capasso, L., Kennedy, K. A. R., & Wilczak, C. A. (1999). *Atlas of occupational markers on human remains.* Teramo: Edigrafital S.P.A.

Ceruti, M. C. (2004). Human bodies as objects of dedication at Inca Mountain Shrines (North-Western Argentina). *World Archaeology, 36*(1), 103–122.

Chapman, R., Kinnes, I., & Randsborg, K. (2009). *The archaeology of death (second printing).* Cambridge: Cambridge University Press.

Charles, D. K., & Buikstra J. E. (2002). Siting, sighting, and citing the dead. In *The place and space of death,* eds. Helaine Silverman, and David A. Small, 13–25. Washington, D. C.: Archeological Papers of the American Anthropological Association.

Conklin, B. A. (1995). "Thus are Our Bodies, Thus was Our Custom": Mortuary cannibalism in an Amazonian society. *American Ethnologist, 22*(1), 75–101.

Darling, A. J. (1999). Review of Man Corn: Cannibalism and violence in the prehistoric American Southwest by Christy G. Turner, II and Jacqueline A. Turner. *Latin American Antiquity, 10*(4), 441–442.

Duncan, W. N. (2005). *The bioarchaeology of ritual violence in postclassic El Petén, Guatemala (AD 950–1524).* Unpublished PhD dissertation, Southern Illinois University, Carbondale.

Ferguson, T. J., Dongoske, K. E., & Kuwanwisiwma, L. J. (2001). Hopi perspectives on southwestern mortuary studies. In D. R. Mitchell & J. L. Brunson-Hadley (Eds.), *Ancient burial practices in the American Southwest* (pp. 9–26). Albuquerque: University of New Mexico Press.

Fiorato, V., Boylston, A., & Knüsel, C. (2000). *Blood red roses.* Oxford: Oxbow Book Publishers.

Fitzsimmons, J. L., & Shimada, I. (2011). *Living with the dead: Mortuary ritual in Mesoamerica.* Tucson: University of Arizona Press.

Formicola, V., & Buzhilova, A. P. (2004). Double child burial from Sunghir (Russia): Pathology and inferences for upper Paleolithic funerary practices. *American Journal of Physical Anthropology, 124*(3), 189–198.

Galloway, A. (1999). *Broken bones: Anthropological analysis of blunt force trauma.* Springfield: Charles C. Thomas.

Hamilakis, Y., & Konsolaki, E. (2004). Pigs and gods, burnt animal sacrifices as embodied rituals at a Mycenaean Sanctuary. *Oxford Journal of Archaeology, 23*(2), 135–151.

Harvig, L., Lynnerup, N., & Amsgaard Ebsen, J. (2012). Computed tomography and computed radiography of Late Bronze Age cremation urns from Denmark: An interdisciplinary attempt to develop methods applied in bioarchaeological cremation research. *Archaeometry, 54*(2), 369–387. doi:10.1111/j.1475-4754.2011.00629.x.

Heller, A. (2003). Archeology of funeral rituals as revealed by Tibetan tombs of the 8th to 9th century. In *Ērān ud Anērān: Studies presented to Boris Ilich Marshak on the occasion of his 70th birthday,* http://www.transoxiana.org/Eran/. Accessed February 15, 2013.

Howey, M. C. L., & O'Shea, J. M. (2006). Bear's journey and the study of ritual in archaeology. *American Antiquity, 71*(2), 261–282.

Isbell, W. H. (2010). *Mummies and mortuary monuments: A postprocessual prehistory of Central Andean social organization.* Austin: University of Texas Press.

Iserson, K. V. (1994). *Death to dust: What happens to dead bodies?* Tucson: Galen Press.

Jacobi, K. P. (2003). The malevolent undead: Cross-cultural perspectives. In C. D. Bryant (Ed.), *Handbook of death and dying. Volume 1: The presence of death* (pp. 96–109). Thousand Oaks: Sage Publications.

Kimmerle, E. H., & Baraybar, J. P. (2008). *Skeletal trauma: Identification of injuries resulting from human remains abuse and armed conflict.* Boca Raton: CRC Press.

Knapp, A. B., & Ashmore, W. (1999). Archaeological landscapes: Constructed, conceptualized, ideational. In W. Ashmore & A. Bernard Knapp (Eds.), *Archaeologies of landscape: Contemporary perspective* (pp. 1–30). Malden: Blackwell Publishers Ltd.

LaFraniere, S. (2011). China curbs fancy tombs that irk poor. The New York Times. Accessed October 15, 2012.

Liénard, P., & Boyer, P. (2006). Whence collective rituals? A cultural selection model of ritualized behavior. *American Anthropologist, 108*(4), 814–827.

Linseele, V., Van Neer, W., & Hendrickx, S. (2007). Evidence for early cat taming in Egypt. *Journal of Archaeological Science, 34*(12), 2081–2090.

Malone, C., Stoddart, S., Bonanno, A., & Trump, D. (2009). *Mortuary customs in Prehistoric Malta: Excavations at the Brochtorff Circle at Xaghra, Gozo (1987–94)*. Cambridge: McDonald Institute for Archaeological Research.

Margolis, M. M. (2007). *The isolated human bone from Grasshopper Pueblo* (AZ P:14:1[ASM]). Unpublished MA thesis, University of Arizona, Tucson.

Martin, D. (1996). On the cultural ecology of sky burial on the Himalayan Plateau. *East and West, 46*(3/4), 353–370.

Martin, D. L. (1998). Owning the sins of the past: Historical trends in the study of Southwest human remains. In A. H. Goodman & T. L. Leatherman (Eds.), *Building a new biocultural synthesis: Political-Economic perspectives on human biology* (pp. 171–190). Ann Arbor: University of Michigan Press.

Martin, D. L., & Akins, N. J. (2001). Unequal treatment in life as in death: Trauma and mortuary behavior at La Plata (AD 1000–1300). In D. R. Mitchell & J. L. Brunson-Hadley (Eds.), *Ancient burial practices in the American Southwest* (pp. 223–248). Albuquerque: University of New Mexico Press.

Martin, D. L., Akins, N. J., Goodman, A. H., & Swedlund, A. C. (2001). *Harmony and discord: Bioarchaeology of the La Plata Valley. Totah: Time and the rivers flowing excavations in the La Plata Valley* (Vol. 242). Santa Fe: Museum of New Mexico, Office of Archaeological Studies.

Martin, D. L., & Frayer, D. W. (1997). *Troubled times: Violence and warfare in the past*. Amsterdam: Gordon and Breach.

Merbs, C. F. (1989). Trauma. In M. Y. Iscan & K. A. R. Kennedy (Eds.), *Reconstruction of life from the skeleton* (pp. 161–199). New York: Alan R. Liss.

Millaire, J.-F. (2004). The manipulation of human remains in Moche Society: Delayed burials, grave reopening, and secondary offerings of human bones on the Peruvian north coast. *Latin American Antiquity, 15*(4), 371–388.

Mitchell, D. R., & Brunson-Hadley, J. L. (2001). *Ancient burial practices in the American Southwest: Archaeology, physical anthropology, and Native American perspectives*. Albuquerque: University of New Mexico Press.

Morey, D. F. (2006). Burying key evidence: The social bond between dogs and people. *Journal of Archaeological Science, 33*(2), 158–175.

Morris, E. H. (1924). *Burials in the Aztec Ruin*. The Archer M. Huntington Survey of the Southwest. New York: Anthropological Papers of the American Museum of Natural History.

Murphy, E. M. (2008). *Deviant burial in the archaeological record*. Oxford: Studies in Funerary Archaeology 2, Oxbow Books.

Mytum, H. C. (1989). Public health and private sentiment: The development of cemetery architecture and funerary monuments from the eighteenth century onwards. *World Archaeology, 21*(2), 283–297.

Mytum, H. C. (2004). *Mortuary monuments and burial grounds of the historic period*. New York: Kluwer Academic/Plenum Publishers.

O'Shea, J. (1981). Social configurations and the archaeological study of mortuary practices: A case study. In R. Chapman, I. Kinnes, & K. Randsborg (Eds.), *The archaeology of death* (pp. 39–52). Cambridge: Cambridge University Press.

Onar, V., Armutak, A., Belli, O., & Konyar, E. (2002). Skeletal remains of dogs unearthed from Van-Yoncatepe Necropolises. *International Journal of Osteoarchaeology, 12*, 317–334.

Pachico, E. (2011). Mexico drug lords live on in Narco-Graveyard. InSight: Organized crime in the Americas. http://insightcrime.org/insight-latest-news/item/1079-mexico-drug-lords-live-on-in-narco-graveyard. Accessed October 15, 2012.

Pearson, M. P. (1999). *Archaeology of death and burial*. College Station: Texas A&M University Press.

Pearson, O. M. (2000). Activity, climate and postcranial robusticity: Implications for modern human origins and scenarios of adaptive change. *Current Anthropology, 41*(4), 569–589.

Pennisi, E. (2004). Burials in Cyprus suggests cats were ancient pets. *Science, 304*(5668), 189.

Pepper, G. H. (1909). The exploration of a burial-room in Pueblo Bonito, New Mexico. In F. Boaz, R. B. Dixon, F. W. Hodge, A. L. Kroeber, & H. I. Smith (Eds.), *Putnam anniversary volume: Anthropological essays* (pp. 196–252). New York: G. E. Stechert and Co.

Perry, E. M., Stodder, A. L. W., & Bollong, C. A. (2010). Animas-La Plata Project: Bioarchaeology. SWCA Anthropological Research Papers No. 10, vol, XV. Phoenix: SWCA Environmental Consultants.

Pickering, M. P. (1989). Food for thought: An alternative to 'cannibalism in the Neolithic'. *Australian Archaeology, 28*, 35–39.

Pinhasi, R., & Bourbou, C. (2008). How representative are human skeletal assemblages for population analysis. In R. Pinhasi & S. Mays (Eds.), *Advances in human paleopathology* (pp. 31–44). West Sussex: Wiley.

Potts, D. T. (2000). *Ancient Magan: The secrets of Tell Abraq.* Mayfair: Trident Press Ltd.

Rakita, G. F. M., & Buikstra, J. E. (2005). Corrupting flesh: Reexamining Hertz's perspective on mummification and cremation. In G. F. M. Rakita, J. E. Buikstra, L. A. Beck, & S. R. Williams (Eds.), *Interacting with the dead: Perspectives on mortuary archaeology for the new millennium* (pp. 97–106). Gainesville: University Press of Florida.

Rakita, G. F. M., Buikstra, J. E., Beck, L. A., & Williams, S. R. (2005). *Interacting with the dead: Perspectives on mortuary archaeology for the new millennium.* Gainesville: University Press of Florida.

Redfern, R. (2008). New evidence for iron age secondary burial practice and bone modification from Gussage All Saints and Maiden Castle (Dorset, England). *Oxford Journal of Archaeology, 27*(3), 281–301.

Robb, J. E., Bigazzi, R., Lazzarini, L., Scarsini, C., & Sonego, F. (2001). Social "status" and biological "status": A comparison of grave goods and skeletal indicators from Pontecagnano. *American Journal of Physical Anthropology, 115*(3), 213–222.

Roberts, C. A. (2009). *Human remains in archaeology: A handbook.* Bootham: Council for British Archaeology.

Saxe, A. A. (1970). *Social dimensions of mortuary practices.* Unpublished PhD dissertation, University of Michigan, Ann Arbor.

Sayer, D., & Williams, H. (2009). *Mortuary practices and social identities in the middle ages: Essays in burial archaeology in honour of Heinrich Härke.* Exeter: University of Exeter Press.

Smith, A. C. T., & Stewart, B. (2011). Organizational rituals: Features, functions and mechanisms. *International Journal of Management Reviews, 13*(2), 113–133.

Sprague, R. (2006). *Burial terminology: A guide for researchers.* Lanham: AltaMira Press.

Tainter, J. A. (1975). Social inference and mortuary practices: An experiment in numerical classification. *World Archaeology, 7*(1), 1–15.

Tainter, J. A. (1978). Mortuary practices and the study of prehistoric social systems. *Advances in Archaeological Method and Theory, 1*, 105–141.

Taylor, T. (2002). *The buried soul: How humans invented death.* Boston: Beacon Press Books.

Trinkaus, E., Churchill, S. E., & Ruff, C. B. (1994). Postcranial robusticity in homo. II: Humeral bilateral asymmetry and bone plasticity. *American Journal of Physical Anthropology, 93*(1), 1–34.

Turner, C. G., II, & Turner, J. A. (1999). *Man corn: Cannibalism and violence in the prehistoric American Southwest.* Salt Lake City: The University of Utah Press.

Ubelaker, D. H. (1999). *Human skeletal remains: Excavation, analysis, interpretation* (3rd ed.). New Brunswick: Aldine Transaction.

Uerpmann, H.-P. (1999). Camel and horse skeletons from protohistoric graves at Mleiha in the Emirate of Sharjah (U.A.E.). *Arabian Archaeology and Epigraphy, 10*(1), 102–118.

Uerpmann, H.-P., & Uerpmann, S. A. J. (2006). *The archaeology of Jebel Al-Buhais, Sharjah, United Arab Emirates, Volume One: Funeral monuments and human remains from Jebel Al-Buhais.* Tübingen: Department of Culture and Information, Government of Sharjah, United

Arab Emirates, in collaboration with the Institut für Ur- and Frühgeschichte und Archäologie des Mittelalters Universität Tubingen, Germany, and in cooperation with Kerns Verlag.

Vigne, J.-D., Guilaine, J., Debue, K., Haye, L., & Gérard, P. (2004). Early taming of the cat in Cyprus. *Science, 304*(9), 259.

Walker, P. L. (1989). Cranial injuries as evidence of violence in prehistoric Southern California. *American Journal of Physical Anthropology, 80*, 313–323.

Walker, P. L. (1997). Wife beating, boxing, and broken noses: Skeletal evidence for the cultural patterning of violence. In D. L. Martin & D. W. Frayer (Eds.), *Troubled times: Violence and warfare in the past* (pp. 145–180). Amsterdam: Gordon and Breach.

Walker, P. L., Johnson, J. R., & Lambert, P. M. (1988). Age and sex biases in the preservation of human skeletal remains. *American Journal of Physical Anthropology, 76*(2), 183–188.

Walker, W. H. (1998). Where are the witches of prehistory? *Journal of Archaeological Method and Theory, 5*(3), 245–308.

Walker, W. H. (2002). Stratigraphy and practical reason. *American Anthropologist, 104*(1), 159–177.

Wang, X., Chen, F., Zhang, J., Yang, Y., Li, J., Hasi, E., et al. (2010). Climate, desertification, and the rise and collapse of China's historical dynasties. *Human Ecology, 38*(1), 157–172.

White, T. D. (1992). *Prehistoric cannibalism at Mancos 5MTUMR-2346*. Princeton: Princeton University Press.

White, T. D., & Folkens, P. A. (2005). *The human bone manual*. Burlington: Academic.

White, T. D., Folkens, P. A., & Black, M. T. (2012). *Human osteology* (3rd ed.). Burlington: Academic.

Chapter 6
Bioarchaeology of Individuals: Identity, Social Theory, and Skeletal Analysis

Bioarchaeological analysis of human remains almost always begins with reconstructing the basic identity of the individuals whose bones have been recovered. Part of what makes bioarchaeology such an interdisciplinary and integrative approach to understanding the past is that it relies on gleaning as much information as possible from the human remains in conjunction with many other cultural and environmental considerations. While context and taphonomy are important (Chaps. 4 and 5), the biological remains have many indicators of what life was like for the person prior to their death. For bioarchaeologists, human remains represent the *only* direct information about human biology. There are many methods for the analysis of human remains and new techniques are regularly advanced in bioarchaeology and forensic journals and books. Methods for the analysis of human remains offer insight into aspects of identity such as the age at death, sex/gender, stature, pathology, trauma, and activity that in turn provide unique and nuanced information about the lived experience of the person who died. Not often, but sometimes, it is even possible to say how and why the person died.

Before reviewing how to reconstruct identity, it is important to discuss what the term identity implies. In today's society identity is the way in which individuals express themselves (e.g., the clothes they wear, their jobs, their ethnicity and age, or having a tattoo on their body), the group or subculture that they associate with (e.g., believing in one religion or another), and the way they perceive and describe themselves (e.g., liberal, middle class, vegetarian). Díaz-Andreu and Lucy (2005:1–2) state that identity is something that people consciously choose and therefore it is never in stasis but rather it changes throughout the life of an individual. Given the complex, abstract, and fluid nature of identity, it is challenging to reconstruct these kinds of self-identifications from human remains. However, people "live" their identities and as such there are huge social/cultural forces that affect the body.

Some theoreticians go so far as to suggest that physical bodies are largely socially created (Lorber and Martin 2011). Another way to think about this is that the physical body is actually an embodiment of the biological, social, and material worlds that people live in (see Chap. 9 for examples of materiality and human remains).

D.L. Martin et al., *Bioarchaeology: An Integrated Approach to Working with Human Remains*, Manuals in Archaeological Method, Theory and Technique, DOI 10.1007/978-1-4614-6378-8_6, © Springer Science+Business Media New York 2013

In some ways this is an extended way of thinking about the biocultural perspective that most bioarchaeologists operate within. Biology can influence social relations, and social relations can affect and impact the body. To separate out biology from culture, or the social from the biological, is simply not a useful way to analyze bodies.

How does skeletal tissue reveal biocultural identities? Bone tissue is a unique organ in the body that has a very limited way of responding to growth, development, maintenance, and stressors of all kinds (Martin et al. 2001). While this both limits the ability of bone to respond in endless ways to different stimuli, it also makes it relatively easy to understand how bones might embody various biocultural influences. These include culturally specific things such as the foods people choose to or are able to eat, the pathogens they are exposed to, the activities they take part in by choice or through coercion, and the amount of violent interactions they encounter. These and other biocultural factors often leave decipherable "signatures" on the skeletal system.

6.1 Theorizing Skeletal Indicators

The reconstruction of biocultural identity involves looking at as many skeletal indicators as possible but with emphasis placed on the contextualization of these indicators. Chapter 3 provided some possibilities for theorizing the work done with human remains, and utilized here to facilitate how to use theory is the idea of humans having simultaneously multiple bodies that reflect the different spheres of influence in which all humans live. Scheper-Hughes and Lock (1987:7–8) suggest that bodies are shaped in three overlapping but distinctive ways. Operationalizing the notion that the body can be described by three different perspectives helps to open up new pathways to thinking about the biological indicators that bioarchaeologists typically use to reconstruct identity.

Starting with individual's body, Scheper-Hughes and Lock suggest that this body is a reflection of each person's lived experiences, which they refer as the "body-self" (Scheper-Hughes and Lock 1987:7). The body-self would include self-identification of individuals by their age, sex and/or gender, ancestry or kinship, congenital anomalies (visible biological or behavioral anomalies that they were born with), and employment or occupation. These characteristics help define the body-self because they reveal biological realities influenced by specific social definitions and customs. All societies have boundaries within which individuals largely inhabit because of these kinds of identifying features. The traditional approach is to argue that these characteristics (such as age, sex and body type) are simply the product of genetics, growth and development, and diet. However, the reality is that the body is never simply a product of genes and biology because social and cultural influences shape the expression of those physical characteristics. The phenotype of humans is very plastic and responsive to cultural forces.

A second type of body that humans have is the social body, which can be thought of as the more symbolic representation of the larger spheres of social influence. Scheper-Hughes and Lock describe the social body as "... a natural symbol with which to think about nature, society, and culture" (1987:7). The idea here is that the social body is less about the physicality of flesh and bones and more focused on the context in which the body is situated. For bioarchaeologists, this encourages a detailed reconstruction of the mortuary context (Chap. 5), as well as body modification (Chap. 9), gender expression (Chap. 4), and other ways in which individuals are embodiments of the larger cultural worldview, ideology, and social institutions in which they live. Analysis of the social body could be undertaken by an examination of all of the ways that society affects biological health and well-being. Inequality, structural violence, and differential access to resources will take the form of nutritional problems, communicable diseases, and other biological indicators of poor health.

The third kind of body is that which is shaped by political forces and involves political identity or the "body politic" as labeled by Scheper-Hughes and Lock (1987:7). The ways that politics and coercive social structures impact humans are perhaps the most important in that there is potential to cause much harm to the body. The body politic is shaped by the ".... regulation, surveillance, and control of bodies (individual and collective) in reproduction and sexuality, in work and leisure, in sickness and other forms of deviance and human difference" (1987:8). It is this last body that is often of most interest because it is these signatures that offer insight into a person's social status, health, and risk of early death. The kinds of political activities found in prestate and state societies in the ancient world include culturally sanctioned violence such as warfare and captive-taking, slavery and indentured servitude, and enforcement of punishment for crimes. All of these kinds of socially created categories can place individuals at extreme risk for trauma, early death, and poor health. This theory also provides a way to think about power and which individuals have it and which do not. One can theorize about the relationship between power and the use of force and how it would differently impact perpetrators of violence, victims of violence, and witnesses to violence. Theorizing about the body politic in this way provides a way to frame questions about social order and social control enacted in ancient societies and what these might mean for individual well-being.

Analyzing human remains as embodiments of different kinds of identities is not without limitations and challenges. The primary limitations of this endeavor are that many of the bony alterations to the skeleton are interrelated and dependent on one another as well as cumulative in nature. The more skill a bioarchaeologist has in human osteology, skeletal analysis, and anatomy, the better the ability to tease out subtle and nuanced changes to the skeleton. Part of the ability to identify aspects of skeletal morphology that are due to individualized activities and disease episodes is in knowing when bone deviates from normal. Learning to distinguish normal from pathological or anomalous bone involves looking at bone markers across the skeleton and also in using a comparative methodology that provides information on the range of variability in expression.

Through detailed bioarchaeological reconstructions of age at death, sex and gender, stature, and indicators of activity, it is possible to reveal important aspects of the biocultural identities of past people. As discussed in the Chap. 3, theory can inform us about general patterns known to exist in human behavior, and this can structure the way that data are collected and interpreted. If the research question is "what is the earliest finding of cancer in the New World?" this would necessarily structure data collection to be focused on looking only for occurrences of cancer in New World human remains. If the research question is "what forms of social control were used to enforce cultural practices such as paying tribute to the rulers and how did this affect health?" then it would be useful to use a theoretical perspective such as Scheper-Hughes and Lock to organize how data are collected. The following organizes various indicators of biocultural identity such as age at death, sex, pathology, trauma, and activity in a way that illustrates how use of social theory about bodies and skeletal indicators can be integrated to produce a more realistic understanding of social systems and human behavior.

6.2   The Individual Body or Body-Self

The individual body is the level of identity that would most closely resemble the biological profile that is most readily deduced from skeletal remains. This includes chronological age at death, the estimation of sex, ancestry, and biological relatedness (i.e., kinship). Provided here is *not* an exhaustive manual for how to analyze skeletal remains; rather it is a very brief overview to provide some of the basic ideas and methods about how to get at the individual body or body-self. Methodologies are provided in many texts (i.e., see Katzenberg and Saunders 2008; DiGangi and Moore 2012 for a discussion of the larger body of analytical techniques currently available).

6.2.1   Shifting Rights and Responsibilities: The Importance of Estimating Age at Death and Sex and Gender

The importance of reconstructing the age at which an individual died and the sex or more importantly the gender they were identified with is that it can reveal a great deal about their roles in society. However, assessing the chronological age and biological sex of an individual at the time of death utilizing only skeletal and dental tissues is not as straightforward as it may seem. While chronological age is the official way that living people track the passage of time on an annual cycle, there is a great deal of individual variation in the process of aging. Psychologists and physical therapists often use developmental or functional age in addition to chronological age to discuss how old a person is.

The process of aging and the assignment of sex are typically thought of as being unaffected by nonbiological processes, but the reality is that culture has a tremendous impact on how these two terms are even defined (see Chap. 3 for a discussion of sex and gender). The two most fundamental age categories are adult and subadult (also called non-adults or juveniles). The term adult refers to a very specific moment in a person's life when they are no longer growing in height due to the fusion and cessation of growth in length of their long bones. In other words, the term adult indicates an individual whose long bones have stopped growing in length. While growth, development, and maintenance of bone continues throughout life, hormones and genetics generally produce this state at an age of 18 for females and 19 or 20 for males.

Assigning the term adult to an individual based on their bone development automatically confers an identity for bioarchaeologists that might be at odds with how the person was categorized by themselves or by others. In some cultures, chronological and biological age is less of a marker of being an adult than is the enactment of a specific ceremony, referred to as a rite of passage. In modern society, being an adult grants certain rights and comes with a range of new obligations and responsibilities, and the same is true of past societies. Thus the age assigned by bioarchaeologists might not be as important of the cultural rights and responsibilities that are assigned when individuals reach some specific age. Once human remains are identified as adult, that individual has been given an identity and comes with social meaning, and this must be factored into the way age categories are ultimately interpreted. During adulthood the degenerative processes are gradual so an individual goes through stages of adulthood over a relatively long period of time. Bioarchaeologists are very efficient at providing an age range for a skeleton based on these slow changes.

The concept of a subadult is similarly problematic. First, the term subadult itself is controversial as several researchers argue that the use of the prefix "sub" implies that individuals in this age category are of less value than adults (Halcrow and Tayles 2008:193; Lewis 2006:2; Sofaer 2006:121). An alternative to the term subadult is the use of the term non-adult because it lacks those negative connotations. Yet, it identifies children by what they are not. An additional problem of the category of subadult is related to the rate of the aging process and the fact that age-related changes are associated with growth and development. Changes among subadults are often very dramatic over a short period of time, such as the exponential growth of the brain and cranium from birth to around 1 year of age (Greenberg and Greenberg 2001). The result is that a subadult that is 15 is very different from a subadult that is only 2 years old, and these differences are biological, social, and cognitive.

There are a number of subdivisions in childhood, all of which are marked by a major change in growth and development, especially in relation to cognition and motor control. The subdivisions of childhood are not necessarily agreed upon, in terms of when they occur and what they are called. There is no consensus among bioarchaeologists about the age categories to use, for example, to define a newborn, an infant, a toddler, or a child. By convention, bioarchaeologists often try and place subadults in meaningful subgroups, but this can vary wildly. Thus, something as seemingly simple as assignment of age at death is fraught with complexities that put biological age markers on the skeleton at odds with social formulations of meaningful age categories.

6.2.1.1 Brief Overview of Methods of Age Estimation and Sex Assignment

Researchers primarily depend on the one or all of the major reference books to assign an age and sex to human remains. These include the *Standards for Data Collection from Human Skeletal Remains* (Buikstra and Ubelaker 1994), *Human Osteology: A Laboratory and Field Manual* (Bass 2005), *Human Osteology* (White et al. 2012), and *Developmental Juvenile Osteology* (Scheuer and Black 2000). While these texts are extremely useful because they pull together a lot of the primary methodological approaches into single texts, the limitation of these texts is that the context in which the original data was collected is often not well known and the perspectives of the original researchers are not represented.

Subadult age is obtained by two methods that include the analysis of tooth formation and eruption, as well as the timing of growth and eventual fusion of cranial and postcranial skeletal elements. The development and eruption of dentition is arguably one of the most accurate methods of assigning age (Cardoso 2007; Hillson 1996). The value of dental development is that although there is variation in the stages, the variation is minimal. Currently, it is very difficult to assign sex to subadults, especially in the youngest age categories.

In regard to the age estimation of adult individuals, the methods are more diverse and take skill and years of training because it is based on alterations to the bone that occur due to the normal processes of aging. In general, identification of how old an individual is at the time of death is based on the process of degeneration (Minot 1907; Franklin 2010). Skeletal degeneration is sometimes difficult to see because of the potentially large number of areas to be assessed, including all the vertebra, ribs, and joints or articular surfaces. An additional concern is that degeneration is not a straightforward process, and there is actually quite a large amount of variation between individuals in how their skeleton responds to age-related degeneration. However, the fact remains that while many factors may contribute to the breakdown of skeletal tissue, the primary cause is the process of wear and tear that occurs with aging. While degeneration of almost every part of the skeleton has been analyzed in order to assess age, the articular surfaces of bones, especially the pubic symphysis (Todd 1920; Brooks and Suchey 1990) and auricular surface of the innominate (Lovejoy et al. 1985; Buckberry and Chamberlain 2002), are the most common and accurate regions.

Sex estimation of adults relies on evaluation of changes in the skeletal anatomy based on shifting hormone profiles related to the development of secondary sexual characteristics and changes to the pelvic girdle related to the reproductive ability of females. Secondary characteristics involve the analysis of differences in the size and shape of cranial morphology or general differences in overall body size. This can include the size and shape of the bones. Changes associated with reproduction in females are related to alterations of the morphology of the pelvic inlet to allow for the birthing of term infants. Changes in the pelvis have additional effects on other anatomical features because a wider pelvic inlet alters the angle and shape of the femur (upper thigh) bone.

6.2.2   Kinship's Role in Marriage, Trade, and Warfare: Assessing Ancestry or Biological Relatedness

A crucial part of any reconstruction of the individual body is determining how and if this individual is related to other individuals. To accomplish this, the bones are measured and observed, and patterns of similarities and differences are highlighted. Traditionally, the goal of assessing the degree of similarity between different sets of human skeletal remains was to place them into categories of "race" (see Chap. 2 for a discussion of this concept). However, recent work in both bioarchaeology and forensic anthropology has moved away from "race" and instead focused on the identification of two general levels of relatedness, ancestral decent group and regionally based biologically similar populations.

The notion of ancestral group differs from "race" because it is not based on phenotypic characteristics such as skin color. Traditional categories (i.e., *Caucasoid*, *Negroid*, and *Mongoloid*) are not useful to capture the full range of human variation in phenotypic traits that exist within and between populations (Steadman 2009). Assignment of ancestry to an individual is based on regional population variability that has to do with geographic and environmental adaptations and with genetic flow between individuals.

6.2.2.1   Brief Overview of Methods of Assessing Ancestry and Biological Relatedness

Two primary methods used to assess ancestry and biological relatedness are the measurement of the skeletal and dental remains (i.e., metric traits) and the observation of morphological changes to the bones and teeth (i.e., nonmetric traits). Metric traits were originally considered to be due to genetic attributes. There is debate now over whether or not metric traits are a good indicator of genetics because there is evidence that environment plays a significant role in determining the size and shape of bones. This was highlighted in *Skeletal Biology in the Great Plains* (Owsley and Jantz 1994) where the researchers looking at metric measures as a means of differentiating biological distance found that it was both genetics and environmental factors that were causing the metric differences between populations inhabiting neighboring portions of the Plains cultural area.

The analysis of nonmetric or observable genetic traits involves recording three essential conditions of small irregularities that are genetic in origin but benign (not pathological). These include the presence or absence of the trait, the expression of the trait, and the degree of development of the trait (Whitehead et al. 2005:231). Nonmetric traits were emphasized as a way to compensate for the limitation of metric traits. This movement toward nonmetric traits was based on work conducted in the late 1960s and 1970s that suggested that "… minor nonmetric traits of the skeleton apparently do indirectly reflect part of the underlying genotype of the population and that biological comparison of related groups is legitimate and accurate using population incidence of these traits" (El-Najjar and McWilliams 1978:119). However, even nonmetric traits are questionable as White et al. (2012) cite several studies that have shown that these measures lack a clear and concise standard for determination, they are often not discontinuous and discrete, and the precise genetic basis is not well known. Analyzing ancestry of known populations from South Africa using nonmetric traits, L'Abbé et al. found that nonmetric features were not very effective and that they were often confounded by sex and age at death (2011).

Recognizing the limitations of both metric and nonmetric traits, the most effective means of identifying biological relationship is to combine these traits with other genetic indicators such as congenital changes to the skeleton. In both bioarchaeology (Stojanowski 2005; Duncan 2012) and forensic anthropology (Birkby et al. 2008), there has been a push to analyze metric and nonmetric traits from a biocultural perspective that considers these factors along with other markers of identity that can be read from the bones. In addition to this, analytical techniques that extract and analyze DNA from bone and teeth are showing promise in providing biological affinities (see Chap. 8).

6.3 The Social Body

The culturally constructed worlds that people live in have an effect on the quality and length of their lives. Depending on one's place in that world, the quality can be better or worse. Analysis of human remains can offer insight into how the body becomes a record of people's place in the world. It is possible to track poor nutrition and disease in the past because these leave patterned alterations and lesions on the skeleton. Bioarchaeologists must look for multiple lines of evidence and carefully piece together ancillary information to be able to interpret bony changes that may be due to things emanating from the conditions in which they are living. This includes the location of the bone alteration, the extent and status of the bone alteration, and how many bones are affected. Known as differential diagnosis, this approach has become the foundation of bioarchaeological approaches to understanding nutrition (Goodman and Martin 2002; Steckel and Rose 2002) and disease (Ortner 2003; Ortner and Putschar 1985; Aufderheide and Rodríguez-Martin 2003).

6.3.1 Veneration of Loved Ones or Disposal of the Deviants: The Importance of Mortuary Context

What is most important to keep in mind when considering mortuary context is an understanding of the site layout, burial position and location, presence and quantity of grave goods, and the comparison of burials to one another (see Chap. 5). The mortuary context is an important indicator of the social body, and it may reveal a great deal about how the society as a whole viewed the individual after death (e.g., grave offerings, ancestor veneration) as well as how the body even after death continues to have agency within the society (trophy taking, secondary burial). Thus information about the social body will necessarily need to take the mortuary context into consideration along with the analysis of the human remains.

6.3.2 Why Isn't Everyone Healthy and Well Fed? What Nutrition Can Reveal About Inequality

Over the last several decades, the analysis of the human remains has proven to be critical to the reconstruction of ancient diets and past nutrition. The importance of recreating nutrition is that the quality and quantity of the diet of past peoples and patterns of better and worse health within groups can reveal a great deal about an individual's social identity. Beyond the implications of social status and access to resources is assessing the morbidity and mortality patterns of a population. Nutrition-related changes to the skeletal anatomy include variations in height or stature, the development of dental defects including enamel hypoplasias and caries, and the presence of conditions such as cribra orbitalia and porotic hyperostosis. There are other more specialized kinds of studies that can be done that are not covered here. Provided here is a very general overview of some of the basic ways that the social body can be used to interpret signs of poor health and nutrition.

6.3.2.1 Stature

The importance of being able to analyze subadult growth and adult morphology from skeletal remains is that patterns in growth are powerful tools for assessing nutritional status. Evidence suggests that subsistence patterns and differential access to quality foods have an impact on the length of the long bones so much so that stature is a reflection of the quality and reliability of an individual's diet (Steckel 2008; Auerbach 2011; Mummert et al. 2011). Variation in size among contemporary groups is almost completely dependent on the environmental conditions that effect an individual's nutrition (Steckel 1995), thus it is possible to discern differences in diet based on its effect on skeletal morphology.

Looking at both adults and subadults provides different insights into how nutrition affects stature. Among subadults, several studies have compared children in precontact societies with children living in the modern industrial world and found that the pattern in stature is not always straightforward to interpret. Variability exists within and between groups suggesting a role for local conditions in growth and development. For example, Owsley and Jantz (1985) noted that children from indigenous groups inhabiting the Plains region of North American were rather close in stature to Euro-American children living in postindustrial America. This is different from what Martin et al. (1991) found when they looked at stature of precontact indigenous groups living in the American Southwest who were much shorter than children living today. The difference between the Plains and the Southwest children is a result of contrasting traditional subsistence activities which translate into different diets and different nutritional quality. Plains diet is much more diverse with the incorporation of a greater range of meat and wild plants compared to the Southwest diet that is more dependent on fewer dietary resources and an especially high reliance on maize.

Nutrition is critical because there are specific periods of bone development when having good nutrition is important for normal growth and development in children. For example, one especially important period in development is around the ages of 2–3 when children are weaning and are becoming more involved in activities such as the gathering of subsistence resources. However, weaning is only one stage at which children's growth can be hampered, and bioarchaeological and ethnographic research has shown that children are at risk at various stages of growth and development (Baustian 2010; Johnston 1962; Little and Gray 1990; Maresh 1955).

The value of analyzing adult stature is that unlike the stature of subadults, they have obtained their optimal growth in height. The result is that adults provide a picture of stature that is not complicated by the issue of growth.

6.3.2.2 Dental Health

Dentition is arguably one of the best indicators of nutrition because the teeth are formed during the most critical periods of growth and development, and they are a great reflection of subsistence because we use them to process our food. There are two indicators on the teeth that can be utilized to measure nutritional stress, linear enamel hypoplasias (LEH), and caries or cavities. LEHs are disruptions in the formation of enamel as a result of physiological disruption, usually complicated by poor nutrition (Goodman and Rose 1991). Unlike bone that remodels repeatedly throughout an individual's lifetime, once teeth are formed there is no process that takes away or adds new enamel, and it is this static nature of the teeth that makes them so useful for assessing nutrition.

One limitation of LEHs is that since the teeth are formed early in life, they only are useful for assessing nutrition in the earliest years of life. However, within the early years of life, infancy through weaning, LEHs are very informative in terms of when the individual experienced nutritional stress. By assessing the location of each hypoplasia in relation to the root of the tooth, it is possible to identify when nutritional stress occurred in that individual's growth and development.

Dental caries or cavities are another change in dentition that is useful for assessing general health and diet. Carious lesions are especially useful because they can occur at any point in an individual's lifetime and as such are not simply an indicator of nutrition during the earliest years of life. One major pattern that has been correlated with the frequency of caries is the shift to an agricultural subsistence (Cohen and Armelagos 1984; Pinhasi and Stock 2011). The reason for this increase in caries is a shift from a relatively diverse diet prior to agriculture to a reliance on low-quality, high sugar and starch diets such as maize, or other grains such as wheat, barley, sorghum, or millet (Eshed et al. 2006; Lukacs 1996; Zvelebil and Dolukhanov 1991).

Utilizing dental changes like LEHs and caries together is an especially useful means of identifying dietary patterns and general health. Recent clinical data looking at the co-occurrence of LEHs and caries on the second molar found that the presence of LEHs as a child significantly increased the likelihood of having caries later in life (Hong et al. 2009). Considering these and other indications of oral health provides an effective means of recreating access to dietary resources and subgroups at risk for poor oral health.

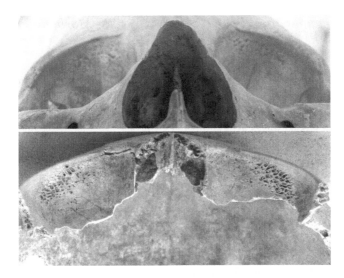

Fig. 6.1 Cribra obitalia: *top*—healed (Pueblo Bonito burial 327.080); *bottom*—active (Pueblo Bonito burial 327.101) (Courtesy of the Division of Anthropology, National Museum of Natural History)

6.3.2.3   Other Indicators of Nutritional Status

Poor nutrition does not only effect the growth of the bones but also the overall health as well. Two diseases in particular have been linked to differences in nutrition, cribra orbitalia (seen in the orbits of the cranium), and porotic hyperostosis (seen on the vault portions of the crania). Cribra orbitalia is the proliferation of bone at the top of the interior of the eye orbits, while porotic hyperostosis involves the development of porosity in the diploe and thinning of external surface of the cranial vault. Researchers often combine these two conditions together as it is assumed they are a result of the same nutritional problem (Fig. 6.1).

Recent research has shown that the cause of these two conditions is not well understood. Walker et al. (2009) argue that iron deficiency is unlikely to be the cause of either condition. They suggest that the conditions have a multifactorial cause that likely involves B12 deficiency during nursing and increased gastrointestinal infections around the time of weaning (2009:119). However, the exact cause is not what is important; what matters is that the presence of these conditions on the skeletal remains indicates that the individual was suffering some sort of dietary deficiency indicative of nutritional stress. Walker et al. also indicate that cribra orbitalia and porotic hyperostosis are disorders that relate to growth and development because children have more trouble maintaining their red blood cell levels and as a result are more susceptible to stress (2009:111). This suggests that these pathological conditions are most useful for assessing nutritional stress during the earliest years of life.

6.3.3   Inequality and Poor Health: Infection, Morbidity, and Early Death

Many of the diseases that bioarchaeologists can identify in the past are those that are chronic in nature. Epidemics in contrast are fast and leave no trace on the bone and are only identifiable in the past by the presence of unusual burial patterns (e.g., mass graves) and in sudden spikes in mortality rates. Chronic diseases are the ones that reveal a great deal of information about an individual's social body because they are indicative of exposure to human or animal waste and garbage, exposure to the elements, poor hygiene related to social status, and a greater risk for injuries to become infected.

Systemic inflammation caused by infection in the body can affect the outer surface (periosteum) of long bones. Bones become implicated in generalized inflammatory responses and develop additional bone growth along the surfaces called periosteal reactions. Looking at periosteal reactions (the irregular bone growth on the surfaces of long bones), Weston (2008) found that the cause of these proliferative changes to the bone was nearly impossible to discern with any accuracy. However, Ortner (2003:181) argues that ". . . The causative organism, in close to 90% of cases [of periosteal reactions], is *Staphylococcus aureus*; the second in frequency is *Streptococcus*, with other infectious agents making up for the remainder."

Until there is more information on the exact etiology of periosteal reactions on bones, the best one can say is that they may be from common transmissible bacterial infections such as staph and strep, but the lesions might also be indicative of some other condition. And, it is likely that as with many diseases, periosteal reactions are multifactorial in nature, so isolating specific agents may not be possible. Yet, infectious diseases have been one of the most significant selective forces in human evolution (Armelagos et al. 2005). With the spread of many new infectious diseases and the reemergence of tuberculosis and malaria (Barrett et al. 1998), understanding who is impacted by disease within a society is critical to understanding the lives of the people that bioarchaeologists study. Most bioarchaeological studies combine all possible indicators of stress and look for multiple confirmations that an individual was morbid at the time of death.

6.4   The Political Body or Body Politic

Even in very ancient societies, individuals had jobs and work to do, and these were often delegated based on cultural decisions based on social arrangements. Individuals may also be coerced into performing certain tasks or they may be born into a class of people whose job is to toil in the fields. The body politic as a theoretical approach to one of the kinds of bodies influenced by social structures invites bioarchaeologists to examine the skeleton for signs of habitual use of the musculoskeletal system and for signs of trauma and violence. Activity-related changes and trauma may be interrelated and contribute to the overall poor health of individuals. Societal institutions and dictates may force individuals to live in a particular way, and bioarchaeologists seeking to understand the body politic can use several lines of evidence to reconstruct the ways different individuals may be mistreated. This is especially true of people who are part of a subaltern group such as captives, slaves, or servants. However, "elites" or higher status individuals may also be forced to play certain roles within a society, and this can be reflected in the bones. For example, warriors may use interpersonal conflict as a means of attaining and maintaining higher status which can reveal itself in the skeletons and mortuary context (Tung 2007; Walker 1989). Through the analysis of markers of activity and evidence of trauma, it is possible to identify how social status and conflict in past societies impacted people physically.

6.4.1 Ancient Occupations: Captives, Warriors, Elites, and Slaves

Reconstructing activity in the past using skeletal remains is an area of intense research (Jurmain et al. 2012). Walker et al. (2012:66) recently pointed out that reconstructions of activity "… continue to strengthen contextual approaches of human remains." Activity-related changes on bone may reveal social and political relations within a particular culture when there are obvious differences in activity levels among various members of the society. The skeleton necessarily responds to any habitual and daily use of specific muscles because one of the functions of skeletons is for anchoring muscles and facilitating movement. When considered with other skeletal changes, such as those due to poor nutrition and disease, activity-related changes can reveal a great deal about an individual's identity.

There are generally three methods for reconstructing activity on human remains. All of the approaches involve the analysis of postcranial elements. The first method examines the degenerative process of the articular surfaces in order to identify patterns that are indicative of overburdening the joints (osteoarthritis and degenerative joint disorder). The second is a method that involves the measurement of the size and shape of the long bones (i.e., robusticity). Finally, a third method evaluates the presence or absence of bony growths at the site of muscle, tendon, and ligament attachment sites. These are called entheses or musculoskeletal stress markers.

6.4.1.1 Osteoarthritis and Degenerative Joint Disease

The term degenerative disease is somewhat ambiguous because all adults from middle adulthood onward show signs of wear and tear at the joints, and this degeneration is part of the natural aging process. However, the changes to the skeletal system that are of interest in reconstructing the body politic are those that are atypical. This is where the patterning of the change is not what is expected because it either occurs earlier in life than is normal or effects the body disproportionately in terms of side of the body or mechanical function of the bone. Osteoarthritis is the term used to describe changes throughout the skeleton including osteophytic growth on the ribs

Fig. 6.2 Osteoarthritis
(Pueblo Bonito burial
327.070) (Courtesy of the
Division of Anthropology,
National Museum of Natural
History)

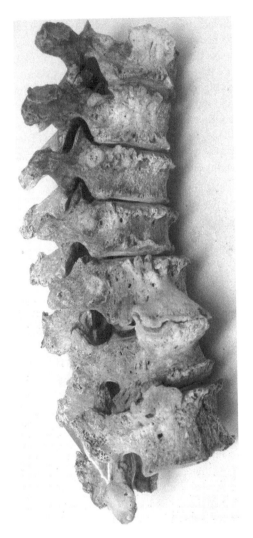

and vertebrae. Degenerative joint disease (DJD) is more specifically related to changes at the articular surface of the joints (Fig. 6.2).

Osteoarthritis and DJD have been associated with disproportionate amounts of activity and used as an indicator of social stratification and inequality (Woo and Sciulli 2011; Merbs 1983). In the La Plata example of captive women in Chap. 5, it was found that females with head wounds and an inconsiderate burial also showed signs of asymmetrical wear and tear on their joints possibly due to hard labor (Martin et al. 2010). "Beaten down and worked to the bone" is the way the bioarchaeologists examining the La Plata female skeletons described what they found. Osteoarthritis that is found in only one location of younger adults can be related to an older traumatic injury that has healed but that has left the joint system weakened or no longer functional. All of these factors taken together are useful for identifying and reconstructing activity patterns of individuals, and they may reveal the social category that these individuals were placed into, likely against their will.

6.4.1.2 Robusticity

The measure of robusticity is essentially an estimate of the overall size and shape of the diaphysis of a long bone. Stock and Shaw state that size and shape are indications of the bone's strength (2007:412). The value of analyzing robusticity is that it measures morphological aspects of specific areas of the long bone as a result of forces applied during biomechanical loading (stress and strain produced by working a set of muscles really hard). This is important to the reconstruction of activity because differences in the level of robustness can provide an indication of the amount or duration of activities a person performed over their lifetime. However, robusticity can also simply reveal the effects of walking long distances and conducting other movements throughout the day.

Robusticity is a measure that can be taken on any of the long bones in the body. However, the elements that are of most interest are the humerus in the upper arm and the femur and tibia of the lower leg. The robusticity of the humerus can reveal patterns of activity that involve repetitive or strenuous use of the arms, which could include a multitude of activities from rowing, grinding corn or grain, weaving, or even arm wrestling. In contrast, because of their function in locomotion and the fact that they are weight bearing, the femur and tibia can reveal a pattern of carrying heavy loads, maintaining a particular posture, or walking great distances and navigating over rugged terrain (Ruff 2008; Pearson 2000).

The traditional method for collecting metric data on robusticity involves taking the external measurements of specific regions of the long bones, such as anterior and posterior diameter or medial and lateral diameter of the midshaft of the femur, and comparing these measures to the physiological length of the bone (Bass 2005; Cole 1994). Another method of obtaining robusticity is to cut cross-sections out of the bone and evaluate cortical thickness (see Chap. 8 for more discussion of this technique). While the latter method is often argued to provide more accurate measures of robusticity, Stock and Shaw (2007) suggest that external measures of the bone are less invasive and that they also provide an accurate means of assessing robusticity.

6.4.1.3 Entheses

Entheses are the points on the bone where ligaments connect bone to bone and where tendons connect muscle to bone (Benjamin et al. 2006). The importance of studying entheses is that it has been argued that over an individual's lifetime, activity puts stress on the ligaments and tendons and that in turn strains the places on the bone where they attach. The consequence of this stress is that the outer cortex of the bone (i.e., the periosteum) is agitated. The response to this insult is the bone reacts by producing new bone which results in the development of ridges or crests of bone that can be observed and recorded (Fig. 6.3).

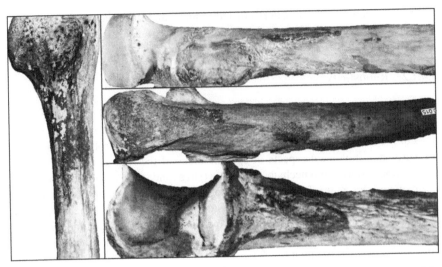

Fig. 6.3 Entheses of the upper arm (Carlin burial 3) (Photo by Ryan P. Harrod, courtesy of the Department of Anthropology, University of Nevada, Las Vegas)

A creative way to use entheses would be to compare groups that were known to work at hard tasks all day long (such as slaves) with those that were not forced into hard labor. The La Plata captives discussed above and in Chap. 5 had very pronounced entheses throughout their body (see photos in Martin et al. 2010).

Entheses that are of the most interest are the attachment sites on the extremities that connect the larger muscles that are more likely to be associated with carrying heavy loads and squatting. Two sources that provide a methodology for recording these changes are Mariotti et al. (2007), who provide a ranking system for each entheses, and Capasso et al. (1999) who record presence and absence of entheses.

The problem of confounding variables such as body size in between-group activity comparison can be greatly reduced by investigating the differences in kind rather than differences in intensity of activity. Not only does this approach remove confounding factors from the analysis, it also makes more sense in the theoretical perspective because most socially relevant differences in activity between groups of people are more likely to be qualitative rather than quantitative. Recent research that looks at entheses has placed a much greater emphasis on comparing individuals within the cultural context and not linking the changes to specific activities (Martin et al. 2010; Stefanović and Porčić 2013; Lieverse et al. 2009; Eshed et al. 2004).

Despite their limitations, differences in the development of entheses can reveal differences among members of the same population in relation to activity. If age and sex are controlled for in the study, it is possible to identify individuals who are performing different kinds or greater amounts of activity within the society.

Fig. 6.4 Perimortem panfacial fracture (Carlin burial 10) (Photo by Ryan P. Harrod, courtesy of the Department of Anthropology, University of Nevada, Las Vegas)

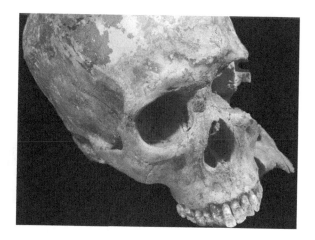

6.4.2 From Domestic Violence to Feuds, Raids, and Warfare: Recognizing Trauma in the Past

Trauma and fractures (healed and unhealed) are an additional data set that can reveal much about the body politic. Whether the result of an accident, occupational hazards, or direct or indirect violence, traumatic injuries offer insight into who within the society is at greater risk of injury, disability, and death.

The timing of trauma on human remains can be estimated based on whether there has been no, some, or complete healing. Differentiating between these three types of fractures is important because the timing of these injuries implies very different implications (for a more detailed description of the difference between antemortem, perimortem, and postmortem trauma, see Chap. 4). In regard to understanding trauma, antemortem (before death) trauma is useful for determining nonlethal violence, while perimortem trauma (around the time of death) often reflects violence that is lethal in nature (Fig. 6.4).

Lethal trauma is critical for understanding the lives of past peoples, but it is often more difficult to use than nonlethal violence as a means of reconstructing the body politic. This is because lethal traumas, unlike nonlethal fractures, are injuries that result at the end of a person's life whereas nonlethal trauma involves injuries that a person sustained during their lifetime (Fig. 6.5).

The experience associated with, and physiological consequence of, the trauma will have an impact on both the individual's lived experience and the groups' as well (see Tilley 2012; Tilley and Oxenham 2011 for an example of one way trauma can have a lasting impact). In some cases, people may have multiple injuries that they accumulate over the life span, which reveals a pattern of reinjury referred to as injury recidivism (Reiner et al. 1990; Judd 2002).

Focusing on the occurrence of nonlethal trauma in conjunction with the other indicators of physiological stress, and inadequate nutrition along with robusticity and activity, may offer an effective way of assessing social hierarchy and

Fig. 6.5 Antemortem nasal fracture (burial 99-3975) of the remains collected by Hrdlička after a brutal Yaqui massacre in June of 1902, at the site of Sierra Mazatan, near Hermosillo, Sonora Mexico

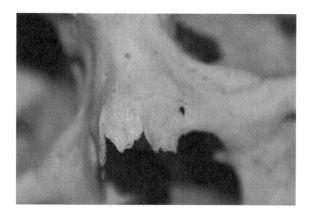

complexity of a culture. For example, Martin et al. looking at a population in the American Southwest (2008, 2010), Tung at a population of Wari women in Peru (2012), Koziol for a group of women at Cahokia (2012), and Wilkinson looking at women in the ancient US northeast (1997) have all been able to show the presence of captive women in past societies based on patterns of trauma and other indicators. As Scheper-Hughes and Lock theorized, the body politic is most shaped by the "control of bodies" in "work and leisure, in sickness and in other forms of deviance" (1987:8). One way to get at individuals who have had their bodies controlled in some way is to be able to identify perpetrators and victims in these kinds of coercive cultural settings.

6.5 Summary

The indicators of identity reviewed here are not an exhaustive survey of every technique for the analysis of human remains. Instead, common indicators of biocultural identity have been organized around questions about the human condition. What are the effects of inequality on health and longevity? Can socially sanctioned violence such as the capture of women who will become slaves be reconstructed for prestate societies? Are warriors more at risk for early death than others in the group? Organizing skeletal indicators of biocultural identity such as age at death, sex/gender, diseases, and trauma using the theoretical framing provided by Scheper-Hughes and Lock was simply a way to demonstrate how these indicators can be used in creative and more integrative ways. Questions about the lives of poor people, prisoners, captives, slaves, and the disabled are as important in the past as they are today. Generating a variety of identities that individuals possess is the first step in the analysis using this perspective.

References

Armelagos, G. J., Brown, P. J., & Turner, B. L. (2005). Evolutionary, historical and political economic perspectives on health and disease. *Social Sciences and Medicine, 61*, 755–765.

Auerbach, B. M. (2011). Reaching great heights: Changes in indigenous stature, body size and body shape with agricultural intensification in North America. In R. Pinhasi & J. T. Stock (Eds.), *Human bioarchaeology of the transition to agriculture* (pp. 203–233). Chichester: Wiley-Blackwell.

Aufderheide, A. C., & Rodríguez-Martin, C. (2003). *The Cambridge encyclopedia of human paleopathology* (reprint edition). Cambridge: Cambridge University Press.

Barrett, R., Kuzawa, C. W., McDade, T. W., & Armelagos, G. J. (1998). Emerging and re-emerging infectious diseases: The third epidemiologic transition. *Annual Review of Anthropology, 27*, 247–271.

Bass, W. M. (2005). *Human osteology: A laboratory and field manual* (5th ed.). Columbia: Missouri Archaeological Society.

Baustian, K. M. (2010). *Health status of infants and children from the bronze age tomb at Tell Abraq, United Arab Emirates*. MA thesis, University of Nevada, Las Vegas.

Benjamin, M., Toumi, H., Ralphs, J. R., Bydder, G., Best, T. M., & Milz, S. (2006). Where tendons and ligaments meet bone: Attachment sites ('entheses') in relation to exercise and/or mechanical load. *Journal of Anatomy, 208*, 471–490.

Birkby, W. H., Fenton, T. W., & Anderson, B. E. (2008). Identifying southwest Hispanics using nonmetric traits and the cultural profile. *Journal of Forensic Sciences, 53*(1), 29–33.

Brooks, S., & Suchey, J. (1990). Skeletal age determination based on the os pubis: A comparison of the Acsadi-Nemskeri and Suchey-Brooks methods. *Human Evolution, 5*(3), 227–238.

Buckberry, J. L., & Chamberlain, A. T. (2002). Age estimation from the auricular surface of the ilium: A revised method. *American Journal of Physical Anthropology, 119*(3), 231–239.

Buikstra, J. E., & Ubelaker, D. H. (1994). *Standards for data collection from human skeletal remains*. Fayetteville: Arkansas Archaeological Survey, Research Series, No. 44. A copy of standards is required in order to fill out these forms accurately. It may be obtained from the Arkansas Archeological Survey, 2475 N. Hatch Ave., Fayetteville, AR 72704, http://www.uark.edu/campus-resources/archinfo/.

Capasso, L., Kennedy, K. A. R., & Wilczak, C. A. (1999). *Atlas of occupational markers on human remains*. Teramo: Edigrafital S.P.A.

Cardoso, H. F. (2007). Environmental effects on skeletal versus dental development: Using a documented subadult skeletal sample to test a basic assumption in human osteological research. *American Journal of Physical Anthropology, 132*(2), 223–233.

Cohen, M. N., & Armelagos, G. J. (1984). *Paleopathology at the origins of agriculture*. Orlando: Academic.

Cole, T. M., III. (1994). Size and shape of the femur and tibia in northern plains. In R. L. Jantz & D. W. Owsley (Eds.), *Skeletal biology in the great plains: Migration, warfare, health, and subsistence* (pp. 219–233). Washington, DC: Smithsonian Institution Press.

Díaz-Andreu, M., & Lucy, S. (2005). Introduction. In M. Díaz-Andreu, S. Lucy, S. Babi , & D. N. Edwards (Eds.), *The archaeology of identity: Approaches to gender, age, status, ethnicity and religion*. New York: Routledge.

DiGangi, E. A., & Moore, M. K. (2012). *Research methods in human skeletal biology*. Oxford: Academic.

Duncan, W. N. (2012). Biological distance analysis in contexts of ritual violence. In D. L. Martin, R. P. Harrod, & V. R. Pérez (Eds.), *The bioarchaeology of violence* (pp. 251–275). Gainesville: University of Florida Press.

El-Najjar, M. Y., & McWilliams, K. R. (1978). *Forensic anthropology: The structure, morphology, and variation of human bone and dentition*. Springfield: Thomas.

Eshed, V., Gopher, A., Galili, E., & Hershkovitz, I. (2004). Musculoskeletal stress markers in Natufian hunter-gatherers and Neolithic farmers in the Levant: The upper limb. *American Journal of Physical Anthropology, 123*, 303–315.

Eshed, V., Gopher, A., & Hershkovitz, I. (2006). Tooth wear and dental pathology at the advent of agriculture: New evidence from the Levant. *American Journal of Physical Anthropology, 130*, 145–159.

Franklin, D. (2010). Forensic age estimation in human skeletal remains: Current concepts and future directions. *Legal Medicine, 12*(1), 1–7.

Goodman, A. H., & Martin, D. L. (2002). Reconstructing health profiles from skeletal remains. In R. H. Steckel & J. C. Rose (Eds.), *Backbone of history: Health and nutrition in the western hemisphere* (pp. 11–60). Cambridge: Cambridge University Press.

Goodman, A. H., & Rose, J. C. (1991). Dental enamel hypoplasias as indicators of nutritional status. In M. A. Kelley & C. S. Larsen (Eds.), *Advances in dental anthropology* (pp. 279–293). New York: Wiley-Liss.

Greenberg, S., & Greenberg, M. (2001). *Handbook of neurosurgery* (5th ed.). New York: Thieme.

Halcrow, S. E., & Tayles, N. (2008). The bioarchaeological investigation of childhood and social age: Problems and prospects. *Journal of Archaeological Method and Theory, 15*, 190–215.

Hillson, S. W. (1996). *Dental anthropology*. Cambridge: Cambridge University Press.

Hong, L., Levy, S. M., Warren, J. J., & Broffitt, B. (2009). Association between enamel hypoplasia and dental caries in primary second molars: A cohort study. *Caries Research, 43*(5), 345–353.

Johnston, F. E. (1962). Growth of long bones of infants and young children at Indian Knoll. *American Journal of Physical Anthropology, 20*(3), 249–253.

Judd, M. A. (2002). Ancient injury recidivism: An example from the Kerma period of Ancient Nubia. *International Journal of Osteoarchaeology, 12*, 89–106.

Jurmain, R., Cardoso, F. A., Henderson, C., & Villotte, S. (2012). Bioarchaeology's Holy Grail: The reconstruction of activity. In A. L. Grauer (Ed.), *A companion to paleopathology* (pp. 531–552). Malden: Blackwell Publishing.

Katzenberg, M. A., & Saunders, S. R. (2008). *Biological anthropology of the human skeleton* (2nd ed.). Hoboken: Wiley.

Koziol, K. M. (2012). Performances of imposed status: Captivity at Cahokia. In D. L. Martin, R. P. Harrod, & V. R. Pérez (Eds.), *The bioarchaeology of violence* (pp. 226–250). Gainesville: University of Florida Press.

L'Abbé, E. N., Van Rooyen, C., Nawrocki, S. P., & Becker, P. J. (2011). An evaluation of non-metric cranial traits used to estimate ancestry in a South African sample. *Forensic Science International, 209*(1–3), 195.e191–195.e197.

Lewis, M. E. (2006). *The bioarchaeology of children: Perspectives from biological and evolutionary anthropology*. Cambridge: Cambridge University Press.

Lieverse, A. R., Bazaliiski, V. I., Goriunova, O. I., & Weber, A. W. (2009). Upper limb musculoskeletal stress markers among middle Holocene foragers of Siberia's Cis-Baikal region. *American Journal of Physical Anthropology, 138*, 458–472.

Little, M. A., & Gray, S. J. (1990). Growth of young nomadic and settled Turkana children. *Medical Anthropology Quarterly, New Series, 4* (3, Steps toward an Integrative Medical Anthropology), 296–314.

Lorber, J., & Martin, P. Y. (2011). The socially constructed body: Insights from feminist theory. In P. Kvisto (Ed.), *Illuminating social life: Classical and contemporary theory revisited* (5th ed., pp. 183–206). Thousand Oaks: Pine Forge Press.

Lovejoy, C. O., Meindl, R. S., Pryzbeck, T. R., & Mensforth, R. P. (1985). Chronological metamorphosis of the auricular surface of the ilium: A new method for the determination of adult skeletal age at death. *American Journal of Physical Anthropology, 68*(1), 15–28.

Lukacs, J. R. (1996). Sex differences in dental caries rates with the origin of agriculture in South Asia. *Current Anthropology, 37*(1), 147–153.

Maresh, M. M. (1955). Linear growth of the long bone of extremities from infancy through adolescence. *American Journal of Diseases of Children, 89*, 725–742.

Mariotti, V., Facchini, F., & Belcastro, M. G. (2007). The study of entheses: Proposal of a standardised scoring method for twenty-three entheses of the postcranial skeleton. *Collegium Antropologicum, 31*(1), 291–313.

Martin, D. L., Akins, N. J., Crenshaw, B. J., & Stone, P. K. (2008). Inscribed on the body, written in the bones: The consequences of social violence at La Plata. In D. L. Nichols & P. L. Crown (Eds.), *Social violence in the prehispanic American southwest* (pp. 98–122). Tucson: University of Arizona Press.

Martin, D. L., Akins, N. J., Goodman, A. H., & Swedlund, A. C. (2001). *Harmony and discord: Bioarchaeology of the La Plata Valley. Totah: Time and the rivers flowing excavations in the La Plata Valley* (Vol. 242). Santa Fe: Museum of New Mexico, Office of Archaeological Studies.

Martin, D. L., Goodman, A. H., Armelagos, G. J., & Magennis, A. L. (1991). *Black Mesa Anasazi health: Reconstructing life from patterns of death and disease.* Carbondale: Southern Illinois University Press.

Martin, D. L., Harrod, R. P., & Fields, M. (2010). Beaten down and worked to the bone: Bioarchaeological investigations of women and violence in the ancient Southwest. *Landscapes of Violence, 1*(1), Article 3.

Merbs, C. F. (1983). *Patterns of activity-induced pathology in a Canadian Inuit population.* Paper presented at the Archaeological Survey of Canada Mercury Series, No. 119, Hull

Minot, C. S. (1907). The problem of age, growth, and death. *The Popular Science Monthly, 71*, 509523.

Mummert, A., Esche, E., Robinson, J., & Armelagos, G. J. (2011). Stature and robusticity during the agricultural transition: Evidence from the bioarchaeological record. *Economics and Human Biology, 9*(3), 284–301.

Ortner, D. J. (2003). *Identification of pathological conditions in human skeletal remains.* London: Academic.

Ortner, D. J., & Putschar, W. G. (1985). Identification of pathological conditions in human skeletal remains. *Smithsonian Contributions to Anthropology, 28*, 1–488.

Owsley, D. W., & Jantz, R. L. (1985). Long bone lengths and gestational age distributions of post-contact Arikara Indian perinatal infant skeletons. *American Journal of Physical Anthropology, 68*, 321–328.

Owsley, D. W., & Jantz, R. L. (1994). *Skeletal biology in the great plains: Migration, warfare, health, and subsistence.* Washington, DC: Smithsonian Institution Press.

Pearson, O. M. (2000). Activity, climate and postcranial robusticity: Implications for modern human origins and scenarios of adaptive change. *Current Anthropology, 41*(4), 569–589.

Pinhasi, R., & Stock, J. T. (2011). *Human bioarchaeology of the transition to agriculture.* Chichester: Wiley-Blackwell.

Reiner, D. S., Pastena, J. A., Swan, K. G., Lindenthal, J. J., & Tischler, C. D. (1990). Trauma recidivism. *American Surgeon, 56*, 556–560.

Ruff, C. B. (2008). Biomechanical analyses of archaeological human skeletons. In M. A. Katzenberg & S. R. Saunders (Eds.), *Biological anthropology of the human skeleton* (2nd ed., pp. 183–206). Hoboken: Wiley.

Scheper-Hughes, N., & Lock, M. M. (1987). The mindful body: A prolegomenon to future work in medical anthropology. *Medical Anthropology Quarterly, New Series, 1*(1), 6–41.

Scheuer, L., & Black, S. M. (2000). *Developmental juvenile osteology.* San Diego: Elsevier Academic Press.

Sofaer, J. R. (2006). *The body as material culture: A theoretical osteoarchaeology.* Cambridge: Cambridge University Press.

Steadman, D. W. (2009). *Hard evidence: Case studies in forensic anthropology* (2nd ed.). Upper Saddle River: Pearson Education, Inc.

Steckel, R. H. (1995). Stature and the standard of living. *Journal of Economic Literature, 33*, 1903–1940.

Steckel, R. H. (2008). Biological measures of the standard of living. *Journal of Economic Perspectives, 22*(1), 129–152.

Steckel, R. H., & Rose, J. C. (2002). *The backbone of history: Health and nutrition in the western hemisphere.* Cambridge: Cambridge University Press.

Stefanović, S., & Porčić, M. (2013). Between-group differences in the patterning of musculo-skeletal stress markers: Avoiding confounding factors by focusing on qualitative aspects of physical activity. *International Journal of Osteoarchaeology, 23*(1), 94–105.

Stock, J. T., & Shaw, C. N. (2007). Which measures of diaphyseal robusticity are robust? A comparison of external methods of quantifying the strength of long bone diaphyses to cross-sectional geometric properties. *American Journal of Physical Anthropology, 134*, 412–423.

Stojanowski, C. M. (2005). *Biocultural histories of La Florida: A bioarchaeological perspective*. Tuscaloosa: The University of Alabama Press.

Tilley, L. (2012). The bioarchaeology of care. *The SAA Archaeological Record, 12*(3), 39–41.

Tilley, L., & Oxenham, M. F. (2011). Survival against the odds: Modeling the social implications of care provision to seriously disabled individuals. *International Journal of Paleopathology, 1*(1), 35–42.

Todd, T. W. (1920). Age changes in the pubic bone. I. The male white pubis. *American Journal of Physical Anthropology, 3*, 285–334.

Tung, T. A. (2007). Trauma and violence in the Wari Empire of the Peruvian Andes: Warfare, raids, and ritual fights. *American Journal of Physical Anthropology, 133*, 941–956.

Tung, T. A. (2012). Violence against women: Differential treatment of local and foreign females in the heartland of the Wari Empire, Peru. In D. L. Martin, R. P. Harrod, & V. R. Pérez (Eds.), *The bioarchaeology of violence* (pp. 180–198). Gainesville: University of Florida Press.

Walker, P. L. (1989). Cranial injuries as evidence of violence in prehistoric southern California. *American Journal of Physical Anthropology, 80*, 313–323.

Walker, P. L., Bathurst, R. R., Richman, R., Gjerdrum, T., & Andrushko, V. A. (2009). The causes of porotic hyperstosis and cribra orbitalia: A reappraisal of the iron-deficiency-anemia hypothesis. *American Journal of Physical Anthropology, 139*, 109–125.

Walker, P. L., Buikstra, J. E., & McBride-Schreiner, S. (2012). Charles "Chuck" merbs: Reconstructing behavior through the bones. In J. E. Buikstra & C. A. Roberts (Eds.), *The global history of paleopathology: Pioneers and prospects* (pp. 60–69). Oxford: Oxford University Press.

Weston, D. A. (2008). Investigating the specificity of periosteal reactions in pathology museum specimens. *American Journal of Physical Anthropology, 137*(1), 48–59.

White, T. D., Folkens, P. A., & Black, M. T. (2012). *Human osteology* (3rd ed.). Burlington: Academic.

Whitehead, P. F., Sacco, W. K., & Hochgraf, S. B. (2005). *A photographic atlas for physical anthropology*. Englewood: Morton Publishing Company.

Wilkinson, R. G. (1997). Violence against women: Raiding and abduction in prehistoric Michigan. In D. L. Martin & D. W. Frayer (Eds.), *Troubled times: Violence and warfare in the past* (pp. 21–44). Amsterdam: Gordon and Breach.

Woo, E. J., & Sciulli, P. W. (2011). Degenerative joint disease and social status in the terminal late Archaic period (1000–500 b.c.) of Ohio. *International Journal of Osteoarchaeology*, 10.1002/oa.1264.

Zvelebil, M., & Dolukhanov, P. (1991). The transition to farming in eastern and northern Europe. *Journal of World Prehistory, 5*(3), 233–278.

Chapter 7
Bioarchaeology of Populations: Understanding Adaptation and Resilience

Human remains represent a uniquely rich data set for a wide range of research questions having to do with the origins and evolution of disease and the limits of adaptability to changing environmental and cultural conditions. Being able to say something about the lived experience and untimely death of individuals who lived in different time periods in faraway lands provides snapshots of historical moments that are invaluable. For example, a recent volume entitled *Bioarchaeology of Individuals* demonstrated the value of applying the interdisciplinary methods of bioarchaeology to persons of interest in the archaeological record (Stodder and Palkovich 2012). The previous chapter provided an overview of how individual identities and social personas are pieced together from a wide variety of skeletal and dental indicators.

Putting individuals who lived (and died) together presents an entirely different and useful way of examining the past. The population-level analysis permits researchers to see patterns and trends at the larger aggregate level. It is only through a population-level analysis that differences in health, activity, trauma, or other indicators of stress can be examined by demographics (age and sex), through time, and by region. This type of analysis allows researchers to ask questions such as: In this population, did females work harder than age-matched males? What is the frequency of nutritional anemia in the children who died before the age of 6? What causes certain individuals by age, sex, region, or social status to die younger than others?

The passage of NAGPRA and NAGPRA-like initiatives in other countries provides an opportunity for the descendants of indigenous people to participate in designing research projects that can answer questions about the state of their ancestor's lives (Silliman 2008). But, there is tremendous diversity in what descendants may want to know (or not know) about their ancestors. Bioarchaeologists need to be prepared to be told that their desire to collect data from ancestral bones is not of interest or value to descendants. Given this reality, researchers should understand how to make their research have value or be of interest to these communities. For example, population-level analyses have revealed important aspects of resilience during periods of enormous change, and these findings are helpful in correcting

D.L. Martin et al., *Bioarchaeology: An Integrated Approach to Working with Human Remains*, Manuals in Archaeological Method, Theory and Technique, DOI 10.1007/978-1-4614-6378-8_7, © Springer Science+Business Media New York 2013

historical narratives of the colonial period that generally tell a story of population disintegration and decline (Stodder et al. 2002). The analysis of the past can even reveal hidden atrocities such as the removal of Yaqui individuals from the battle field so they could be shipped back to a museum for study (Pérez 2010) or even clear up historical accounts about indigenous groups that are false like the century old account of how the Paiute killed the settlers at Mountain Meadows in Utah, when it was local Euro-American residents that were actually the culprits (Novak 2008). As was mentioned in Chap. 3, these are examples of how bioarchaeology done in collaboration with indigenous groups can bring about restorative justice (Colwell-Chanthaphonh 2007). The former director of the National Museum of the American Indian (a Native American himself) has suggested that human skeletal remains are a valuable source of information that offers a way to bring important information to Native Americans (West 1993). He also suggests that working with indigenous groups can promote new research projects and agendas for bioarchaeologists. In an attempt to rectify certain misunderstandings (on all sides) of what the potential of biological data are, bioarchaeology has a mandate to demonstrate that it is a relevant field of study and population analyses are one way to accomplish that.

7.1   Relevance of Population Data

While working with human remains, bioarchaeologists are often asked by onlookers some version of this question: Why is it important to use skeletal remains to document patterns of health and disease for people who are long gone, especially since the modern condition is arguably more pressing? Why not concentrate scientific efforts on people living today because the need there is so great? One answer is that often the ultimate cause of poor health is not proximally located, rather it is an "upstream" manifestation of a situation displaced temporally and/or spatially (Farmer 2003). Bioarchaeologists have the methods to extract information about the past that encompass environmental, cultural, and biological factors along with sequential time periods. Disease can be located in time and space, and an examination of the interrelatedness of ecological, behavioral, and biological variables is possible. These kinds of data are crucial to the understanding of the impact of violence or disease on human resilience. No other medical or biological discipline can carry out these kinds of holistic and interdisciplinary studies.

Another aspect of data on the human condition derived from skeletal remains is that information is gleaned from the whole skeleton (see Chap. 6). Physicians often only see radiographs, CT scans, or MRI scans from one part of the body. Biological information is collected from the whole skeleton (or the elements that preserved intact). For example, from a single individual (such as the female burial portrayed in Chap. 5, Fig. 5.4), the analysis revealed that there was a healed cranial depression fracture and trauma to the top of her head and her left hip showed a more recent trauma that was not healed. Putting together the multiple indicators of stress and degeneration across all of the bones of the body is something that physicians rarely do or have the ability to do. The fact that bioarchaeologists can put together information on individuals that looks at the accumulation of stressors and how they have impacted health is a unique offering to the understanding of human adaptation. Bioarchaeological data provide a way to document human resilience by examining healed and healing pathologies, as well as those that occurred around the time of death. Thus, the collective data from the entire skeleton can reconstruct individual profiles that are distinctive and quite complete.

An even more important benefit of bioarchaeology is that in addition to being able to construct profiles of individuals, researchers often have access to large numbers of individuals from a population. It is at this population level that analyses are most relevant to modern-day problems. While physicians or epidemiologists may have access to population data on basic demographic parameters, they do not always have information on the complete body, on populations through time, or on populations from different cultural contexts. This non-reductionist approach to human biology and culture is one of the unique features of anthropological inquiry—analytical units are necessarily linked, broad based, and dimensional.

It has only been through the archaeological and bioarchaeological record that anthropologists and historians have come to understand how changes over time in environment, political and economic structure, subsistence and diet, and settlement patterns can and do have profound effects on population structure and rates of morbidity (sickness) and mortality (death). A now classic and still relevant set of examples for this can be found in the volume *Paleopathology at the Origins of Agriculture* (Cohen and Armelagos 1984) that focuses on the changes in health at the population level related to shifts in subsistence economy in many different locales around the world. Lambert (2009) points out that in addition to the biocultural perspectives on subsistence change provided in studies such as those in the aforementioned volume, bioarchaeologists can also employ an evolutionary perspective. Looking at fitness and the ability to reproduce at a rate that keeps populations stable or in a growth mode is equally as important. Other edited volumes similar to this now proliferate the literature and have paved the way for a more systematic approach to the collection and analysis of population-level demographic, fitness, and disease data.

Another necessity for empirical data on health and disease is to reevaluate the early and entrenched ideas that life was brutal and harsh on the one hand (Hobbes 1651/2003) or simple and bucolic on the other hand (Rousseau 1762/2008). Data from human skeletal remains have refuted both of these simplified notions about life in the past demonstrating diversity in capabilities to adapt. Silliman (2008) working collaboratively with tribal New England groups has shown other ways that ideas about the past have been entrenched. He discusses the impact of the colonial process on native people as generally being thought to be about the disruption and changing of a relatively stable set of adaptations. Notions of change and stasis also can be challenged with data from human remains by showing the vicissitudes within individual lifetimes, as well as age- and sex-related changes in material culture and other variables. In the absence of empirical data about the effects of environmental change or colonial expansion on morbidity, fertility, and mortality, it is difficult to study the limits of human resiliency in the face of change. Environment, resources, diet, and disease have all greatly affected the course of human history, and all of these factors and others likely played a major role in the rise and fall of populations in different parts of the world at different times. Anyone wishing to predict the future of human adaptation and human resilience to the coming climate change or increasing secterian warfare needs to build on these kinds of data from the past.

7.1.1   Adaptation of Populations to Stress

The fundamental biological needs of humans have not changed in thousands or hundreds of thousands of years. However, the means for expressing and satisfying those basic needs continue to vary greatly from culture to culture. Bioarchaeology examines patterns of variability because patterns are very instructive about the capacity of humans to change, modify, adapt, and alter their cultural and behavioral responses to meet their needs. Much of our future survival may depend on our ability to recognize the limits of human responses and coping mechanisms, especially in adverse and extreme conditions of environmental catastrophe, malnutrition and famine, endemic warfare and strife, and rapidly changing ecological, political, and economic conditions.

Bioarchaeologists have used health as one measure of human adaptability, particularly during stressful periods of rapid change or instability. Generally the body can be seen as an agent equipped with mechanisms to protect itself from harm. And, human societies throughout time have developed systems of health-care and ethnopharmacological resources to aid in treating diseases and trauma. When a culture has been around in one place long enough to produce an archaeological record, it is likely due to the fact that their ability to ward off disease and stay relatively healthy was successful. Ethnographic studies of cultures far and wide support the notion that health and disease are both integrated into the cultural system and are part of the adaptive complex (McElroy and Townsend 2009:105).

Of recent interest are questions concerning how human populations respond to stress emanating from interacting cultural and biological spheres of influence. In a review of the literature on biocultural approaches to stress and adaptation as used in bioarchaeological analyses, Zuckerman and Armelagos (2011) argue that disease is a state of disrupted biobehavioral functioning in which the effects of stressors have overridden the capacity of individuals to respond effectively. Disease states compromise individual responses but also can have an impact on activities at the household and population or community levels. Thus, analysis of health and disease can serve to link biological and social consequences of change and adaptation in human groups. The linking of biological and cultural processes is essential for understanding how human groups cope with stress.

The adaptation of human populations is enhanced by a cultural system that buffers the population from environmental stressors. The technology, the social organization, and even the ideology of a group provide a filter through which environmental stressors pass. In most instances, those buffers can attenuate the stressors, thereby lessening their impact on the individual or population. However, in some instances, the buffers are inadequate, and the stressors will then have their full impact on the individual or population.

There is another source of stress that is often overlooked in the analysis of ancient disease. Although culture can act as a buffer (e.g., with a health-care system and use of medicinals), there are instances in which the stressors originate from the cultural system itself and not directly from the environment. There are many examples of culturally induced stressors: For example, the use of fires in closed shelters can produce toxic particles that are inhaled by the occupants, the change in subsistence strategies to a single crop can lead to nutritional deficiencies, or an increase in population size and density can provide the necessary medium for the transmission of contagious infectious pathogens. Those are just a few of the potential culturally mediated stressors that may be responsible for the observed patterns of stress and disease in ancient populations.

The impact of the stressor depends on its strength and duration. An unusually strong stressor that is short in duration may have relatively little effect. The unavailability of food for a few days can usually be tolerated by adults but may be dangerous for infants. A stressor that is relatively minor in the short run (such as a low-level toxin) may create a significant problem for survival if it persists. If stress is long lasting, severe, and uncontrolled, it may have devastating effects. It will be reflected in an increase in morbidity and mortality and a decrease in productivity and reproduction.

Certain segments of the population may be at greater risk because their biological requirements are not matched by biological resources. Newborns, for example, are born with very immature immune systems. They must rely on immunity conferred during their time *in utero* and transferred via breast milk from the mother. Because of their state of biological immaturity, infants are frequently unable to rally from stressors that have only mild effects on a more mature individual. Mortality is particularly high during the first year in many marginal communities.

Thus, the study of disease and maladaptation in archaeological populations always begins with the evaluation of individual skeletons (see Chap. 6). However, it is critically important to move to a population level to understand the full impact of diseases and other stressors on health, longevity, and fitness at the population level.

7.1.2 Chronologies of Pain: Reconstructing Health and Disease Profiles Over Long Periods of Time

Studies of disease in ancient times provide an important dimension to understanding the life struggles of a largely unknown past. Leslie Poles Hartley started the novel *The Go-Between* (1953/2011:17) with the now famous and iconic quote, "The past is a foreign country: they do things differently there." This is good way to think about what it means to reconstruct past events relating to human lifestyles, motivations, biological conditions, and life histories. Using the most scientific and systematic approach to the collection of empirical data from the human remains, bioarchaeologists will still fall short in fleshing out how past people lived and died (especially of course in the absence of any written records). Yet, in this foreign country that we call the past, information on the health status of the ancestors provides long chronologies of health problems that can span for over thousands of years in some regions. This is an invaluable piece of historical information that *only* the human biological remains can provide.

Although physicians and anatomists began publishing observations on unusual cases of pathology in the mid-1800s (e.g., Matthews et al. 1893), the more technical and anthropological analyses began in the 1930s with the now classic works of paleopathologists such as Moodie (1923), Hooten (1930), Hrdlička (1908, 1935), Wells (1964), Jarcho (1966), and Brothwell and Sandison (1967). Wells in particular was an important force in the study of ancient disease, and his holistic approach is best captured in the following quote: "The pattern of disease or injury that affects any group of people is never a matter of chance. It is invariable the expression of stresses and strains to which they were exposed, a response to everything in their environment and behavior. It reflects their genetic inheritance, the climate in which they lived, the soil that gave them sustenance and the animals or plants that shared their homeland. It is influenced by their daily occupations, their habits of diet, their choice of dwelling and clothes, their social structure, even their folklore and mythology" (1964:17). With this integrated approach, the next several generations of paleopathologists and bioarchaeologists developed approaches to disease in the past which emphasized population-level analyses using modern epidemiological methods (see Buikstra and Roberts 2012 for many examples of this).

Bioarchaeologists are at the forefront of contributions to population-level analyses. For several regions in the United States, there are health chronologies spanning hundreds of years. For example, Lambert and Walker (1991), Walker and Johnson (1992), Lambert (1993), and Erlandson et al. (2001) have documented health and dietary reconstruction for the Chumash Indian groups living in southern California going as far back as the archaic period right on up to colonization and historic times. These data highlight the diversity of adaptations to coastal environments. Using a multi-methodology approach involving analysis of a number of skeletal lesions and detailed reconstruction of the environment, Walker demonstrated that Native Americans living in marginal island environments (ca. 800 BC to AD 1150) show greater evidence of health problems than those who lived on the mainland where food was more abundant and diverse. The islanders were shorter in stature (160 cm vs. 162 cm) and had more lesions indicative of iron deficiency anemia (75% vs. 25%). In addition to clarifying the relationship between resources and health conditions, Walker also showed that there were changes over time, with health conditions worsening (increases in infectious disease from 20 to 30%) due to contaminated drinking sources and diarrheal disease.

Other regions of the United States for which there are large skeletal series with hundreds of years of time depth have likewise been studied. Larsen (2001) has focused exclusively on health patterns for inhabitants of the Spanish borderlands from precontact periods through to the colonial period. Meindl et al. (2008a, b) provide paleodemographic and paleopathologic analyses of health conditions for the large and well-preserved Libben site in Ottawa County, Ohio. The American Southwest has likewise provided relatively large skeletal collections from numerous sites, and health conditions from these have been summarized by Martin (1994) and Martin et al. (1991).

When reconstructing health profiles for early indigenous populations living in America, it is clear that many of the health problems facing people worldwide today were endemic and vexing difficulties for early groups as well. For example in the American Southwest, Navajo and Hopi infants and children are plagued with otitis media (middle ear infections) at a rate higher than non-Indian infants and children (Martin and Horowitz 2003:136). Similarly, precontact populations in the Southwest also show high rates of otitis media (up to as high as 80% for infants and children), and this is a problem that dates back to at least 300 BC. The bioarchaeological data are important because it demonstrates that ear infections are not a result of recent historical changes in lifestyle, technology, or health-care behaviors. Prevailing conditions in the Southwest (sand, dust, and wind) facilitate the severe expression of ear infections. The chronic and endemic patterns need to be understood within a biocultural framework. Today antibiotics and interventions such as drainage tubes are available, but unfortunately not to Native Americans living in rural and remote parts of their reservations.

This is one example of how a bioarchaeological study of population-level health can be used to increase the understanding of a problem today and aid in formulating a successful intervention. A temporal population perspective on disease affords a more complete view of health problems. A bioarchaeologist could argue that Hopi and Navajo children are at high risk of ear infections because of the complex interplay of their environment and long-standing cultural behaviors.

Taken together, these examples of bioarchaeological research in understanding chronologies of pain at the individual level and responsive communities at the population level speak to both the resiliency of human groups as well as to what happens when they are pushed beyond their capacity to adapt and survive. Burial populations offer a glimpse at some of the collective pain and suffering that segments of the population experienced. Using age at death as a starting point, the demographic features of populations are used to measure and understand patterns of non-survivorship. It may seem ironic that to understand the living, there is a focus on the dead, but demographic features of populations reveal much about the general nature of the population.

7.1.3   Bioarchaeology of Disability and Community Care

It is tempting to interpret signs of hardship directly from the skeleton through western notions of the disease process and western ideas about pain and disability. Reading the paleopathology literature often invokes interpretations of the quality of life for ancient people that may include ideas about pain, discomfort, suffering, grief, misery, sorrow, sadness, and agony. Caution must be used in doing so because just as bioarchaeologists cannot say that skeletons without pathological lesions on the skeleton were healthy (and therefore happy, fortunate, etc.), bioarchaeologists cannot assume that severe skeletal pathology necessarily caused disability and pain (Dettwyler 1991).

Tilley (2012) presents a way of thinking about caretaking of sick or diseased individuals within the population or community setting. She uses a very productive methodological approach that she calls the Ladder of Inference (2012:40). When human remains are discovered that have a disease that is quite advanced and that obviously would have affected the individual of taking care of themselves and surviving with the disease, the bioarchaeology of care can be carried out using her four-stage method. The first stage involves a detailed diagnosis of the pathology with observational and metric analyses. The second stage involves utilizing medical and clinical literature on the pathology to deduce the range of functional impairments the individual may have suffered. Tilley suggests that the goal of this stage is to clarify if the individual could have survived on his or her own, or if they would have needed care by others in the community. The third stage involves using ethnographic and other cultural information to contextualize the individual within their own community. What kinds of care might have been available and given? How many caretakers would have been necessary?

These kinds of questions are used to interrogate the cultural context of the individual. The final stage puts together information gathered in the other three stages to be able to say something about caregiving in that population. Tilley (2012:40) states that "... while each case of care is unique, there is a fundamental principle to be observed in all cases of health-related care: recognition that care is the product of agency." Thus, going from an individual with a disability to thinking about the dynamic, complex and interrelated notion of caregiving at the population level can link research from an individual to behaviors at the community and population level.

It is important to understand and acknowledge the important dynamic that exists between the individual burial and the population of individuals located within the assemblage. A research design that starts with individuals and moves through to population-level inquiries will provide the most comprehensive and useful conclusions. Studies that focus only on individual case studies may have their place in contributing a small piece of biomedical or paleopathological data, and a solely population-level analysis may contribute to seeing patterns across age, sex, and other subdivisions within the collection. But combining the two perspectives (individual and population) while challenging is ultimately more valuable in what it can reveal about human adaptation and resilience.

7.2   Paleodemography

In recent decades, there have been a number of controversies over the accuracy of estimation of the age at death of an individual skeleton and of estimating mortality patterns in archaeological populations. However, paleodemography remains one of the primary and crucial sets of data for analysis of the general health of populations (Hoppa and Vaupel 2002). Subadults can be aged with precision based on the patterns of tooth calcification and eruption, and new techniques for aging and sexing adults are regularly employed with up to 97% accuracy when the full skeleton is available for study (Uhl 2012).

What bioarchaeologists generally analyze is a sample of individuals who died. It is difficult to know how representative that sample is of the living population. Milner and colleagues capture the challenges of this kind of research: "The linkage between a past population and a skeletal collection is long and torturous. The bones we might examine are the ones that survived a complicated winnowing process that might be summarized by the following sequence: Living $\rightarrow$ Dead $\rightarrow$ Buried $\rightarrow$ Preserved $\rightarrow$ Found $\rightarrow$ Saved" (Milner et al. 2008:571). Another way of thinking about this is that bioarchaeologists have "a sample of a sample of a sample" (Roberts and Manchester 2005:9). We might add that another winnowing occurs because not every skeleton saved actually gets analyzed. Thus, bioarchaeologists must always be careful to document the ways that the collection may be biased, from excavation and taphonomic issues to preservation and curation problems.

7.2.1 Age at Death and Life Table Analyses

Mortality data for skeletal populations derive from assessment of individual ages at death. Traditional presentations of mortality data involve use of survival (from one age to another) to graph survivorship curves. Life expectancy as a function of survival has also been used in many studies. Through paleodemography, population parameters can be generated to examine trends in morbidity and mortality by age group (Gage 2010).

Because mortality data for skeletal populations derive from assessment of individual ages at death, paleopathology and other analyses of health, diet, and disease depend on utilizing those age and sex categories in order to characterize trends at the population level, more commonly referred to as paleoepidemiological analyses. Thus, paleodemography is critical to the establishment of population parameters that can be used to look at trends in morbidity (illness) and mortality by age groups (and by sex for adults).

Traditional presentations of mortality data involve either the direct estimation of life expectancy at birth (based on the mean age at death) or the construction of life tables. In addition to estimating life expectancy at birth, probability of dying and survivorship for all age classes can be estimated. At this level of analysis, the data are completely derivative and removed from the actual skeletal remains because the estimates involve statistical manipulation and only use numbers of individuals dying in different age categories (see Milner et al. 2008 for methods on the construction of life tables and other demographic parameters).

The statistical manipulation of age and sex profiles and the construction of life tables are actually the easier part of paleodemographic analyses. The difficulty lies in *interpretation* of the data. In an important study, Johansson and Horowitz (1986) succinctly summarize the range of unknowns and assumptions that must be taken as a given when reconstructing demographic patterns from skeletal populations. For example, they present four phases or levels of analytical factors that need to be acknowledged when dealing with skeletal populations "whose demographic characteristics are unknown . . . (who) buried their dead over an unknown period of time according to unknown rules" (1986:234). Phase one, according to Johansson and Horowitz, involves archaeological recovery that can present biases in age, sex, class, and other spheres because of mortuary practices and differential preservation. Phase two includes the anatomical and paleopathological analysis of the recovered skeletal material for age and sex. Problems encountered at this level include the margin of unknown error involved in estimations of age and sex, the small sample sizes in age and sex categories, and the inability to determine cause of death in the vast majority of the cases.

Phase three involves demographic analysis of mortality and fertility. Here Johansson and Horowitz point out that it is extremely difficult to prove whether a population was stable, stationary, and closed (assumptions that must be made in order to compare prehistoric age and sex distributions with model life tables derived from living populations). Finally, phase four involves historical reconstruction and theory building based on the mortality estimates. Problems at this level include the inability to establish how valid the derived mortality/fertility estimates are. Errors in phase one or two can greatly affect the validity of the data generated in successive levels of analysis.

In spite of all the potential problems, there is ample evidence that paleodemography is central and critical to our understanding of prehistoric life and death and should be undertaken in all possible cases (e.g., Jackes and Meiklejohn 2008; Storey 2009; Trinkaus 2011). In addition, Roksandic and Armstrong (2011) present a new way to construct life tables. Traditionally life tables use 5- or 10-year-age categories in their calculations. Because methods for providing exact chronological age at death are not yet available, the use of clearly delineated developmental life stages might help get around this problem. For example, in a traditional life table, there may be an age category of new birth to 5 and another category of age 6–10. Current aging techniques do not permit being able to distinguish a 5-year-old from a 6-year-old, and so knowing where to put some individuals in traditional age categories could be difficult and produce inaccuracies. However, Roksandic and Armstrong suggest that using age categories that correspond to developmental stages may be an even better way to divide age categories. For example, they suggest that using eight highly recognizable developmental stages be used: infancy, early childhood, late childhood, adolescence, young adulthood, full adulthood, mature adulthood, and senile adulthood. They suggest that these broader categories map well on to the methods available for assigning age to individuals.

Bioarchaeological analyses that can incorporate these higher-level statistical analyses into their overall research design should do so with as much care as possible. While life tables can reveal a great deal about population structure and demographic features of the population, collaborating with scholars highly trained in life table analyses is recommended. This method provides one way to analyze data at the population level, but it is dependent on careful utilization of the methodologies.

7.3 Modeling the Effects of Stress and Change at the Population Level Using Skeletal Remains

Skeletal tissue represents a durable record of biological and cultural history in ways that cannot be obtained from other archaeological resources. If the skeletal remains represent major elements of the body (the head, arms, pelvis, spine, legs), bioarchaeologists can reconstruct not only the sex of an individual but also the age at death, standing height, level of muscularity, and a range of pathologies and ailments, including trauma as well as other data (see Chap. 6). This allows bioarchaeologists to reconstruct the differential treatment and experiences of subgroups of people in the past (Larsen 1997). Sex differences in habitual activities, patterns of violence and inequality, rates of mortality and morbidity, and other consequences of the lived experience of males and females have all been made within and between groups from across various cultures, regions, and time periods (recently reviewed in Hollimon 2011).

Contextual analyses of skeletal populations have increasingly used multiple lines of data to reconstruct the differential treatment of the sexes in more nuanced ways. The use of multiple lines of skeletal, archaeological, and cultural data has enabled scholars to identify evidence of domestic violence, the capture and enslavement of women, and evidence of individuals who lived as third gender persons. Such scholarship relies entirely on the plasticity of the human skeleton which is shaped by, and thereby reflective of, multiple aspects of an individual's social and biological identity. In this way, human skeletal material provides scholars interested in sex and gender a large body of data through which to generate hypotheses for evaluations that draw together the mutual influence of culture on human biology and vice versa.

Understanding physiological disruption and the impact of stress on any population feeds directly back into the understanding of cultural buffering and environmental constraints. Auerbach (2011) suggests that skeletal populations have great potential for providing information concerning variation in adaptation to shifting availability of food. It is extremely important to understand how disease, nutrition, and early death affect the functional and adaptive consequences for any community. For example, poor health can reduce work capacity of adults without necessarily causing death (Gagnon 2008; Leatherman 2001). Decreased reproductive capacity may occur if maternal morbidity and mortality is high in the youngest adult females (Population Reports 1988). Individuals experiencing debilitating or chronic health problems may disrupt the patterning of social interactions and social unity and may strain the system of social support.

The documentation of patterns of disease in ancient times should ultimately be channeled back into the discussion of contemporary health problems. In modern society, health of infants and children is delicately linked to the function of mothers, families, and communities. We can assume similar dynamics for all human groups, and these interrelated issues must be explored for prestate communities because it supplies a much needed time depth to understanding the history of disease. The bioarchaeologist is in a unique position to monitor the dynamics between changes in the ecological and cultural environment and changes in human response.

If one were to address these hypotheses concerning health dynamics, the demographic and biological impact of stress could be measured by skeletal indicators of growth disruption, disease, and death. Pathological alterations on bone are assessed primarily thorough the systematic description of lesions. Patterns of growth and development provide information on stress. Demographically, a great majority of the human remains recovered from precontact archaeological sites are under the age of 18, and growth and development of children using dental and skeletal data from critical stages could be compared to contemporary groups living in similarly marginal areas. Identifiable, age-specific disruptions in growth yield important information on patterns of childhood developmental disturbances and physiological disruption. The distribution and frequency of specific diseases (nutritional, infectious, degenerative) is also an essential part of the health profile. The patterning and frequencies of nutritional diseases such as iron deficiency anemia are documented for many precontact populations and have obvious implications for understanding adequacy of diet. Infectious diseases likewise well documented for many skeletal series provide an indicator of demographic patterning, population density, and degree of sedentism.

The linking of demographic, biological, and cultural processes within an ecological context is essential for dealing with the kinds of questions that are of crucial interest to health practitioners and demographic researchers the world over. For example, understanding the relationship between political centralization and illness, the impact of population reorganization or collapse on morbidity and mortality, and the relationship between social stratification, differential access to resources, and trauma are all useful in knowing the groups who are most at risk for poor health and early death. These kinds of problems demand a multidimensional approach because they cross over numerous disciplinary boundaries.

The following section provides three case studies that demonstrate different ways that the analysis of human remains at the population level can reveal patterns of past lives and this contributes to our understanding of the past and dispels many of the erroneous stories that have been perpetuated about past cultures.

7.3.1 Case Study: The People of the Mounds—Status and Inequality at Cahokia

The Mississippi River Valley is the home of arguably one of the most complex societies in North America; this region is at the heart of the cultural tradition early Europeans referred to as the "mound builders." Characterized by the creation of large earthen works and mounds, the various cultures that were categorized together as the mound builders were complex agricultural societies. These mounds range in size but include the largest constructed features in the New World north of the Aztec and Incan Empires. Given their complexity, when early colonist first unearthed them, they thought that these elaborate human-made structures were too complex for the indigenous people to have made, so they had to be evidence of an early European presence in North America that predated the "natives" inhabiting the continent (e.g., the Lost Tribes of Israel, Viking explorers, or Phoenician sailors). These racist ideals persisted until the advent of scientific archaeology (Jefferson 1787/1955; Thomas 1884) which clearly demonstrated that these earthwork structures and mounds were not constructed by some unknown mound builder society but by the ancestors of the modern indigenous peoples of North America.

While the stories of marauding Vikings, globe-trotting Phoenicians, and wandering "Lost Tribes" faded from the discussion of the mounds, scientific investigation of the region continued to grow. The result of America's early and ongoing fascination with the mound builder culture led to the development of a long tradition of archaeology in this region. Eventually the Mississippi River Valley would serve as the birthplace of bioarchaeology with pioneers like Buikstra (1972) and George Armelagos with his students Lallo (1973) and Rose (1973) developing some of the earliest demographic reconstruction of the past cultures of North America. To this day, this region remains a major center of bioarchaeology research.

Looking specifically at one site in the Mississippi River Valley that developed during the last stage of the so-called mound builder cultural tradition, illustrates how population-level analysis in bioarchaeology can inform our understanding of the past. Cahokia is a site in Illinois near modern-day St. Louis, Missouri, and it is a large complex site with anywhere between a hundred (Moorehead 2000) and two hundred mounds (Pauketat 2009). In fact, based on estimated population size and the constructed features at the site, it is believed that Cahokia was one of the most complex societies in North American prehistory.

With complexity comes social stratification and inequality, which are two things that bioarchaeological analysis can reveal. Social stratification and inequality can be revealed through the analysis of site complexity, mortuary context, and general health of the human skeletal remains. The trade system in Cahokia was vast with artifacts and exotic materials coming in from and going out to distant locations. However, according to Milner (1998), it appears that the chiefs of Cahokia were receiving more trade goods than they were redistributing and that reduced reciprocity is a further sign of greater complexity and more power in the hands of the rulers. This is evident in the analysis of the grave goods and more specifically in the presence of exotic grave goods. For example, tools have proven to be especially important in determining the complexity of Cahokia. These are the copper hand axes and chipped-stone hoes. Copper axes are only found in the graves of high-ranking individuals in the society making them one of the most significant status identifiers in Cahokia, while chipped-stone hoes reinforce the agrarian nature of the society.

Mound 72, a large and elaborate burial mound at Cahokia, offers a perfect case study for understanding social stratification and inequality. Excavated in the late 1960s through the early 1970s by Fowler (1991), this mortuary context reveals the presence of both elite individuals as well as captives or slaves (Rose 1999; Koziol 2012). The individuals are identified as captives based on how they are buried and the presence of traumatic injuries (Koziol 2012). They are identified as nonlocals based on markers of genetic relatedness using dentition (Rose 1999) as well as dietary reconstructions using isotopic data (Ambrose et al. 2003). Support for captives at Cahokia is also confirmed by evidence that violence in the form of raiding and warfare was present in the Mississippi River Valley before and after Cahokia (Milner 2007; Milner and Ferrell 2011). All of these factors taken together present a system of social inequality and control within the site of Cahokia.

The importance of understanding how bioarchaeology can reveal social stratification and inequality in the Mississippi River Valley is that it sheds light on how complex these past societies were. By understanding the complexity within this one site, it is easy to see how these societies were able to build such elaborate earthen works and mounds more than 500 years before the arrival of Christopher Columbus and the European explorers and settlers that followed him.

7.3.2   Case Study: The Pueblo—Hardship and Poor Health in the American Southwest

The record of human habitation in the American Southwest is both long and continuous. Although many Native American groups in historical times were decimated by disease, killed in battle, or forcibly removed from their original homelands, inhabitants of the American Southwest represent largely intact and *in situ* communities in spite of the genocidal policies and colonial pressures exerted in that area for over 500 years. One particularly widespread group, the Pueblo Indians, is the continuation of precontact cultural traditions recorded by archaeologists as the Anasazi cultural traditions but referred to in a variety of more appropriate ways by today's Pueblo Indians such as the ancestral Pueblo people. Criticism of the label "Anasazi" (a Navajo word) is justified since it has little relevance to contemporary Pueblo people (Ladd 1991). Use of the phrase ancestral Pueblo emphasizes the fact that the people being studied by archaeologists are the ancestors of the Pueblo people living there today. The term Pueblo was applied by the Spanish conquistadors to American Indians living in adobe houses and farming villages near the Rio Grande River in the 1500s and is used today by Pueblo groups to refer collectively to the large and heterogenous Indian populations that lived there long before contact, during contact, and today (see chapters in Downum 2012).

Archaeological investigations coupled with analyses of the human remains suggest viable and highly adaptive agricultural communities that used innovation and flexibility in order to survive and thrive in a largely marginal and seemingly hostile environment for agriculture. Stodder (2012) provides an overview of the historical trajectory of studies based on human remains from the American Southwest. She shows how the methodological orientations in general have changed over time, from descriptive synchronic studies to more synthetic diachronic overviews. She documents the ways that research questions have changed over time as well. Theories about social and political processes have become more prominent in the approach to interpreting data from the human remains in this region and that has opened up new ways of thinking about the biocultural adaptation of these groups over hundreds of years.

The biological costs of living in a desert environment included endemic nutritional anemias and early mortality for some segments of some populations. Trauma and violence were sporadic but significant suggesting periodic strife either within or among groups. Massacres and warfare have been documented throughout the precontact period. Regional and temporal variability in frequencies of disease demonstrates that factors such as population density and environmental fluctuations often affected the ability of groups to positively respond to the challenges of desert living. Strategies such as frequent migrations, extensive alliances, and creative uses of the local ecology provided a buffer allowing population growth and widespread influence, particularly during the twelfth and thirteenth centuries. Patterns of health and disease examined in conjunction with archaeological data are useful in reconstructing the past in ways that reflect the complexities and multidimensionality of ancient lifeways.

Swentzell (1993:141), a scholar and writer who was born and raised on the Santa Clara Pueblo, has ruminated in various publications about the work that bioarchaeologists and archaeologists do, intent as they are on reconstructing the lifeways and life histories of her ancestors. She looks at the archaeological ruins and the artifacts and sees something quite different worth noting: "The 'old ones' did not live according to an elaborate and formalized ideology of absolute truth. . . They lived knowing that this place, this time, is all that there is. . . This place is where it all happens—happiness, sadness, pain, obligation, responsibility, and joy. Human life, in the traditional Pueblo world, is based on philosophical premises that promote consideration, compassion, and gentleness toward both human and nonhuman beings. . . At death, cycles and transformation are honored" Swentzell (1993:141).

Part of the problem regarding the analysis of human remains has been the inability of the research community to precisely articulate the ways that information from the past can inform the present in the Southwest. With respect to health, during the 1950s, reports began to surface that demonstrated that American Indian rates of infant mortality and adult morbidity were alarmingly high and disproportionate to the rates for the general US population (Moore et al. 1975). These data were not looked at historically in order to understand the impact and effects of colonization and economic and racial oppression related to the placement on reservations. Further, because knowledge of Indian health in America is limited to a few geographic areas and is largely synchronic, it is not surprising that no clear picture of precontact and historic trends in health has been fully documented. With over 40,000 years of rich and diverse culture history, it is disheartening to note the lack of attention to and detailed treatment of health status in broad and encompassing ways. A systematic analysis of indigenous health before and after colonization and changes in health status resulting from colonization and reservation life would fill the void that currently exists. Today, a crisis exists in Indian communities over access to good health-care and critical resources such as jobs, water, and land. In order to contextualize and create solutions to these problems, a better understanding of historical trends would be extremely useful.

Today, there are a number of traditional Pueblo villages in Arizona and New Mexico. Continuities of these communities with precontact groups are evident in material culture, subsistence patterns, religious and ideological behaviors, and biology.

What is unique about the Pueblo people is this unbroken cultural continuity existing into the present. The Pueblos, as descendants of New World explorers who settled in Mesoamerican and North America thousands of years ago, have maintained a persistent hold on traditional values by adapting to novel and changing economic, ecological, political, and cultural conditions (Eggan 1979).

All Pueblo Indians share historical ties to the Southwest, as well as characteristic patterns of social organization and ceremonial cycles. However, there is much cultural (including political and ideological) and linguistic diversity evident both today and in the precontact and historic past. Ortiz presents one key to understanding Pueblo longevity in the Southwest when he states that "the Pueblo peoples have shown a genius for maintaining that which is essential to their lives while also receiving, absorbing, and reinvigorating. . . other ways of life" (1979:3).

Bioarchaeological studies and the American Southwest do have much to contribute to our understanding of population adaptation and resilience. Stodder (1989:145) suggests that it is often assumed that environmental perturbations affecting poor agricultural yield result in subsistence stress and increased health problems. In testing this hypothesis with temporal health data from Mesa Verde in southeastern Colorado, she was able to show that increases in disease were more immediately attributable to sedentism and population aggregation than to the effects of poor agricultural yield and subsistence stress. Stodder points out that diet or subsistence stress are only one of several factors that directly affect morbidity and mortality. Cultural processes such as restricted access to resources, political organization, settlement patterns, food processing and storing techniques, trade relations, and a host of other culturally influenced behaviors most likely helped to mediate the negative effects of subsistence stress through technological, social, and ideological systems.

In an examination of these same communities over time, Stodder (1989) also noted that there are more active nonspecific infectious lesions (versus healed lesions) in the later Mesa Verde samples. Stodder presents the archaeological context as reconstructed for ecological and subsistence factors, and then provides a demographic study of patterns of mortality, growth disruption, nutritional problems, and other indicators of morbidity. Using methods in paleoepidemiology, she notes a trend toward increasing morbidity (illness) and higher mortality (deaths) in younger ages. Tying this in to climatic, ecological, nutritional, and political changes occurring at the end of the later period, Stodder's interpretation is one that suggests that community health became increasing compromised over time.

In summarizing the changes in health from early to late periods, Stodder presents compelling data that health declined. The early sample subadults exhibit a gradual increase in the probability of dying, in the prevalence of nutritional anemias, and in the frequency of developmental defects from birth to age 5. Peaks in childhood morbidity at ages 2–3 suggest weaning stress. In the late sample, the probability of dying is highest at the age of 1 year, and peaks in childhood morbidity occurring later (at ages 4–5) suggesting delayed weaning. Nutritional anemia and developmental defects are more prevalent in this group as well. Stodder also shows that late adult females have more indicators of stress than early females.

Stodder's work is important because it emphasizes the importance of population-level analyses which take many factors into consideration in the interpretation of community health. Human skeletal remains, when used as part of the archaeological database, can provide time depth and geographic variability to the understanding of short- and long-term consequences of living in marginal environments. Studies on the health and disease of southwestern groups have incorporated skeletal remains to address health status over time and have provided indisputable data that the ancestral Pueblo people were "stressed" to varying degrees in different areas at different times. But we are left with many questions, and the mechanisms and buffering afforded by different groups are by no means yet fully understood.

The inventory of diseases suffered by prehistoric Southwest people is relatively long but very incomplete. While published data support the ubiquity of infections, anemia, dental disease, developmental problems, and trauma, paleopathologists cannot provide a detailed scenario of how these diseases actually played out at the group level. The various indicators of stress most certainly have overlapping etiologies, but the pattern of these morphological changes confirms that stress in the Southwest was primarily chronic; it affected infants and children to a degree not seen in adults, and most likely contributed significantly to morbidity if not mortality.

Health then for the ancient Southwest could be summarized in the following manner: there were major and persistent nutritional deficiencies resulting from a largely maize diet; crowded and unsanitary living conditions enhanced the chances of picking up communicable diseases such as gastroenteritis; dental problems including caries and periodontal disease were a major concern; most adults had arthritis and spinal degeneration from carrying heavy loads; parasites such as lice and helminths were common; and infant and childhood mortality was high. With respect to trends over time, a continuum of health problems suggests that there were changes in the patterns with an increase in diseases associated with large and aggregated populations (Martin 1994).

7.3.3   Case Study: The Calusa—Complex Hunter-Gatherers of Spanish Florida

The term Spanish Florida is used to describe a region in the southeastern portion of the United States in present-day Georgia and Florida where the Spanish established a number of missions. This region, also known as La Florida, is important because it is a focal point for early contact and missionization in the New World.

Within La Florida, there has been an extensive amount of research conducted on indigenous populations via the analysis of human remains recovered from sites associated with the missions. The result of this analysis is that a great deal of information about indigenous populations has been produced expanding our understanding of demography before, during, and after contact. While numerous articles and books have been published on this region, the volume *Bioarchaeology of Spanish Florida* edited by Larsen (2001) is arguably the most comprehensive. The authors of each chapter utilize biocultural approaches to ask very specific questions about the lives of the indigenous groups inhabiting this region, including diet, quality of life, genetic relationships, and exposure to pathology (Harrod 2009).

What bioarchaeological research has illustrated is that the impact of colonialism was not homogeneous in nature as the diet, health, and mortality among the different indigenous groups in the region were affected to varying degrees by the establishment of the Spanish missions. Using a range of methodological approaches including more advanced methods like the ratio of stable isotopes of nitrogen and carbon within the bones and microscopic evidence of wear on the teeth, as well as more traditional methods including robusticity, dental health, and hypoplasia prevalence, and the analysis of pathological conditions, bioarchaeologists in this region over the last several decades has generated a fairly complete picture of life in La Florida during the protohistoric period (Stojanowski 2005a, b). The biocultural approach to bioarchaeology is evident in the four-field approach that typifies research in this region.

Bioarchaeology has helped to expand our understanding of one rather unique population in La Florida, the Calusa people. This culture is interesting for a number of reasons. First, they were the indigenous group to make initial contact with Ponce de León on his quest to find the fountain of youth (Widmer 1988:223). They were such a formidable group that they actually drove him off the first time they encountered him and fatally wounded him when he later returned. "When Ponce de Leon first set foot on Florida soil in 1513, eighty war canoes of the Calusas were forced to retreat after a day-long battle. When he returned 8 years later, a Calusa arrow wounded him so badly that he died in Cuba a short time later" (Brown and Owens 2010:33–34). Second, they differ from other groups in La Florida because they are not agriculturalists but were complex hunter-gatherers located along the coast in the southwestern portion of the region. The Calusa show no evidence of being engaged in agriculture (Thompson and Worth 2011). In fact, the culture was so opposed to agriculture that Arnold states the following: "When presented with opportunities to cultivate plants by the Spanish in the early Historic era, Calusa males refused, indicating that scratching in the dirt was beneath their station in life" (Arnold 2001:8).

Similar to what is found along the Pacific Coast among the Northwest Coast cultures and southern California among the Chumash, the Calusa appear to be complex, semisedentary hunter-gatherers that were exploiting the abundant fishing and marine resources for their subsistence (Johnson and Earle 2000). At the time of colonial contact in La Florida, the Calusa had established a large political domain in the region that relied on a system of exploitation where they raided the neighboring agricultural groups on the interior of the continent for resources (Widmer 1988). The result of this exploitation eventually led to the development of a system of tribute where the neighboring cultures produced excess resources that were traded to the Calusa to prevent raiding. An account of the Calusa during this period of Spanish Florida describes the social status these chiefs hold by describing the material objects in the possession of one chief. This individual wore strings of beads on his legs and a very visible gold ornament was on his forehead, which clearly signified his higher social status (Hann 2003).

Hann (2003) discusses the complexity of the Calusa by equating the power and prestige held by the leaders of this group of coastally adapted hunter-gatherers to that of Charlemagne. The reason for the association is that right around AD 800 when Charlemagne was attempting to unify and control Europe, these cultures were developing large cities and constructing massive earthwork structures. In fact, the complexity of the Calusa chiefdom was much greater than what was found among the Pacific Northwest cultures. "In the small pantheon of complex hunter-gatherers the Calusa of Florida stand clearly at the top" (Arnold 2001:7). The Calusa were more complex because the leaders possessed more authority than what was present among the Northwest Coast leaders and had a long history of interaction with the agricultural chiefdoms and prestate societies found throughout the southeastern United States (Hann 2003; Widmer 1988).

It is through bioarchaeological analyses that we are able to reveal the demography of cultures in the past and provide support for or against the narratives that exist about these people. For example, support for preexisting stories of the Calusa is provided by recent analysis of the remains by Hutchinson (2004:155–156) that suggest no real evidence of warfare but a prevalence of nonlethal violence prior to circa AD 800. The implication is that the type of conflict was likely raiding, which matches the ethnographic accounts of the Calusa. In contrast, bioarchaeological work by Kelly et al. (2006) suggests that the narratives cannot always be trusted. Given the complexity of this large stratified society, there are researchers who, citing ethnographic examples, argue that the Calusa did practice agriculture. However, recent stable isotope analysis of the burials culturally identified as Calusa indicates that maize was not a part of the diet (Kelly et al. 2006:259).

Unlike populations in the Southwest, little is known about the fate of the Calusa. There are theories about the effect of colonization and the epidemics that it brought with it or that small groups of the Calusa migrated south to Cuba as they were pushed out of the region. However, more may be known in the future as more bioarchaeological research is conducted. For example, Stojanowski (2011) looking at biodistance data and utilizing an evolutionary approach informed by ethnographic, linguistic, and archaeological evidence found that many of the northern cultures were not wiped out. Instead he argues that while the indigenous groups in this region were affected by colonization, skeletal indicators suggest that some of the groups lived on and integrated with the modern-day Seminole people. Perhaps future research will reveal that the descendants of the Calusa live on in a living group in the Southeast United States or perhaps Cuba.

7.4   Summary

Bioarchaeologists interested in studying populations of ancient human remains have a long checkered past in their relationship with North American indigenous peoples. Historically, the "other" has been prodded, X-rayed, measured, and treated as a scientific specimen, devoid of a cultural context. Bioarchaeologists applied their trade with virtually no recognition or acknowledgement that there might be a social, political, religious, or ethical realm to what they did, and this arrogance was challenged with the passage of NAGPRA in 1990. The real and perceived damage done to indigenous peoples by reductionist scientific studies of ancient biology and health can be countered with bioarchaeology that recognizes and responds to the ethical and cultural dimensions inherent it its practice.

Pueblo scholar and writer Naranjo (1995) suggests that archaeologists too often look for exclusive and universal truths in the archaeological data, but oral histories reflect multiple truths and thus multiple meanings. "The myths, stories and songs describe a world in which a house or structure is not an object, as such, but part of a cosmological worldview that recognizes multiplicity, simultaneity, inclusiveness and interconnectedness" (Swentzell 1993:142). Bioarchaeologists are not routinely encouraged to include oral histories and ethnographic research into their studies, and superficial use of these data sets from the living would not change the criticisms currently directed at them from the Native American communities.

Although much can be learned from an admittedly rudimentary outline of basic relationships for past peoples, a more textured and layered understanding would come with the inclusion of a wider range of knowledge. For example, spatial arrangements, architecture, material cultural, birth, illness, and death are major cultural and biological events captured in both the empirical archaeological record and in traditional oral history, and both of these avenues can lead to a more authentic understanding of how and why people get sick. Anyon et al. (1996:14) states that oral tradition and archaeology present two overlapping ways of understanding the past and that the "real history" of a population as revealed in oral tradition is the "same history that archaeologists study."

Understanding cycles of life and death has long been used by bioarchaeologists as one measure of human adaptability, particularly during periods of rapid change or instability. Yet, it is imprudent to think that, as scientists, we can look at the physical remains of a people long gone and reconstruct all of the important events that made up their lives. Without the textured layering of oral tradition and the voices of those most closely related to the people we study, we are destined to create a series of scenarios that, although grounded in theoretical modeling and empirical observations, are wanting in relevance and authenticity (Martin 1998). It is incumbent upon the scientific community to use information that has been collected over the years from artifacts and human remains in a way that does not trivialize or diminish the lives of the living descendants of those being studied. If scientists are to continue to make inquiries into the past, they must make relevant and link those findings directly to solving pressing problems today. For example, representatives from the Hopi Nation have worked closely with archaeologists and have deemed some biological studies of ancient remains to be valuable: "Some Hopis are also interested in the genetic affinity between different tribes in the Southwest and what this means for prehistoric migrations. In addition to affinity, the age, sex, and pathologies of disinterred human remains are deemed to be important variables…" (Ferguson et al. 1993:33).

The biological past does hold clues to solving health problems and preventing violence in the present and for future generations. Tracking patterns of variability in health is important because these instruct in important ways about the capacity of humankind to change, modify, adapt, innovate, and alter their behaviors to meet their needs. Future survival may depend on our ability to recognize the limits of human adaptability and coping mechanisms, especially in adverse and extreme conditions of political and economic oppression, environmental catastrophe, malnutrition and famine, and narrowing ecological resources.

References

Ambrose, S. H., Buikstra, J. E., & Krueger, H. W. (2003). Status and gender differences in diet at Mound 72, Cahokia, revealed by isotopic analysis of bone. *Journal of Anthropological Archaeology, 22*(3), 217–226.

Anyon, R., Ferguson, T. J., Jackson, L., & Lane, L. (1996). Working together: Native American oral traditions and archaeology. *SAA Bulletin, 14*(2), 14–16.

Arnold, J. E. (2001). The Chumash in the world and regional perspectives. In J. E. Arnold (Ed.), *The origins of a Pacific Coast chiefdom: The Chumash of the Channel Islands* (pp. 7–8). Salt Lake City: The University of Utah Press.

Auerbach, B. M. (2011). Reaching great heights: Changes in indigenous stature, body size and body shape with agricultural intensification in north America. In R. Pinhasi & J. T. Stock (Eds.), *Human bioarchaeology of the transition to agriculture* (pp. 203–233). Chichester: Wiley-Blackwell.

Brothwell, D. R., & Sandison, A. T. (1967). *Disease in antiquity.* Springfield: Charles C. Thomas.

Brown, V. P., & Owens, L. (2010). *The world of the southern Indians: Tribes, leaders, and customs from prehistoric times to the present (reprint).* Montgomery: NewSouth Books.

Buikstra, J. E. (1972). *Hopewell in the lower Illinois river valley: A regional approach to the study of biological variability and mortuary activity.* Unpublished Ph.D. dissertation, University of Chicago, Chicago.

Buikstra, J. E., & Roberts, C. A. (2012). *The global history of paleopathology: Pioneers and prospects.* Oxford: Oxford University Press.

Cohen, M. N., & Armelagos, G. J. (1984). *Paleopathology at the origins of agriculture.* Orlando: Academic.

Colwell-Chanthaphonh, C. (2007). History, justice, and reconciliation. In J. L. Barbara & A. S. Paul (Eds.), *Archaeology as a tool of civic engagement* (pp. 23–46). Lanham: AltaMira.

Dettwyler, K. A. (1991). Can paleopathology provide evidence for "compassion"? *American Journal of Physical Anthropology, 84*(4), 375–384.

Downum, C. E. (2012). *Hisat'sinom: Ancient peoples in a land without water.* Santa Fe: School for Advanced Research Press.

Eggan, F. (1979). Pueblos: Introduction. In A. Ortiz (Ed.), *Handbook of North American Indians, Volume 19, Southwest* (pp. 224–235). Washington, DC: Smithsonian Institution Press.

Erlandson, J. M., Rick, T. C., Kennett, D. J., & Walker, P. L. (2001). Dates, demography, and disease: Cultural contacts and possible evidence for Old World epidemics among the protohistoric Island Chumash. *Pacific Coast Archaeological Society Quartelrly, 37*(3), 11–26.

Farmer, P. (2003). *Pathologies of power: Health, human rights, and the new war on the poor.* Berkeley: University of California Press.

Ferguson, T. J., Dongoske, K. E., Jenkins, L., Yeatts, M., & Polingyouma, E. (1993). Working together: The roles of archaeology and ethnohistory in Hopi cultural preservation. *Cultural Resource Management, 16,* 27–37.

Fowler, M. L. (1991). Mound 72 and early Mississippian at Cahokia. In J. B. Stoltman (Ed.), *New perspectives on Cahokia and views from the periphery* (pp. 1–28). Madison: Monographs in World Archaeology, No. 2, Prehistory Press.

Gage, T. B. (2010). Demographic estimation: Indirect techniques for anthropological populations. In C. S. Larsen (Ed.), *A companion to biological anthropology* (pp. 179–193). Chinchister: Wiley-Blackwell.

Gagnon, C. M. (2008). Bioarchaeological investigations of pre-state life at Cerro Oreja. In L. J. Castillo, H. Bernier, G. Lockard, & J. Rucabado (Eds.), *Arqueología mochica nuevos enfoques*. Lima: Fondo Editorial de la Pontificia Universidad Católica del Perú e Institut Français d'Études Andines.

Hann, J. H. (2003). *Indians of central and south Florida* (pp. 1513–1763). Gainesville: University Press of Florida.

Harrod, R. P. (2009). Book Review: Bioarchaeology of Spanish Florida: The impact of colonialism. In: Larsen, C. S. (ed.). Gainesville: University Press Florida (2001). *Southeastern Archaeology, 28* (1):126–127.

Hartley, L. P. (2011). *The go-between*. New York: The New York Review of Books. (Original work published 1953)

Hobbes, T. (2003). *The Leviathan*. Bristol: Thoemmes Continuum. (Original work published 1651)

Hollimon, S. E. (2011). Sex and gender in bioarchaeological research: Theory, method, and interpretation. In S. C. Agarwal & B. A. Glencross (Eds.), *Social bioarchaeology* (pp. 149–182). Malden: Wiley-Blackwell.

Hooten, E. A. (1930). *Indians of Pecos Pueblo: A study of their skeletal remains*. New Haven: Yale University Press.

Hoppa, R. D., & Vaupel, J. W. (2002). *Paleodemography: Age distributions from skeletal samples*. Cambridge: Cambridge University Press.

Hrdlička, A. (1908). Physiological and medical observations among the Indians of the southwestern United States and northern Mexico. In *Bulletin, No. 37, Bureau of American Ethnology*, 103–112. Washington, DC: Smithsonian Institution Press.

Hrdlička, A. (1935). Ear exostoses. *Smithsonian Miscellaneous Collections, 93*, 1–100.

Hutchinson, D. L. (2004). *Bioarchaeology of the Florida Gulf Coast: Adaptation, conflict, and change*. Gainesville: University Press of Florida.

Jackes, M., & Meiklejohn, C. (2008). The paleodemography of central Portugal and the Mesolithic-Neolithic transition. In J.-P. Bocquet-Appel (Ed.), *Recent advances in palaeodemography: Data, techniques, patterns* (pp. 209–258). Dordrecht: Springer.

Jarcho, S. (1966). The development and present condition of human paleopathology in the United States. In S. Jarcho (Ed.), *Human paleopathology* (pp. 3–30). New Haven: Yale University Press.

Jefferson, T. (1955). *Notes on the state of Virginia* (W. Peden, Ed.). Chapel Hill: University of North Carolina Press. (Original work published 1787)

Johansson, S. R., & Horowitz, S. (1986). Estimating mortality in skeletal populations: Influence of the growth rate on the interpretation of levels and trends during the transition to agriculture. *American Journal of Physical Anthropology, 71*(2), 233–250.

Johnson, A. W., & Earle, T. (2000). *The evolution of human societies: From foraging group to agrarian state* (2nd ed.). Stanford: Stanford University Press.

Kelly, J. A., Tykot, R. H., & Milanich J. T. (2006). Evidence for early use of maize in peninsular Florida. In: J. Staller, R. Tykot, & B. Benz (Eds.), *Histories of maize: Multidisciplinary approaches to the prehistory, linguistics, biogeography, domestication, and evolution of maize* (pp. 249–261). 2009: Left Coast Press.

Koziol, K. M. (2012). Performances of imposed status: Captivity at Cahokia. In D. L. Martin, R. P. Harrod, & V. R. Pérez (Eds.), *The bioarchaeology of violence* (pp. 226–250). Gainesville: University of Florida Press.

Ladd, E. (1991). On the Zuni view. In *The Anasazi: Why did they leave? Where did they go? Sponsored by the Bureau of Land Management*. Dolores, Colorado: A panel discussion at the Anasazi Heritage Center. Sponsored by the Bureau of Land Management, Natural and Cultural Heritage Association, Albuquerque.

Lallo, J. W. (1973). *The skeletal biology of three prehistoric American Indian societies from Dickson Mounds*. Unpublished Ph.D. dissertation, University of Massachusetts, Amherst.

Lambert, P. M. (1993). Health in prehistoric populations of the Santa Barbara Channel Islands. *American Antiquity, 58*(3), 509–521.

Lambert, P. M. (2009). Health vs. fitness: Competing themes in the origins and spread of agriculture? *Current Anthropology, 50*(5), 603–608.

Lambert, P. M., & Walker, P. L. (1991). Physical anthropological evidence for the evolution of social complexity in coastal Southern California. *American Antiquity, 65*, 963–973.

Larsen, C. S. (1997). *Bioarchaeology: Interpreting behavior from the human skeleton.* Cambridge: Cambridge University Press.

Larsen, C. S. (2001). *Bioarchaeology of Spanish Florida: The impact of colonialism.* Gainesville: University Press of Florida.

Leatherman, T. L. (2001). Human biology and social inequality. *American Journal of Human Biology, 13*(2), 292–293.

Martin, D. L. (1994). Patterns of health and disease: Health profiles for the prehistoric Southwest. In G. J. Gumerman (Ed.), *Themes in Southwest prehistory* (pp. 87–108). Santa Fe: School of American Research Press.

Martin, D. L. (1998). Owning the Sins of the Past: Historical Trends, Missed Opportunities, and New Directions in the Study of Human Remains. Building a new biocultural synthesis: political-economic perspectives on human biology. The University of Michigan Press, Ann Arbor, 171–90.

Martin, D. L., Goodman, A. H., Armelagos, G. J., & Magennis, A. L. (1991). *Black Mesa Anasazi health: Reconstructing life from patterns of death and disease.* Carbondale: Southern Illinois University Press.

Martin, D. L., & Horowitz, S. (2003). Anthropology and alternative medicine: Orthopedics and the other. *Techniques in Orthopaedics, 18*(1), 130–138.

Matthews, W., Wortman, J. L., & Billings, J. S. (1893). Human bones of the Hemenway collection in the United States Medical Museum. *Memoirs of the National Academy of Sciences, 7*, 141–286.

McElroy, A., & Townsend, P. K. (2009). *Medical anthropology in ecological perspective* (5th ed.). Boulder: Westview.

Meindl, R. S., Mensforth, R. P., & Owen Lovejoy, C. (2008a). The libbon site. In R. Pinhasi & S. Mays (Eds.), *Advances in human paleopathology* (pp. 259–275). West Sussex: Wiley.

Meindl, R. S., Mensforth, R. P., & Owen Lovejoy, C. (2008b). Method and theory in paleodemography, with an application to a hunting, fishing, and gathering village from the eastern woodlands of north America. In M. A. Katzenberg & S. R. Saunders (Eds.), *Biological anthropology of the human skeleton* (2nd ed., pp. 601–618). Hoboken: Wiley.

Milner, G. R. (1998). *The Cahokia chiefdom: The archaeology of a Mississippian society.* Washington, DC: Smithsonian Institution Press.

Milner, G. R. (2007). Warfare, population, and food production in prehistoric eastern north America. In R. J. Chacon & R. G. Mendoza (Eds.), *North American indigenous warfare and ritual violence* (pp. 182–201). Tucson: University of Arizona Press.

Milner, G. R., & Ferrell, R. J. (2011). Conflict and death in a late prehistoric community in the American Midwest. *Anthropologischer Anzeiger, 68*(4), 415–436.

Milner, G. R., Wood, J. W., & Boldsen, J. L. (2008). Advances in paleodemography. In M. A. Katzenberg & S. R. Saunders (Eds.), *Biological anthropology of the human skeleton* (2nd ed., pp. 561–600). Hoboken: Wiley.

Moodie, R. L. (1923). *Paleopathology, an introduction to the study of ancient disease.* Urbana: University of Illinois Press.

Moore, J. A., Swedlund, A. C., & Armelagos, G. J. (1975). The use of life tables in paleodemography. In A. C. Swedlund (Ed.), *Population studies in archaeology and biological anthropology: A symposium* (pp. 57–70). Washington, DC: Memoirs of the Society for American Archaeology 30.

Moorehead, W. K. (2000). The Cahokia mounds. In J. E. Kelly (Ed.), *Classics in southeastern archaeology series.* Tuscaloosa: The University of Alabama Press.

Naranjo, T. (1995). Thoughts on migration by Santa Clara Pueblo. *Journal of Anthropological Archaeology, 14*, 247–250.

Novak, S. A. (2008). *House of Mourning: A biocultural history of the Mountain Meadows Massacre*. Salt Lake City: The University of Utah Press.

Ortiz, A. (1979). Introduction. In A. Ortiz (Ed.), *Handbook of North American Indians, Volume 19, Southwest* (pp. 1–4). Washington, DC: Smithsonian Institution Press.

Pauketat, T. (2009). *Cahokia: Ancient America's great city on the Mississippi*. New York: Viking-Penguin.

Pérez, V. R. (2010). From the Singing Tree to the Hanging Tree: Structural violence and death with the Yaqui Landscape. *Landscapes of Violence* 1 (1): Article 4.

Population Reports. (1988). *Mother's lives matter: Maternal health in the community*. Reports Issues in World Health Series L, No. 7.

Roberts, C. A., & Manchester, K. (2005). *The archaeology of disease* (3rd Aufl.). Ithaca: Cornell University Press.

Roksandic, M., & Armstrong, S. D. (2011). Using the life history model to set the stage(s) of growth and senescence in bioarchaeology and paleodemography. *American Journal of Physical Anthropology, 145*(3), 337–347.

Rose, J. C. (1973). *Analysis of dental micro-defects of prehistoric populations from Illinois*. Unpublished Ph.D. dissertation, University of Massachusetts, Amherst.

Rose, J. C. (1999). Mortuary data and analysis. In M. C. Fowler, J. C. Rose, B. Vander Leest, & S. R. Alher (Eds.), *The Mound 72 area: Dedicated and sacred space in early Cahokia* (pp. 63–82). Springfield: Reports of Investigations, No. 54, Illinois State Museum.

Rousseau, J.-J. (2008). *The social contract*. New York: Cosimo. (Original work published 1762)

Silliman, S. W. (2008). Collaborative indigenous archaeology: Troweling at the edges, eyeing the center. In S. W. Silliman (Ed.), *Collaborating at the trowel's edge: Teaching and learning in indigenous archaeology* (pp. 1–21). Tucson: University of Arizona Press.

Stodder, A. L. W. (1989). Bioarchaeological research in the basin and range region. In A. H. Simmons, A. L. W. Stodder, D. D. Dykeman, & P. A. Hicks (Eds.), *Human adaptations and cultural change in the greater Southwest* (pp. 167–190). Wrightsville: Arkansas Archaeological Survey Research Series, no. 32.

Stodder, A. L. W. (2012). The history of paleopathology in the American Southwest. In J. E. Buikstra & C. A. Roberts (Eds.), *The global history of paleopathology: Pioneers and prospects* (pp. 285–304). Oxford: Oxford University Press.

Stodder, A. L. W., Martin, D. L., Goodman, A. H., & Reff, D. T. (2002). Cultural longevity in the face of biological stress: The Anasazi of the American Southwest. In R. H. Steckel & J. C. Rose (Eds.), *The backbone of history: Health and nutrition in the western hemisphere* (pp. 481–505). Cambridge: Cambridge University Press.

Stodder, A. L. W., & Palkovich, A. M. (2012). The bioarchaeology of individuals. In C. S. Larsen (Ed.), *Bioarchaeological interpretations of the human past: Local, regional, and global perspectives*. Gainesville: University Press of Florida.

Stojanowski, C. M. (2005a). The bioarchaeology of identity in Spanish colonial Florida: Social and evolutionary transformation before, during, and after demographic collapse. *American Anthropologist, 107*(3), 417–431.

Stojanowski, C. M. (2005b). *Biocultural histories of La Florida: A bioarchaeological perspective*. Tuscaloosa: The University of Alabama Press.

Stojanowski, C. M. (2011). Social dimensions of evolutionary research: History in colonial Southeastern U.S. *Evolution: Education and Outreach, 4*(2), 223–231.

Storey, R. (2009). An estimation of mortality in a pre-Columbian urban population. *American Anthropologist, 87*(3), 519–535.

Swentzell, R. (1993). Mountain form, village form: Unity in the Pueblo world. In S. H. Lekson & R. Swentzell (Eds.), *Ancient land, ancestral places: Paul Logsdon in the Pueblo Southwest* (pp. 139–147). Santa Fe: Museum of New Mexico Press.

Thomas, C. (1884). Who were the mound-builders? *The American Antiquarian, 6*, 90–99.

Thompson, V. D., & Worth, J. E. (2011). Dwellers by the Sea: Native American adaptations along the southern coasts of North America. *Journal of Archaeological Research, 19*, 51–101.

Tilley, L. (2012). The bioarchaeology of care. *The SAA Archaeological Record, 12*(3), 39–41.

Trinkaus, E. (2011). Late Pleistocene adult mortality patterns and modern human establishment. *Proceedings of the National Academy of Sciences of the United States of America, 108*(4), 1267–1271.

Uhl, N. M. (2012). Age-at-death estimation. In E. A. DiGangi & M. K. Moore (Eds.), *Research methods in human skeletal biology* (pp. 63–90). Oxford: Academic.

Walker, P. L., & Johnson, J. R. (1992). The effects of European contact on the Chumash Indians. In J. W. Verano & D. H. Ubelaker (Eds.), *Disease and demography in the Americas* (pp. 127–139). Washington, DC: Smithsonian Institution Press.

Wells, C. (1964). *Bones, bodies, and disease: Evidence of disease and abnormality in early man.* London: Thames and Hudson.

West, W. R., Jr. (1993). Research and scholarship at the National Museum of the American Indian: The new "inclusiveness". *Museum Anthropology, 17*(1), 5–8.

Widmer, R. J. (1988). *The evolution of the Calua: A nonagricultural chiefdom on the southwest Florida coast.* Tuscaloosa: The University of Alabama Press.

Zuckerman, M. K., & Armelagos, G. J. (2011). The origins of biocultural dimensions in bioarchaeology. In S. C. Agarwal & B. A. Glencross (Eds.), *Social bioarchaeology* (pp. 15–43). Malden: Wiley-Blackwell.

Chapter 8
Special Applications in Bioarchaeology: Taking a Closer Look

There has been an explosion in new techniques and methodological approaches in bioarchaeology. Several factors contribute to why more methods than ever before have been developed. First, there have been major scientific breakthroughs in other scientific disciplines such as chemistry, biology, geology, and physics that have provided techniques that can be applied to the study of human skeletal remains. Second, as the field has increasingly incorporated methods from other disciplines, funding and interest have increased for projects that emphasize shared heritage, ancient migrations, and biological affinities. Finally, the recent growth and interest in bioarchaeology and forensic anthropology have fueled a use of new methodological approaches (Martin and Harrod 2012). What these new methods have in common is that they provide a specialized approach for the analysis of bone and teeth that offer ways of using those tissues to get information that is simply not obtainable through more traditional techniques.

The focus here is to provide a very brief synopsis of what some of these special applications can provide in terms of additional information about identity, ancestry, and migration. Ethical issues abound in this area because it entails genetic and biomedical research using tissues from human remains, and this comes with all the attendant complexities of informed consent and weighing the costs and benefits (O'Rourke et al. 2005). Also, these techniques are defined by the scientific community as destructive, in that tissue is lost when the analyses are done. For example, two increasingly popular special applications include stable isotope analysis and the reconstruction of DNA profiles (or genetic analysis), both of which can destroy several grams of skeletal or dental tissues. Both of these methods provide bioarchaeologists and forensic anthropologists with the ability to provide data on biological identity and affiliation beyond what can be obtained at the surface level with the more traditional analyses such as those discussed in Chap. 6. Isotopic analysis allows for a better and more precise identification of past diets, as well as a means of understanding the relatedness of individuals within populations. These special applications permit researchers to move from the macroscopic to the microscopic and on to the molecular levels. These include histological and cellular (cortical bone thickness and osteon analysis) and molecular (isotopic or elemental analysis and genetic analysis) techniques.

D.L. Martin et al., *Bioarchaeology: An Integrated Approach to Working with Human Remains*, Manuals in Archaeological Method, Theory and Technique, DOI 10.1007/978-1-4614-6378-8_8, © Springer Science+Business Media New York 2013

8.1   Under the Surfaces of Bone and Teeth

Bone and teeth are called hard connective tissues, and although they may appear on the surface to be weathered and fragmentary, underneath the surface there may still be remnants of a protein (collagen), the organic component of these tissues. The larger component is composed of calcium and phosphate salts (hydroxyapatite), and that is what gives these tissues their hardness. Both of these components, collagen and apatite, when extracted from ancient bones, can be analyzed using various kinds of instrumentation. In the process of extracting and preparing collagen and apatite for analysis, the ancient bone specimens are forever altered, and in most cases, tissues are lost in the process. While these methods permit bioarchaeologists to take a much closer look at what bones and teeth can reveal about a person, they also present unique ethical, analytical, and interpretive challenges.

One limitation of most of these techniques is that they are considered by the scientific community to be destructive in that some amount of bone or tooth tissue is dramatically altered or destroyed during the analysis. Identifying these special applications as destructive has led to a variety of concerns from tribal representatives. Concerns have been raised because to carry out these studies a small amount of bone or tooth is destroyed. In addition to this, many Native Americans are wary of biomedical, genetic, and scientific studies on them or their ancestors. Harry (2009) provides a very thorough overview of issues raised for Native Americans when nonnative scientists and bioarchaeologists conduct research. She demonstrates with multiple examples that "Research has historically been a top-down, outside-in process, with Indigenous peoples serving merely as research subjects, not partners, without any meaningful participation or potential to benefit from outcomes of the research" (2009:147).

However, the kinds of data that these special applications can provide have also led some tribes to participate in DNA studies because they do see some potential benefits resulting. These include legal proceedings where cultural affiliation, ancestry, and ancestral migrations can be proven and thus work in the tribe's favor. O'Rourke et al. (2005:237) who have been partially successful in obtaining permission from some indigenous groups to conduct research on ancient DNA (aDNA) summarize the future of special applications this way: ". . . inferences drawn from aDNA studies will be used for legal purposes, possibly including water and land right issues based on prior occupancy arguments, definitions of tribal identities, and other applications."

8.1.1   Destruction, Alteration, and Transformation: Ethical Considerations

What do these terms (destructive and altered) mean to different constituents who have an interest in ancient human remains? In almost all cases, the human remains have already been altered and destroyed prior to being discovered and excavated. It was discussed in Chap. 4 that natural and cultural (both ancient and modern) taphonomic processes destroy, alter, and damage bones in a multitude of ways. Roots, carnivores, insects, water, and other bioturbations can reduce the 206 bones of the human body to a small number of fragmentary bones that barely resemble a complete skeleton. Ancient practices of cutting, defleshing, dismembering, burning, and shaping human bone and teeth as discussed in Chap. 9 can also render human remains diminished and fragmentary. Often small bones of the body such as ribs or fragile bones of the body such as cranial bones (or infant bones in general) may be lost altogether or represented by small wafer-thin pieces.

The terms destructive and invasive are poor descriptors for what is actually involved in these new techniques. Katzenberg (2001) provides a very thoughtful rumination on the nature of conducting work on ancient human remains that destroys material in the process. She points out that prior to NAGPRA and NAGPRA-like legislation and considerations, she and her colleagues were cautious about destroying ancient bone out of consideration for future researchers and the possibility that removing bone tissue may prohibit some new innovative technique from being carried out. She notes that researchers in the 1980s were not concerned about what the descendants might think about bone destruction and that this sentiment would now be unacceptable. But she points out that bioarchaeologists have always had some trepidation about doing anything to reduce the amount of ancient bone available for study. In framing the question regarding what constitutes destructive analysis, Katzenberg points out that some analytical techniques such as atomic absorption spectrometry require samples to be destroyed in the process, but with other techniques such as X-ray fluorescence (XRF), bone is reduced to a powder and analyzed, but the powder remains available for future analyses (or repatriation). The first method is considered destructive, and the second is not.

Pfeiffer (2000) suggests that the terminology that bioarchaeologists typically use to describe their work to non-bioarchaeologists is unnecessarily reduced to the terms destructive or nondestructive techniques. For her work in histological analyses of bone to determine age and health parameters, she prefers to use the term transformative. She discusses the making of bone thin sections as a transformation rather than destruction of bone tissue. This is not just a semantic exercise to make the research seems less problematic. It is a very different way of conceptualizing the complexity of what happens to bone tissue in a way that invites a different kind of thinking.

Bioarchaeologists have been limiting their opportunities to communicate with nonscientists and tribal representatives by using restrictive and narrow jargon. In our own experience, we have found that providing alternative ways to think about how much tissue is necessary, how much is lost, and what the procedures entail can open up rather than shut down discussions (Martin et al. 2001). Explaining methods in a way that suggests opportunities and alternative ways of thinking about what is gained by these methods may provide non-bioarchaeologists with a broader and more nuanced way of thinking about histological, biochemical, and molecular studies.

8.1.2 Complex and Varied Perspectives of Descent Communities

Although the amount of tissue destroyed is typically quite small (0.5–10 g), some tribes have a "no destructive analyses" clause in their policies and deliberations over the types of studies that they will permit or sanction (Harry 2009:162). However, some tribes are permitting these special applications so that information on ancestry can be used by scientists and others to better understand the origins of indigenous people in the New World. Malhi et al. (2007) obtained permission to analyze DNA from two 5,000-year-old individuals from the Canoe Creek, Soda Creek, and Dog Creek bands. The results of this study resulted in broadening existing views on the first migrations of people into the New World.

Tallbear (2003) discusses cases where DNA analysis has been used by tribes to prove their ancient ancestry so that the US government will declare them as federally recognized tribes. However, Harry (2009) summarizes two cases (Kennewick Man and Spirit Cave Man) where DNA analysis has been used to block tribes from repatriating ancestral remains. Thus, the use of something like DNA analysis can cut both ways, permitting tribes to prove their ancestry as well as blocking tribes from proving their rights to their ancestor's remains. The issues become even more complex if taken in the larger prospective of the hegemonic colonial policies (Marks 2005) and of biomedical studies conducted on indigenous people without their knowledge or consent (Drabiak-Syed 2010). Zimmerman (2001:169) takes this further with an exploration of what he calls "scientific colonialism." He argues that the history of archaeological (and by extension bioarchaeological) studies and indigenous people is one of arrogance and the taking of data from peoples who are marginalized and subordinated. These cautionary tales reveal that there is more at stake for indigenous people than just the studies bioarchaeologists conduct on human remains.

Thus, this is a very complex, rapidly changing, and relational area that bioarchaeologists need to constantly be aware of and sensitive to. There is no easy ethical stance to take that will fit all cases and all groups. With over 565 federally recognized tribes in the United States and many tribes seeking federal recognition, there are likely to be many different ways that special application studies are and will be viewed in the future. While there is the potential to create many more collaborative projects involving molecular and biochemical analyses with indigenous people, there will also likely be much opposition to these studies. And, not all of the opposition rests on whether or not the analyses are destructive but more on whether the tribes can utilize the information in ways that benefit them and that do not compromise their integrity. Sacrificing a wafer-thin fragment of an ancestor's rib bone may be something that tribes are willing to do to obtain information of value to them or it may not.

One of the many uses of DNA data derived from human remains is in the area of understanding the peopling of the New World and the origins of Native Americans. In a review that captures the scientific fascination with this topic, "The Human Genetic History of the Americas: The Final Frontier," O'Rourke and Raff (2010) provide a detailed overview of the interdisciplinary nature of these studies.

Combining DNA data from modern and ancient indigenous groups, archaeological data, and demographic modeling involving scientists from a dozen disciplines, the review enthusiastically demonstrates how valuable these data are to the question of indigenous origins. To many Native Americans, this information is less than useful and even frivolous because many oral histories provide answers to those kinds of questions to their satisfaction (Deloria 1997).

Other criticisms of the scientific approach to Native American origins are that there is no integration of native scholarship, oral histories, or indigenous voices in any of this literature. Echo-Hawk (2000) presents a case study using the Arikara, an indigenous group found archaeologically and historically in what is now North Dakota. He presents a compelling case that oral tradition *and* the archaeological record both describe Arikara origins and migrations and that the two together would be a far better approach to presenting an understanding than either alone. Likewise, Whiteley (2002) takes a case study using the Hopi to demonstrate that combining archaeological data with oral tradition provides a *more* rigorous and expanded interpretation of the past. His study makes a good case for why oral histories must be viewed as primary sources of evidence for interpretation of the past and not relegated to mythical narratives lacking credibility.

Another complicating factor in thinking about the possibilities of these special applications such as DNA-based studies in reconstructing biological identities and affinities is that at the precise moment that scholars are suggesting that identity is fluid, relational, and dynamic, DNA studies tend to make biological identity something rooted in the DNA. Brodwin (2002:323) elegantly captures this in his critique of genetic studies: ". . . essentialist identities grow ever more powerful and seductive. New genetic knowledge, for example, adds the cachet of objective science to the notion that one's identity is an inborn, natural and unalterable quality. Rapid advances in sequencing and analyzing the human genome have strengthened essentialist thinking about identity. . ." This highlights some of the complexities involved in how aDNA information will be interpreted by different groups.

A final case study that reveals the interest of tribes in special application studies comes from an ongoing study of ancestral Omaha human remains. An Associated Press archive of a 1991 breaking story reported that at the request of Omaha Indians, bioarchaeologists analyzed 40 Omaha burials from the late 1700s. It was initially thought that infectious disease from colonial encounters had killed large numbers of Omaha Indians. Tribal officials permitted a special application involving isotopic analysis of lead in the historic human remains. Lead isotopes were found to be in very high levels in half of the skeletal remains tested. This suggested that a good proportion of the deaths may have been due to lead poisoning from a variety of trade items available to Indians at this time. These included pottery, paint, casks, and bullets. Dennis Hastings, the tribal historian at the time, stated that ". . . the skeletal remains of our ancestors are speaking to us through science" (Secter 1991). Reinhard and Ghazi (1992) did a follow-up study and tested for lead in various soils and artifacts. The study found that lead was being mined in the Missouri region at that time and used in making a wide variety of paints and pottery. This study confirmed that through use of trade items, individuals were at risk for lead poisoning.

8.2   Synopsis of Select Special Applications in Bioarchaeology

Techniques for the analysis of bone and teeth that get below the anatomical surface to utilize preserved collagen and apatite have become central to understanding aspects of ancestry, kinship, health, diet, disease, growth, and development from ancient, historic, and forensic human remains.

8.2.1   Histological and Cellular Analyses

At the histological level, bone is arranged in multicellular units called osteons. Analysis of bone histology generally looks for normal and abnormal features of osteons at the histological and cellular level (Fig. 8.1).

Too many, too few, or poorly mineralized osteons are all signals of some kind of physiological problem. The number and size of osteons changes predictably as humans age, and so quantifying osteons can be used to provide an approximate age for individuals. Histological analysis of bone can provide information on age at death, some pathologies, and biomechanical properties. The size of the specimen necessary to conduct microscopy on bone is approximately 100 micra (about the thickness of a piece of paper). However, bones must be cut with a saw so that a cross section of the bone is obtained. The amount of tissue destroyed is quite small, and destruction of bone can be minimized by taking a 1 cm "plug" (weighing approximately 3 g) from preselected points from the midshaft of the femur, the middle third of the sixth rib, and the distal portion of the radius. This technique is used in hospitals when removing a small piece of tissue for a biopsy from patients. These sites are clinically significant and have been used by other researchers in studies of bone remodeling (Robling and Stout 2008). The bone samples are then wafer cut, embedded, and analyzed using a high-powered microscope with camera and computer attachment.

8.2.1.1   Cross-Sectional Analysis

Cutting out a piece of the bone to analyze the osteons is a highly effective technique for estimating the age of the individual, measuring the loss of bone with age (osteoporosis), or determining loading stress placed on the bone during locomotion. Agarwal and colleagues looking at a population at the site of Çatalhöyük in Turkey found that while there was no difference in cortical bone density as seen in the cross section among males and females in particular, there was a difference in the life history and loss of bone between males and females. Interestingly, the men of the community lost bone late in life, while the women lost bone in the middle of their lives and then maintained that level throughout old age (2011:9). The result is that by old age, the both have similar amounts of bone loss, which differs significantly from what is seen among modern populations. This suggests that perhaps other factors, such as culture and gender-based divisions of labor, contribute to the loss of bone or osteoporosis (2011:14).

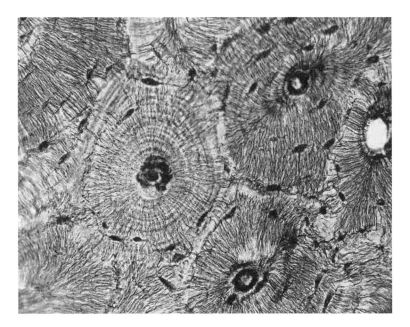

Fig. 8.1 Cross-section illustrating osteons (Courtesy of Wikimedia Commons)

Over the last several decades, scholars such as Ruff (2008) have explored the ways that cross-sectional analysis of the bone can reveal information about activity, especially locomotion. Using cross-sectional analysis, Ruff has shown that populations traveling long distances or navigating rugged terrain tend to have thicker cortical bone.

8.2.2   Molecular Analysis

The tissues of the human body are mostly composed at the molecular level of water, carbon, nitrogen, oxygen, hydrogen, calcium, and phosphorus. The human body has trace amounts of many kinds of elements and minerals in tissues of bone and teeth, such as strontium, zinc, copper, fluorine, iron, lead, and others. The amount of some of these trace elements in the human body reveals information about the kinds of food regularly consumed because different foods (e.g., terrestrial vs. marine animals and desert vs. prairie grasses) provide different amounts of these trace elements.

Collagen is not always adequately preserved in ancient human remains, and it is only through assessment of the collagen that it can be determined whether or not a study is possible. Isotopic analyses are conducted on the collagen fraction of bone because it maintains its isotopic integrity (unlike the inorganic or mineral portion of bone). While any bone can be used, many researchers utilize bones that are already fragmentary and broken, and often ribs are sacrificed for this invasive analysis.

Samples of bone can be taken from any bone. Approximately 1 g (0.035 oz) of bone powder is necessary (about the size and weight of 1 raisin) and can be taken from any already broken rib or femur fragment. Only the inner portion of the cortical bone can be used, and so to extract 1 g of uncontaminated bone powder, usually about 2–3 g of bone is necessary. The sample is demineralized, the calcium salts are dissolved, and carbon dioxide is released, recovered, and purified. Through distillation and filtration, approximately 20 mg of dry gelatin is obtained, and mass spectrometric analysis of the gel is performed.

8.2.2.1 Isotopic and Elemental Analysis

Analysis of stable isotopes such as carbon, oxygen, and nitrogen (nonradioactive atoms of the same element but with different atomic masses) in skeletal and dental tissue can reveal geographic location as well as specific dietary components. Strontium, nitrogen, oxygen, and carbon are stable isotopes that provide different values that reflect particular geographic locations. Humans consume plants, marine resources, meat, and other products produced in specific geographic locations, and these products lock in various stable isotopes at particular (known) rates. These are referred to as chemical signatures. The analysis of stable carbon isotopes from bone collagen can reveal the relative proportions of different kinds of plant foods that individuals ate prior to death.

Schwarcz and Schoeninger (1991) provide a very good overview of how these kinds of analyses work. For example, stable carbon isotopes analysis can differentiate the amount of plants that have a C3 vs. a C4 photosynthetic pathway because these different plants metabolize the two stable isotopes of carbon differently. C4 plants include maize, amaranth, chenopodium, portulaca, many common grasses, sorghum, agave, yucca, and prickly pear. C3 plants include nuts, beans, wheat, rice, and tubers. Humans who eat quite a bit of maize or other C4 plants would have a higher (or less negative) 13C/14C value than those eating more of some other foods. Individuals eating primarily a C4 plant diet would have a delta13C of around −7.5 0/00. Isotope ratios are expressed as per mil (using the symbol 0/00) deviations from the corresponding ratios of widely recognized standards. The deviations are provided as delta values. Bone collagen has a carbon turnover rate of approximately 10 years, so an isotopic value provides an average for the diet over a long period of time. Seasonal changes or short-term food shortages are not easily determined using this method.

Atomic absorption spectrophotometry with graphite furnace is performed as another way to measure trace elements in the bone. Two grams of bone are necessary for the analysis of major and minor elements. The bone is digested in nitric acid and dried. Strontium, calcium, zinc, magnesium, barium, copper, sodium, and lead have all been used in studies to differentiate dietary components. Major elements (reported as a percentage) and trace elements (reported as parts per million or micrograms per gram of bone ash or whole bone) both have been used in this type of analysis. An alternative to this method is the use of XRF, which works by measuring the loss of electrons from specific molecules as a result of concentrated radiation being shot at the sample. The energy from the radiation knocks electrons loose from different elements at different levels. The XRF device is capable of capturing and recording this process. Problems associated with using this technique include the possibility that elements are on the bone as a result of postmortem taphonomic processes. However, researchers such as Swanton et al. (2012) are indentifying methods of compensating for surface elements.

The value of any of these approaches is that elements can reveal a lot about the life of an individual. Strontium and oxygen isotopes can reveal place of birth and migration in bone and tooth tissues that form at different ages. Dental enamel forms in early life and so isotope composition can reveal early childhood residence, whereas bone reveals residence patterns in the last decade of life (Katzenberg 2001). Strontium alone is very indicative of the amount and kind of plants consumed in the diet, and the ratio of strontium/calcium in bone is an important indicator of the trophic level of the plants consumed. For example, isotopic data from the cave burial site Alepotrypa along the coast in Greece dating to the Late Neolithic (ca. 5000–3200 bc) found that contrary to what was expected, this group was not consuming a marine-based diet but was already reliant on a primarily agricultural diet (Papathanasiou et al. 2000). Schurr (1998) found that in children, weaning can be discerned in children by looking at nitrogen isotopes, while Blakely (1989) has demonstrated that strontium/calcium ratios are elevated in pregnant and lactating females.

The goal of isotopic analysis is to identify characteristics of individuals that contributes to building their individual identity and to the understanding of population demography. For example, it has been suggested that isotopes should be considered in relation to one another. Copper and zinc have been successfully used to estimate the amount of red meat in the diet. In a combined analysis of samples with low levels of strontium and magnesium but high levels of zinc might provide an indication that the individual consumed more meat. In contrast, high concentrations of strontium, zinc, and magnesium indicate seafood consumption. Interpretations regarding protein resources can be cross-tested by comparison with the nitrogen isotope values. This research looks at multiple isotopes and their relative levels in relation to one another. For example, Coltrain et al. (2007:317) conducted elemental studies on male burials from the large ceremonial site from Chaco Canyon, New Mexico (900–1150 ad). The study revealed that the elaborately buried males from Chaco Canyon were eating significantly more meat than other people throughout the region.

Similarly, by integrating carbon isotope values with element levels, it is possible to discriminate between different kinds of plant food in the diet. A diet characterized isotopically as being in the beans, tubers, and nuts family and elementally by bone levels high in strontium and magnesium and low in zinc would imply a dependency on nuts (such as pinyon) rather than tubers and beans. Lead is becoming increasing of interest in health studies because of its highly detrimental effect on growing children. Calculation of lead levels in precontact groups will be essential in reconstructing historic and contemporary lead toxicities.

8.2.2.2 Genetic Analysis

Analytical techniques for extracting DNA (genetic material located in every living cell) from fossils and ancient archaeological remains provide another destructive method that is of interest to both the scientific community as well as some tribal representatives. Although much of the research has been conducted on specimens so well preserved that mummified soft tissue is still available, a handful of researchers are working on DNA extraction from ancient dry bone. To date, these analyses are very preliminary, and there are many technical problems associated with the interpretation of the data. The technique is in the process of being perfected and involves polymerase chain reaction which is a very sensitive DNA-amplification technique that can analyze a single molecule of DNA. Although this technique is in its infancy (primarily in laboratories in Europe at the moment), it suggests that only very small samples of intact material will be needed for analysis. These techniques are not limited to human remains but are also being used on archaeological faunal and thousands of years old ancient fossils as well.

DNA from different kinds of analyses yields information on ancestry, biodistance, kinship, migration, and genetic diseases and can be helpful for identifying the biological sex in unknown adults and children. There have been so much research conducted on genetics in anthropology that it has led to the development of a whole separate discipline called molecular anthropology (Zuckerkandl 1963).

A recent study by Kemp and Schurr (2010) that looked at mitochondrial DNA (mtDNA) throughout the New World has important implications for the peopling of America. What they found was that the genetic diversity present suggests two things. First, that the initial population of the New World was from a population that entered prior to the opening of the ice-free corridor. Second, that the populations today came from a single population not multiple populations as has been suggested based on morphometric features of the skeleton (2010:14). The implication of these findings is that they argue against the Clovis-first hypothesis and against recent claims made about the Kennewick skeletal remains. The Clovis-first hypothesis states that the first Native Americans came from a population utilizing big game hunting lithic technology (Clovis tools) and that they entered through the ice-free corridor as they followed the big game herds. The Kennewick Man argument claims that he is not related to indigenous North Americans but rather that he is a descendant of Polynesian populations.

8.2.3 Photographs and X-Rays

Even photographs and X-rays may be considered both nondestructive and destructive. Photographs are helpful for documentation of human remains, and handheld X-ray instruments are increasingly being utilized in the field and lab. Most bioarchaeologists would consider these practices nondestructive to human remains. But these analytical tools must also be cleared with tribal representatives, and sometimes there is an unwillingness to permit photographs and X-rays for a variety of reasons that include the destruction and disturbance of the soul and spirit of the remains. Or there may be mistrust that the photos and X-rays will be used for purposes that counter tribal values.

Traditionally, nondestructive analyses take advantage of the surface morphology (also called the gross anatomical level) by observing and measuring features, landmarks, normal and pathological developments, anomalies (unusual characteristics), and other visible aspects. Being able to see some of the internal structures of bone has been important in tracking patterns of growth and development of ancient infants and children, and so with the advance in portable radiology instruments, X-rays have proven invaluable. X-rays have also provided information on adult bone dimensions, robusticity, strength, and thickness. When X-rays have been permitted, they have the added advantage of providing permanent information on human bones below the surface, inviting an analysis of the internal structures that make up bone.

8.3 Summary

While these kinds of special applications are on the rise, they raise many complex issues that are likely to continue into the future. Early on, many of these methods were discussed in terms of being destructive applications, and this limited how these were perceived by nonscientists. Less controversial and more generally accepted in the past were scientific observations and measurements (as discussed in Chap. 6) taken on human remains. In these studies, bone is not diminished, altered, or damaged during the analysis. However, the notion that these new methods are somehow worse because they are destructive is a false binary. There are other ways to conceptualize what actually is happening to bone and teeth, and bioarchaeologists would do well to not fall in the trap (that they created) by forcing techniques into either/or categories. What these words (destructive and nondestructive) convey is neither accurate nor helpful in presenting the more nuanced and gray areas of what these special applications entail and what they provide.

Most scientists would not consider the excavation and removal of bones to a laboratory, or the touching, photographing, and X-raying of bones, to be destructive, invasive, or in any way altering the physical properties of the bone. Yet for some descendants, these analytical procedures applied to their ancestors (e.g., retrieval, curation, making scientific observations, measuring, photographing, X-raying) are highly problematic. Bioarchaeologists cannot assume that their standard notions about what it means to perform a nondestructive or a destructive technique on human remains are universal. This binary way of dividing scientific analytical techniques obscures the potential utility of individual procedures to tribal and nontribal people who have an interest in ancient remains. And it works both ways. Tribal representatives may put X-rays into the destructive category because in a metaphysical sense, the releasing of radioactive heavy metal ions that pass through the bones may be considered as invasive and destructive as cutting the bone and reducing it to ash.

With NAGPRA legislation, ancient burials cannot be excavated or analyzed without permission from tribal authorities. In some situations when ancient burials are encountered in the US, there is a desire by tribal consultants and representatives to not even remove the remains for analysis. Excavating and removing human remains by some tribal people is considered highly dangerous, disturbing, destructive, and invasive. These notions form part of an ideology that informs how human remains are to be understood, just as researchers have their own ideology about what constitute proper protocols and laboratory practices. However, there are also situations when burials are encountered and tribal authorities do permit analysis. Often these are *in situ* analyses where bones are exposed for analysis but are not ever removed from the ground. This is a challenging method for bioarchaeologists but one that respectfully provides a compromise between the wishes of descendant communities to not cause destruction or disturbance to the remains and the desire of bioarchaeologists to obtain information to reconstruct the identity and the lived experience of the ancestors. In the future, more collaboration, more dialogue, and more compromise will strengthen the science of bioarchaeology to be an integrative, engaged, relevant, and ethical approach to dealing with human remains in the United States and beyond.

References

Agarwal, S. C., Glencross, B. A., & Beauchesne, P. (2011). Bone growth, maintenance and loss in the Neolithic community of Çatalhöyük, Turkey: Preliminary results. In *Archaeological research facility laboratory reports* (pp. 1–33). Berkeley: Archaeological Research Facility, UC Berkeley.

Blakely, R. L. (1989). Bone strontium in pregnant and lactating females from archaeological samples. *American Journal of Physical Anthropology, 80*(2), 173–185.

Brodwin, P. (2002). Genetics, identity, and the anthropology of essentialism. *Anthropological Quarterly, 75*(2), 323–330.

Coltrain, J. B., Janetski, J. C., & Carlyle, S. W. (2007). The stable- and radio-isotope chemistry of western basketmaker burials: Implications for early Puebloan diets and origins. *American Antiquity, 72*(2), 301–321.

Deloria, V., Jr. (1997). *Red earth, white lies: Native Americans and the myth of scientific fact.* Golden: Fulcrum Publishing.

Drabiak-Syed, K. (2010). Lessons from Havasupai Tribe v. Arizona State University Board of regents: Recognizing group, cultural, and dignitary harms as legitimate risks warranting integration into research practice. *Journal of Health and Biomedical Law, 6,* 175–225.

Echo-Hawk, R. C. (2000). Ancient history in the new world: Integrating oral traditions and the archaeological record in deep time. *American Antiquity, 65*(2), 267–290.

Harry, D. (2009). Indigenous peoples and gene disputes. *The Chicago-Kent Law Review, 84,* 147.

Katzenberg, M. A. (2001). Destructive analyses of human remains in the age of NAGPRA and related legislation. In *Out of the past: The history of human osteology at the University of Toronto.* Ontario: CITD Press. Retrieved from http://tspace.library.utoronto.ca/citd/Osteology/Katzenberg.html. Accessed February 15, 2013.

Kemp, B. M., & Schurr, T. G. (2010). Ancient and modern genetic variation in the Americas. In B. M. Auerbach (Ed.), *Human variation in the new world: The integration of archaeology and biological anthropology* (pp. 12–50). Carbondale: Occasional Papers, No. 38, Center for Archaeological investigations, Southern Illinois University.

Malhi, R. S., Kemp, B. M., Eshleman, J. A., Cybulski, J. S., Smith, D. G., Cousins, S., et al. (2007). Haplogroup M discovered in prehistoric North America. *Journal of Archaeological Science, 34*, 642–648.

Marks, J. (2005). Your body, my property: The problem of colonial genetics in a post-colonial world. In L. Meskell & P. Pels (Eds.), *Embedding ethics: Shifting boundaries of the anthropological profession* (pp. 29–46). New York: Berg.

Martin, D. L., Akins N. J., Goodman A. H., & Swedlund A. C. (2001). *Harmony and discord: Bioarchaeology of the La Plata Valley. Totah: Time and the rivers flowing excavations in the La Plata Valley* (Vol. 242). Santa Fe: Museum of New Mexico, Office of Archaeological Studies.

Martin, D. L., & Harrod, R. P. (2012). New directions in bioarchaeology, special forum. *The SAA Archaeological Record, 12*(2), 31.

O'Rourke, D. H., Hayes, M. G., & Carlyle, S. W. (2005). The consent process and aDNA research: Contrasting approaches in North America. In T. R. Turner (Ed.), *Biological anthropology and ethics: From repatriation to genetic identity* (pp. 231–240). Albany: State University of New York Press.

O'Rourke, D. H., & Raff, J. A. (2010). The human genetic history of the Americas: The final frontier. *Current Biology, 20*(4), R202–R207.

Papathanasiou, A., Larsen, C. S., & Norr, L. (2000). Bioarchaeological inferences from a Neolithic ossuary from Alepotrypa Cave, Diros, Greece. *International Journal of Osteoarchaeology, 10*, 210–228.

Pfeiffer, S. (2000). Palaeohistology: Health and disease. In M. A. Katzenberg & S. R. Saunders (Eds.), *Biological anthropology of the human skeleton* (pp. 287–302). Hoboken: Wiley.

Reinhard, K. J., & Ghazi, A. M. (1992). Evaluation of lead concentrations in 18th-century Omaha Indian skeletons using ICP-MS. *American Journal of Physical Anthropology, 89*, 183–195.

Robling, A. G., & Stout, S. D. (2008). Histomorphometry of human cortical bone: Applications to age estimation. In M. A. Katzenberg & S. R. Saunders (Eds.), *Biological anthropology of the human skeleton* (2nd ed., pp. 149–182). Hoboken: Wiley.

Ruff, C. B. (2008). Biomechanical analyses of archaeological human skeletons. In M. A. Katzenberg & S. R. Saunders (Eds.), *Biological anthropology of the human skeleton* (2nd ed., pp. 183–206). Hoboken: Wiley.

Schurr, M. R. (1998). Using stable nitrogen-isotopes to study weaning behavior in past populations. *World Archaeology, 30*(2), 327–342.

Schwarcz, H. P., & Schoeninger, M. J. (1991). Stable isotope analyses in human nutritional ecology. *Yearbook of Physical Anthropology, 34*, 283–321.

Secter, B. (1991). A Tribe Reveals Its Deadly Secret From the Grave: Sophisticated new tests indicate Omahas were decimated in the last century by poisoning caused by handling of lead. Los Angeles Times. Accessed October 15, 2012.

Swanton, T., Varney, T., Coulthard, I., Feng, R., Bewer, B., Murphy, R., et al. (2012). Element localization in archaeological bone using synchrotron radiation X-ray fluorescence: Identification of biogenic uptake. *Journal of Archaeological Science, 39*(7), 2409–2413.

TallBear, K. (2003). DNA, blood, and racializing the tribe. *Wicazo Sa Review, 18*(1), 81–107.

Whiteley, P. (2002). Prehistoric archaeology and oral history: The scientific importance of dialogue. *American Antiquity, 67*, 405–415.

Zimmerman, L. J. (2001). Usurping Native American voice. In T. L. Bray (Ed.), *The future of the past: Archaeologists, Native Americans, and repatriation* (pp. 169–184). New York: Routledge.

Zuckerkandl, E. (1963). Perspectives in molecular anthropology. In S. L. Washburn (Ed.), *Classification and evolution* (pp. 243–272). Chicago: Aldine.

Chapter 9
Body as Material Culture

The human body is more than the bone, tissue, and blood that give it its corporeal features. It is the ultimate symbol of social interactions and cultural ideology and as such is laden with meaning. It is the physical vehicle that contains human consciousness and moves people through their world, and it is what people leave behind after they die. The body is examined here in terms of how it is utilized to symbolize the worldview of the people within a particular culture. It is critical to view the body in ways beyond the simple categories of age, sex, stature, and presence or absence of nutritional deficiencies, disease, and trauma. Moving beyond this traditional definition motivates researchers to consider the life history of the individual. It is critical for researchers to remember that these individuals assumed multiple identities throughout their lifetime and that they lived in dynamic and relational social environments that continually influenced their body.

The body must been seen as the study of the interrelatedness of material culture and society. However, more than that, it must be understood in terms of how it coexists and changes in relation to the shared methodological systems and socioeconomic organization of the society it operates in. Sofaer (2006) argues that bioarchaeology must struggle against the structural binary that is perceived to exist between osteology (science) and identity or anthropological archaeology (humanism). Bioarchaeologists tend to either study the bones as clinical specimens or to ask questions about human adaptation and culture that can be answered with empirical data derived from the skeleton. On the one hand, they study the manner by which different diseases and skeletal markers appear on specific elements, and they see each indicator as an isolated phenomenon. On the other hand, they sometimes start by asking questions about the various characteristics of the society. These approaches are a good start but they will provide an incomplete picture of the lived experience of the individual they are not integrated with theory.

It is only through the examination of cultural constructions that it is possible to tease out the different meanings the body can take on through its materiality and reflection of societal norms. These societal norms are part of the formation of the body, in life and death, as a site of constant negotiation. Because it is subject to specific cultural conditions starting even before birth, the body's "objective" qualities cannot be understood without considering this context.

D.L. Martin et al., *Bioarchaeology: An Integrated Approach to Working* 213
with Human Remains, Manuals in Archaeological Method, Theory and Technique,
DOI 10.1007/978-1-4614-6378-8_9, © Springer Science+Business Media New York 2013

9.1 Ritual and Ritualized Behavior

The body is a model which can stand for any kind of bound system (Douglas 1966/1992:115).

Social theorists and cultural anthropologists have long understood the social and cultural significance of the human body. It is a repository for each individual's collection of lived experiences. Bodies are the ultimate form of material culture in that they are the material objects through which culture and biology are synthesized into a holistic human experience. As such, bodies consist of some of the most valuable sources of information for reconstructing past lifeways and behaviors. Archaeology and specifically bioarchaeology has only recently come to see the importance of viewing the body in terms of material culture.

The concepts of ritual and ritualized behavior are explored here. These concepts are extremely important to understanding and interpreting the "bounded system" that is expressed on the body. Ritual is a specific way of producing a pattern of behavior that is centered on inelasticity in performance and is replicated in culturally sanctioned ways with specific themes that illicit particular feelings of compulsion (Liénard and Boyer 2006). The themes surrounding ritualized behavior can include everything from birth rights to fraternal bonding within warrior societies. These rituals can often leave evidence on the skeleton in terms of bone modification or alterations to its size and shape. Complex relationships between body practices and practices of representation are explored through several types of modification in the following sections.

9.1.1 Body Modification and Ornamentation

The body becomes a symbol through the ways in which it is modified. By modifying the body in meaningful ways, human beings establish their identity and social status. The understanding of body modification as a symbol of social identity in the archaeological record has begun to be recognized by bioarchaeologists as a fundamental goal of research on the lived experience of past people. The modification of the body is the most pronounced and visual way to see the expression of social identity. As such, it has captured the imagination of researchers since they first began to try to understand and explain the various cultures of the world (Flower 1881).

Body ornamentation is complex in nature because it is the means for conveying a message, but it also has specific functions for the individual in expressing who they are. Modifications of the body carry information that is simultaneously symbolic as well as signifying a shared vocabulary. However, they are also functional and serve a material purpose. Thus, material culture such as decorative pottery, weapons, or clothing, along with body ornamentation, can express in complex ways the social identity of the person. This is because the human body acts as an interface between the personal and the social. Through body modification or ornamentation, individuals can actively distinguish themselves from others. Body modification accomplishes several things: It provides a marker of personal identity, and it also can demarcate cohesion within groups. These are examples of "cultural bodies" in that they are given distinctive meaning based on the modifications constructed and the symbolic and real boundaries that are created (Blom 2005).

There are some that argue that body modification, while an expression of culture and ideology, is also a canvas for signaling reproductive (i.e., evolutionary) fitness. The argument is that these cultural explanations for why humans modify the body are a proximate explanation but that the ultimate cause is reproductive success (Carmen et al. 2012). While intriguing, all of human body ornamentation and modification cannot be reduced to sexual selection. Humans are no longer constrained by biology alone, and are not simply motivated by food and sex. The biocultural approach argues that this relationship is much more complex, and to understand complex rituals like body modification, it is crucial to integrate biology with culture and ideology.

9.1.1.1 Artificial Cranial Deformation

Cranial deformation is the intentional or unintentional reshaping of the cranial vault of infants as a result of constant pressure being applied to the skull. Among Native American groups, this is most commonly seen as a flattening of the back of the skull as a result of infants being bundled to a cradle board during their first years of life. Artificial or cultural cranial deformation is the process of intentionally reshaping the head to obtain a desired size and shape.

The more elaborate examples of this practice involve forcibly altering the shape of the cranium. Typically, this is achieved by compressing both the back of the head (often in a cradle board) and the front of the head. This can be achieved by binding a board against the frontal bone (the forehead) to create a long sloped head as is seen with the Northwest Coast cultures or wrapping the head in bindings as is the case with the precontact Peruvian cultures.

However, there are numerous different types of cranial deformation. Often the deformation is described by the bone that it affects the most (some examples include fronto-lambdoidal, fronto-occipital, lambdoidal, and occipital). The variation in the type and location of cranial deformation varies widely from culture to culture, and researchers have been describing these patterns since shortly after osteology as a field of study came into being. Some of these early studies document with great detail cranial deformation in Eastern (Neumann 1942) and Southwest (Stewart 1937) regions of the United States.

Blom (2005) carefully analyzed 412 skeletons from several archaeological sites encompassing several different regions in the southern Andes (AD 500–1100). She found very distinctive patterns that helped to identify individuals as living in certain regions. For example, in outlying regions to the governmental and ceremonial center at the site of Tiwanaku, head modification was practiced widely, but there appeared to be specific rules governing whether the head was modified in the fronto-occipital areas or if there was annular (circumferential) modification. Although these look similar, there are important differences that would be readily seen to a trained eye. However, at Tiwanaku, there were both head form styles represented suggesting that diverse groups from neighboring regions migrated to the Tiwanaku capital. Cranial shape modification served as a symbolic means to maintain boundaries and identity of two distinctive regions.

Artificial cranial modification has been practiced for thousands of years by societies throughout the world. Daems and Croucher (2007) discuss how the practice has been performed throughout the Ancient Near East and specifically outlined the phenomena in a late prehistoric Iranian site. Their research blends an analysis of the skeletal remains with the materiality of the site by integrating the study of figurines found in the region to provide insight into the way people modified their bodies as part of their personal and social identities. There is widespread evidence for artificial cranial modification in the Near East. Some Neanderthal crania have demonstrated evidence of the practice, and skulls from Shanidar 1 and 5 in Northern Iraq have shown signs of cranial modification (Trinkaus 1982:198–199; Meiklejohn et al. 1992:84). The practice increases in the Late Neolithic and so too does the increase in figurines mimicking the cranial modification. The work of Daems and Croucher (2007) illustrates how intricate and complex relations between material culture (materialities and bodies) and social identity/ethnicity are made possible when considered together.

Tiesler (2012) conducted an exhaustive overview of Mesoamerican head shaping techniques. She explored how this practice was employed cross-culturally and temporally to express identity, ethnicity, beauty, status, and gender. The archaeological remains at Maya sites have been used to study social structure, political complexity, economic foundation, and cosmology. For example, iconographies from several locations throughout Mesoamerica have shown the tools used for cranial shaping. These representations suggest how human appearance was physically manipulated to conform to behavior that was seen as aesthetically pleasing based on resemblance or iconic relations. The crania was compressed anteroposterior with the infant bound to a round compression board with free tablets placed on the front of the head to compress the crania (Tiesler 2012:14).

Bodies are always marked by their social and physical existence. People transform their bodies in many different small ways, and the practice of altering the cranium leaves a permanent record and can be studied and contextualized. Thus, this bioarchaeological approach allows for social identity to be inferred through which value is created and accumulated in different social forms (Tiesler 2012; Blom 2005; Torres-Rouff 2002).

9.1.1.2 Other Forms of Body Modification and Ornamentation

There are many other examples of the ways that humans modified and altered their bodies as a way to communicate and embody ideals of their societies. Chinese foot binding was practiced for over a thousand years. Similar to cranial deformation, but with more implications for biological damage, the toe bones of young girls were broken and tightly bound so that as the feet grew, the binding forced the toes to curl under the foot and for the mid-arch area to buckle over on itself. This torturous

Fig. 9.1 Dental modification (Pueblo Bonito burial 327.099) (Courtesy of the Division of Anthropology, National Museum of Natural History)

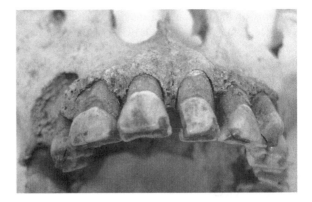

technique was said to symbolize beauty and high status, but it essentially rendered women helpless and hobbled. Thus, the gendered identity that foot binding created was tied to a larger ideology of patriarchy and female submission (Cummings et al. 1997).

Cohen (2008) discusses the revitalization of the ancient practice in parts of what was historically called Burma and today's Myanmar and Thailand of women adding brass rings sequentially to their neck starting as young girls. The weight and restriction of the rings on the neck pushed the clavicles (collar bone) and ribs downward, giving the illusion that these women had especially long necks. While the practice was outlawed in historic times, women are once again reviving this ancient tradition of neck rings because tourists will pay to see the "giraffe women." This practice of body modification and ornamentation has huge significance regarding traditional notions of beauty as well as gender roles. The rings made it difficult for the women to move their head and upper body freely, and often the rings on the neck were coupled with leg and arm rings that further restricted motion.

Dental modification includes the alteration or removal of teeth. This can take the form of filing or drilling teeth to inlay stone or other material (see Ichord 2000 for a pictorial guide to dental modification). The practice of dental modification has been found worldwide and throughout history (Scott and Turner 1988; Milner and Larsen 1991; Alt and Pichler 1998), but there are serious risks associated with this practice (Fig. 9.1). The exposure of dentine through the destruction of the enamel can put an individual at serious risk of infection that can lead to death (Logan and Qirko 1996). In spite of these possible health consequences, dental modification provides people with a highly visible means of establishing identity or gaining social status.

9.2 Ceremonial Warfare, Blood Atonement, and Revenge

Violence in the ancient world is often not straightforward, and levels of violence in groups are highly dynamic, shifting both in time and by location. The context in which violence occurs is crucial for understanding the logic behind the act. The use of violence in human societies is extremely ubiquitous yet challenging to understand in terms of motivation and meaning. There are many different perspectives on violence. Violence is often (erroneously) seen as a human imperative based on innate behaviors of aggression brought about by selective pressures and passed on through evolution. Or the violence is (erroneously) examined as a manifestation of a unique cultural phenomenon that is quintessentially non-western. Explanations for such behaviors are often (erroneously) represented in terms that produce a "mythic other." This exotic or dangerous group enters into the lay epistemology by fulfilling a critical social function in defining conventional morality and behavior through its comparison with "normal" society. Mel Gibson's 2006 movie *Apocalypto* functioned in this capacity by serving as a conduit between quasi-scientific analysis of Maya human sacrifice and ritual body treatment and the public's desire for blood, horror, and chaos.

Violence is a complex behavior that demands a nuanced approach using more recent theoretical frameworks for analysis. Bioarchaeology, for example, has made great strides in addressing the complexity of accessing what is and is not violence based on teasing apart perimortem violence from ritual posthumous body manipulation. Martin et al. (2012) suggest that ancient human remains and the mortuary contexts in which they are located provide a productive way to understand violence in early non-state groups. Creating links between ideology, social relations, trauma, and other indicators of violence on the human remains, bioarchaeology can reveal the complex ways that ritualized violence is embodied.

9.2.1 Ceremonial Warfare

Ceremonial or ritual warfare is a complex topic because it can be nearly impossible to see archaeologically. This is especially evident in the famous documentary "Dead Birds" (Gardner 1963) that chronicles the ritual fighting and ceremonial warfare among the Dani of West New Guinea. In the video, the audience is told about how violence is rampant among the Dani. There is evidence of this violence in their ritual battles, their nomadic lifestyle, and in the construction of watch towers. Dani men keep constant vigilance over any potential threats from the "other"—groups in the surrounding valleys that are not considered to be Dani.

Despite the focus of the movie being on ritual warfare and arguing that violence is such an important part of the Dani ideology, what the viewer actually sees is that violent encounters that result in death or even injury are not the norm. Instead when ritual warfare is actually conducted against their neighbors, it is typically nonlethal and actually ends when someone gets hurt or in rare cases dies. The organization of the battle is that the two rival groups stand on opposites of an area within the a no man's land and throw spears at one another. Though death does occur, it is more common for people to be wounded. These violent encounters would likely not show up archaeologically.

This model of ceremonial violence or ritual warfare was likely common in the past as it offers a way of settling disputes, establishing and increasing status, as well as maintaining lower levels of violence among competing groups in a particular region. Throughout Peru and the rest of the Andean portion of South America, there is a form of ceremonial warfare called *tinku* which is a ritual battle conducted with slings and stones. Prior to being altered by the conquest of the Inka and later the arrival of the Spanish, which shifted the focus from ritual fighting to more lethal violence, this was a nonlethal (most of the time) means of maintaining power relations in the region (Chacon et al. 2007; Tung 2007; Gaither 2012).

The most prominent example of how ceremonial warfare can act as a substitute for conflict is the Northwest Coast, where a major part of the culture was warfare-related activities for thousands of years until the development of the potlatch. Warfare is evident in the emergence of social stratification, the establishment of large trading networks, and the appearance of weapons in the archaeological record and in ethnographic descriptions. Trade was especially important because it provided access to rare or exotic items that could be utilized to establish social hierarchies. Part of this emphasis on trade is the development of a cycle of raiding for slaves (Maschner 1997). The archaeological record includes the appearance of weapons (e.g., lances, war clubs, daggers, and barbed projectile points), fortified houses and villages, the development of a separate warrior class, and increased lethal and nonlethal violence and trophy taking on both skeletal and extant populations (Ames and Maschner 1999; Boas 1966). Intergroup competition and warfare were influenced by the emergence of social status. Also, status was necessary to maintain at the level of individuals, lineages, and villages. A consequence of increased violence was the development of ceremonial warfare that offered an alternative means of competing for status. Driven by the expanding trade and raiding for slaves, a system of gift giving was developed known as the potlatch. Similar in nature to the "moka" of the Big Man systems in Papua New Guinea, the goal is to establish and maintain social status by gifting your rival (Codere 1990). It was not about obtaining wealth as much as it was about display. "Chiefs set out to permanently shame or vanquish their rivals, at least in the eyes of the public" (Miller 2000:85).

9.2.2 Human Sacrifice

Bones and corpses and coffins and cremation urns are material objects that can offer insight into the cultural dimensions of both past and present populations. For example, the Celtic Gauls created sanctuaries in northern France during the late third century BC that were designed to celebrate victory in war and intimidate potential enemies while pleasing the gods of the underworld, whom they believed made them great warriors. After disarticulating the bodies of their enemies, priests crushed their bones to expose the marrow. After breaking the bones, the priests dumped them into a chamber in the ossuaries where they were burned. These sanctuaries may have also served as a communal war trophy depository. The sanctuaries were adorned with long bones and skulls, and some had racks of decapitated warriors left to decompose in the elements (Rives 1995).

Examining the theoretical paradigms of ritual sacrifice and mortuary behavior is key for understanding the motivations behind human sacrifice. Tiesler and Cucina (2007), for example, accomplish this by surveying human bone assemblages evidencing perimortem and posthumous cultural processing from the Classic and Postclassic Maya in order to analyze the complexity, variability, and ambiguities surrounding the manipulation of human remains. Through an integrated interdisciplinary investigation of human remains within a biocultural framework, a broader appraisal of mortuary and non-mortuary behavior of Classic and Postclassic Maya sites can be obtained. As Ashmore and Geller (2005) have discussed, the spatial arrangement of the dead and their mortuary contexts can carry social meaning. Using various Maya mortuary spaces as their examples, they contend that the form and position of funerary monuments can serve as sites of commemoration and social reproduction and as points of orientation in the universe. Positioning of remains within burials may also have symbolic meanings. It is only by addressing the complexity of Maya sacrifice as it relates to conflict, violence, and warfare that the classification of specific types of sacrificial violence is possible.

The identification of violence and trauma requires a nuanced and detailed analysis of both material culture and human remains present at an archaeological site. Thus, it is important to have a comprehensive understanding of the variables that can alter the appearance of skeletal material. Tiesler (2007) outlines new theoretical concepts regarding the distinction between funerary and non-funerary practices based on combined archaeological, osteological, and taphonomic analysis. Through the use of ethnohistoric and iconographic data, Tiesler argues for the existence of alternative sets of taphonomic signatures that can be used to identify funerary and non-funerary practices at Maya archaeological sites.

Lucero and Gibbs (2007) offer a compelling example of how careful analysis of dissimilar remains from two cave sites can differentiate witch killings from ancestral remains and sacrifices. Here again, it is the careful use of taphonomy, agency, and ritual behavior that allow for the recognition multiple depositional practices. Pearson (1993:203) notes that the dead are "manipulated for the purposes of the survivors" and that such funeral ceremonies are the result of "political decisions." The placement of the bodies in caves which the Classic Maya saw as portals to the underworld suggests that these sacrificial victims were placed there to hurry their souls to the underworld. "The Maya dispose of the remains in a nonfunerary manner by placing killed witches (and sacrificial victims) in openings in the earth, especially caves—a long-standing tradition and likely one with prehispanic roots" (Lucero and Gibbs 2007:50).

Witchcraft persecution has long been considered one of the possible explanations for the perimortem modifications of human skeletons of the American Southwest and has become a controversial debate among anthropologists. From AD 900 to 1250, skeletal remains from the Southwest show a variety of injuries (Lambert 1999, 2000; Martin 1997); however, the most inexplicable form of trauma from this period is the dismembered corpses with cutmarks, extensive perimortem fracturing, percussion scars, and burning (Billman et al. 2000; Turner and Turner 1999). On the southern piedmont of Sleeping Ute Mountain in southwestern Colorado, corpse mutilation, cannibalism, and community abandonment around AD 1150 strongly suggest that serious intergroup violence was important in the formation of at least some of these assemblages.

Approximately 40 sites containing human remains with an MNI ranging between 1 and 35 have now been identified as evidencing disarticulated and culturally modified human remains and most date between AD 900 and 1200 (see summaries in Baker 1990; Billman et al. 2000; Turner and Turner 1999; White 1992). There is general agreement that most of these assemblages are the result of pre-Hispanic violent behavior (Darling 1998:747). Ogilvie and Hilton (2000) also have suggested that ritualized violence associated with the destruction of witches could account for such remains and offer an example from a late Pueblo II assemblage (ca. AD 900–1100) from northwest New Mexico.

The mortuary record of the Ancestral Pueblo demonstrates a wide range of variability and ambiguities in the skeletal assemblages. This is particularly true of the culturally modified human remains that have been found in archaeological contexts dating from AD 900 into the historic period. As discussed above, although the disarticulated remains from these assemblages are often thought to be the result of cannibalism and related activities, there are other hypotheses that are currently being debated as well. Darling (1998) presents a compelling hypothesis regarding witchcraft execution that is supported by detailed ethnographic data on the endemic nature of witch killings in Pueblo history. Sacrificial witch killing has a long and complex history that spans nearly every continent and extends back thousands of years. In almost all of these incidents, the accused witches were tortured and murdered in a culturally specific and highly ritualized manner.

Violence reconfigures its victims and the social environment in which it occurs. It must never be viewed as a transitory punctuated event that leaves only a memory with no lasting effects. This is because violence becomes the determining factor that shapes future realities for both individuals and cultures. Thus, it is important to look at cultural realities as they are brought into existence by their daily practice and not as some static entity. There is not a single cultural system but many systems. These systems will continue to morph as the situations around the community change. It should be the goal of the violence researcher (or any anthropologist for that matter) to not search for a single event that delineates and homogenizes a systematic function of a group (sacrifice, violence, or warfare) but rather try to understand how people are bound by events and processes that allow for a fluidity of responses to multiple stimuli.

There are limitations to identifying sacrifice in the past. For example, looking at child sacrifice, especially in the past, often is hard to differentiate from other cultural practices that may lead to early death, such as infanticide, child abuse, or even health-related death. Differentiation of child sacrifice and other child deaths may be possible however, using a bioarchaeological perspective. Bioarchaeology offers a method of understanding the world in which children inhabit and the cultural processes that are not only subjected on, but also created by, them as they navigate their world. This holistic approach to analyzing trauma among children can help differentially diagnose cases of child abuse from infanticide or child sacrifice. The inability to identify infanticide is due to the fact that it is considered as just another form of neglect and as such may be masked by the diagnosis of child abuse. The problem with this approach is that the act of child abuse, especially neglect, though difficult to distinguish from indirect infanticide (Brewis 1992) has very different implications culturally. Calling infanticide child abuse is problematic because it masks a deeper sociopolitical as well as ideological practice. Children designated for sacrifice would have a different bioarchaeological signature based on the ideology surrounding who is sacrificed and how the sacrifices are carried out. For example, they may have been protected and well fed in preparation for being sacrificed (Wilson et al. 2007), or the children selected for sacrifice might present evidence of genetic disease that predisposed them to becoming offerings (Weyl 1968). Arguably, if the child has a higher social status or is selected earlier on because of a birth defect, then one might not expect to see the pattern of injury associated with a long history of abuse.

9.2.3 Captivity and Torture

Bioarchaeological studies can be used to differentiate different kinds of violence on the body as it may be related to captivity, bondage, and torture (Martin and Osterholtz 2012; Blondiaux et al. 2012). Chapter 5 presented the unusual mortuary configuration of the La Plata Ancestral Pueblo adult females. Integrating skeletal analysis, mortuary context, and archaeological reconstruction, multiple lines of evidence were obtained that all pointed to captivity and hard labor. A subgroup of women showed injury recidivism, that is, repeated trauma and injury over the course of a lifetime (see Judd 2002, for one of the first studies linking injury recidivism to violence in ancient societies). Indicators included healed cranial depression fractures likely due to blunt force trauma obtained during raiding and abduction of females. These women also had a variety of healed fractures on the lower body, as well as localized trauma to the joints (e.g., dislocated hip joint). These may be the result of punishment or harsh treatment. These women also had indicators of poor health (infections, nutritional problems). Months or years of hard labor resulted in pronounced muscle markers, traumatic osteoarthritis, and trauma-induced pathologies in these women. They were recovered from burial contexts different from individuals who did not have bodies wracked with trauma and pathology. In this case, the women seem to have been placed without any intentionality or grave offerings and in abandoned pit structures.

Osterholtz (2012) has pioneered methodologies for the analysis of torture and executions. She has found that there are key patterns that are revealed on the skeletal remains. The assemblage at Sacred Ridge, Colorado, for example, is made up of the remains of at least 33 people who were killed, dismembered, and placed in a pit structure around AD 800. Examination of the foot bones of these individuals shows a pattern of injury that is consistent with hobbling and torture, which would have been a performative aspect during the massacre. Individuals would have been forced to watch their kin being hobbled by blows and cuts to the sides of the feet and tortured by beating the soles and tops of the feet. Peeling of the bony tissue as well as cutmarks, buckling of the bone, and other marks consistent with torture and hobbling are present on adult remains of both sexes. Hobbling would have made it impossible for the individual to physically move or flee; this has both physical and psychological effects. Hobbling is visible through the damage to the sides of the feet, caused by both blows and cutting of the ligaments that stabilize the foot for walking and running. Torture through beating the soles of the feet has a long and diverse history worldwide. Torture cements the social control of a captive by literally giving the aggressor power to inflict pain (or to stop the infliction of pain). These types of injury have no utility after death, and so must have been perpetrated prior to death.

Still, there are many challenges in reconstructing captivity, slavery, bondage, and torture. Pain is notoriously difficult to document and even more difficult to objectively score since each individual will feel pain at differing intensities. Equally difficult for bioarchaeologists, and even more important when examining concepts such as torture in a performative light, is the impact that another person's pain has on a witness. In some ways, being forced to watch someone you care for in pain may be as powerful as being subjected to such pain yourself. Not only is someone you care for in distress, but you have no power to mitigate the situation. Pain is inherently relatable, so the examination of pain is a way to humanize work such as that seen at Sacred Ridge, where the scale of the massacre has a tendency to overwhelm individual observations. Through an understanding of the collective studies of these types of violence, it is possible to get an idea of what the individual felt and imagined about themselves and their families in a similar situation.

9.2.4 Massacre

The presence of dead bodies sends a strong social message. The absence of the body or its disarticulation and/or mutilation also creates social trauma because death is often accompanied by particular complex rituals. The body, both alive and dead, is a canvas that allows for social expression and social contest to be recorded. It is the flesh and bone that reveals social subjection, exploitation, and mass murder. The San Juan Basin during the tenth through the thirteenth centuries offers insight into the complexities of violence in the Ancestral Pueblo groups of the American Southwest (Stein and Fowler 1996). In general, most scholars agree that the arid environment and particularly the increasingly impoverished environmental conditions served as a stimulus for a range of manifestations of violence and massacres. There is archaeological evidence for fortified sites, palisades, defensive architecture, aggregation of communities, and structures such as watchtowers (Wilcox and Haas 1994; LeBlanc 1999). Warfare (which in the literature is described as raiding, ambush, intercommunity violence, and intra-ethnic or tribal clashes) and fear of attack are provided as the most likely reasons for the defensive architecture used in the time periods leading up to the tenth century (LeBlanc 1999:119).

Sand Canyon Pueblo in the Northern San Juan Region (AD 1250–1285) represents the full range of burial types with ten formal burials, burials that appear to have been informally placed, and isolated bones (Kuckelman 2007). One individual from an informal burial was a male who was aged 45, sprawled on the floor of a room. At the time of death, he had a healed cranial depression fracture on the left side of his head and a fatal perimortem fracture on the front of his head. Another individual who died violently at this site was a 15-year-old who was located in a collapsed wall. He had a blow to the base of his skull and the nose and several teeth of the maxillary region were broken around the time of death. An additional young person, aged 12–15, was found facedown on the floor of a kiva. A large cranial depression fracture on the back of the head along with additional fractures in the skull suggests blows to the head were the likely cause of death.

Kuckelman et al. (2002) present an interpretation of a massacre rich with disarticulated bone assemblages from the site of Castle Rock (in Colorado). A series of testable hypotheses were formulated utilizing information from cross-cultural, ethnographic, archaeological, and historical sources. Expectations regarding what should be found for each hypothesis were thoroughly discussed. The skeletal elements from the site were not analyzed as a single assemblage; rather, small clusters of bone labeled sequentially as human remains occurrences (HRO) were first analyzed separately as to possible causation and then examined vis-à-vis all of the other HRO for the site as a whole. By destroying elements of materiality "hidden" in a particular identity of a specific community, the perpetrators create not only the physical destruction of a people but also the ideological massacre of their belief system.

Although the hypothesis regarding acts of anthropophagy (a more accurate term for cannibalism) is supported for bones found in one context within the site, it was not supported for all of the fragmentary remains. Using a bioarchaeological methodology that incorporated materiality, the team tested a number of hypotheses and found support for accepting both warfare activities and anthropophagy (or trophy taking in conjunction with anthropophagy). Based on an analysis of the variable patterns of the bone deposits in conjunction with the data from mostly complete individuals and burials, they demonstrate that Castle Rock was the site of a massacre. Some individuals were left on the ground in some places, some individuals may have been torn apart by carnivores after they died, some remains may represent secondary burials that occurred days after the massacre by returning family members, and anthropophagic (or trophy taking) activities may have taken place in a specific location shortly after the massacre. In other words, a careful focus on the full range of subtle differences between and among the broken, chopped, burned, and abraded bones provided a means to determine more specific and exact explanations for a range of simultaneous activities. This type of bioarchaeological approach to the classification and interpretation of disarticulated human remains bridges the chasm between biology and the social and environmental dimensions of the communities being studied (Pérez 2012b).

The osteological record supports large-scale village massacres in places such as Castle Rock (Kuckelman et al. 2002) and Cowboy Wash (Billman et al. 2000). However, the assemblages and burials found at these sites are not comprised simply of dead bodies struck down while fighting. There is a remarkable range of variability in the treatment of corpses (by both the perpetrators of the attack and possibly returning survivors), rituals for burial of the dead that are unique to this time period, and cases of violent deaths. In addition, there is skeletal evidence documenting victims of violent interactions who escaped death. Healed (nonlethal) traumatic injuries and head wounds are present at many Ancestral Pueblo sites but seem to increase during the twelfth and thirteenth centuries (Martin 1997; Martin et al. 2010).

The pre-Hispanic Pueblo landscape is scattered with the occasional occurrence of human skeletal assemblages that are disarticulated, broken, chopped, sometimes burned, and often dismembered. These collections (which include both children and adult males and females) have been variously interpreted to represent cannibalism (Turner 1993; White 1992), witchcraft retribution (Darling 1998), warfare (Wilcox and Haas 1994), and ritualized dismemberment (Ogilvie and Hilton 2000). Whatever the motivation behind these presumably violent deaths and perimortem alterations of the victims' bodies, the deaths and alterations suggest some evidence for violent action directed against subgroups in most cases that seem to be demographically representative (i.e., infants, children, men, and women).

9.2.5   Cannibalism

The concept of cannibalism has captured humankind's imagination for thousands of years. Anthropology and specifically bioarchaeology have been major contributors to this topic through serious academic exploration. Herodotus, who is often cited as the first recorder of other cultures, mentions in the fifth century BC of some mysterious people far beyond the realm of civilization who eat the flesh of man (Arens 1979:10). Humans who partake of these practices have always been found on the outer fringes of the so-called civilized world. These people, though considered barbarians, were never labeled inhuman, for this would serve the purpose of reducing their behavior to an animalistic level, thus naturalizing the act and removing the abomination and dread associated with the ultimate taboo of anthropophagy (Hulme 1986:14). The term anthropophagy described the act of eating human flesh until the end of the fifteenth century.

"Anthropophagi" is a Greek word which was created by combining two preexisting words from the Greek language, "eaters/of human beings." This word was then used by the Greeks to refer to a group of people who were thought to have inhabited the land beyond the Black Sea (Hulme 1986:15). Christopher Columbus coined the term "cannibals," a non-European word, to refer to a group of indigenous people known to exist. These people, known as the Caribs, through a Spanish mispronunciation, became canibs and eventually cannibals (Arens 1979:44).

The first appearance of the word cannibal in a European text came on November 23, 1492. In his journal, Christopher Columbus recounted the approach of an island which the Arawaks had referred to as "Bohio." The following passage is taken from Columbus' journal and refers to a statement made by the Arawaks: "[they] said that this land was very extensive and that in it were people who had only one eye in the forehead, and others whom they called 'canibals'. Of these last, they showed great fear, and when they saw that this course was being taken, they were speechless, he says, because these people ate them and because they are very warlike" (Hulme 1986:16–17).

Few research topics rival cannibalism in the amount of attention and curiosity that this subject matter can invoke. Archaeological assemblages evidencing possible cases of cannibalism range from the Lower Pleistocene to present-day populations. Carbonell et al. (2010) have suggested that the postcranial remains of *Homo antecessor* from the TD6 level of Gran Dolina (Sierra de Atapuerca, Burgos) at more than 800,000 years are the oldest known case of cannibalism. Jiménez-Brobeil et al. (2009) looked at sites between 5500 and 3000 BC in the SE Iberian Peninsula and argued that there are at least four sites evidencing cannibalism.

Anthropologists have defined cannibalism to include a taxonomy which contains three district categories: "(1) endocannibalism, which refers to eating a member of one's own group; (2) exocannibalism, indicating the consumption of outsiders; and (3) autocannibalism, signifying ingesting parts of one's own body" (Arens 1979:17).

Reed reexamined cannibalism in the archaeological record of the American Southwest in 1948. With the exception of two cases, Reed was skeptical and unconvinced that any evidence for cannibalism existed (Nickens 1975:284). Since 1948, a number of researchers working in the Four Corners region of the southwest have reported evidence of possible cannibalism in the archaeological record. In his book *Prehistoric Cannibalism at Mancos 5MTUMR-2346*, White denotes 19 occurrences of possible cannibalism in the Four Corners region of the southwest. Some but not all of the abovementioned sites are listed in his synopsis (1992:36–39).

There is a remarkable range in the types of violence found during the occupation of the San Juan Basin by Ancestral Pueblo people. There is no doubt that violence and warfare played a role in Ancestral Pueblo life. The challenge lies in finding parsimonious data sets that support a scenario for violence that explains the variability. This diversity in individuals with nonlethal injuries, recidivistic individuals, and individuals with health problems suggests that violence could play out differently. The range of variability in the disarticulated and extremely processed material also makes suspect the notion that one overarching activity, such as mini-armies carrying out public executions and cannibalism, could account for the differences in placement, type of modification, degree of breakage, and pattern of cut marks.

As discussed in Chap. 4 and Sect. 9.2.1 of this chapter, it is more than possible that some of the disarticulated assemblages have nothing to do with violence and everything to do with burial rites, veneration, or consecration. Burial rites (unrelated to violence per se) may be occurring simultaneously with acts of violence and intimidation, and it will take careful examination of each assemblage as part of the total site reconstruction to see the difference. Because of this overlooked variability, it is imperative that researchers do not jump to the conclusion of cannibalism when dealing with disarticulated and culturally modified human remains.

With regard to osteological evidence used to support the claim of cannibalism in the American Southwest, the data provides important and ample evidence that there was a high degree of social complexity and mortuary variability in the eleventh and twelfth centuries. This reluctance to consider other viable explanations for the disposition of these assemblages has led to missed opportunities. For example, too many researchers are engaged in what some might call "checklist" osteology. That is, they relied on a data collection form (Turner and Turner 1999:489) that considers only the presence or absence of taphonomic information, and in doing so, they focused on a descriptive account of the processed human remains, largely ignoring and glossing over the many categories of human behavior that produce such assemblages.

9.3   The Body as an Agent in the World

There is often the belief or unconscious thought in modern Western societies that after death, a body has little meaning. However, if this were true, there would not be cemeteries or places like Arlington National Cemetery. These places are significant because they represent dynamic spaces where agency, memory, and cultural meaning are reinvented overtime. In many cultures, the dead body is not simply an object but instead is an entity that has agency in the world. In almost every society in the world, there is evidence that people venerate or memorialize their ancestors (e.g., the Great Pyramids of Egypt, the Catacombs of Rome, and Arlington National Cemetery). If the body could no longer communicate or hold power in particular ways, there would not be so many national displays of the dead in highly visible places. Explored here are the concepts of ancestor veneration or the celebration of those that have died. Behaviors such as keeping the bones of dead people as trophies are shown to be related to the act of taking the power of the enemies that were vanquished. By exploring signatures of ancestor veneration and trophy taking, along with highly ritualized aspects of warfare and raiding, the dead body is seen to hold both symbolic and real power over the living.

9.3.1   Ancestor Veneration

All human societies have their own rituals, beliefs, and customs surrounding death. The way that people dispose of the dead and the meaning associated with death varies from culture to culture and even from neighborhood to neighborhood within the same town. The meaning of the ritual, its timing, and those expected to participate varies enormously. These death rituals serve a variety of functions including helping the spirit or soul move on to its next life, separating the body from the soul, and constructing boundaries between the living and the dead (Hubert 2000:209).

Death ceremonies are comprised of three dramatists, the corpse, the soul, and the mourners, all of which are sent on a ritual journey at death. The marking of this passage occurs through the corpse because the corpse is a transitional object linking the living to the afterlife. Human remains are not simply physical objects, just as death is not simply a biological phenomenon. Both are as culturally constructed as the death rituals inscribed to deal with them (Heinz 1999:155). Thus, the body can be used as a "natural symbol" through which societies and cultures may be interpreted (Heinz 1999:156).

From the earliest human ancestor, disarticulated bone assemblages evidencing perimortem cultural processing have been part of the history of the human species (Pickering et al. 2000). Although it is difficult to establish specific reasons for such behavior, plausible explanations include mortuary practices, ritual destruction, mutilation, cannibalism, and violence. In cultures that practice corpse dismemberment, parts of bodies come to represent whole bodies. Displayed or otherwise memorialized parts of bodies are often considered comforting and have powerful symbolic messages about how to remember and obtain power from the dead.

The practice of dismemberment and/or secondary burial was common in medieval England from AD 1066 to 1555. For example, the difficulty of transporting bodies back to Europe from the Crusades was solved by boiling the corpse to separate flesh from bone. The flesh was then burned and the bones were carried back to Europe for Christian burial (Quigley 1996:82). Thus, when the body of Philip III was boiled in wine and water, his disarticulated skeleton became a concentrated version of the body and was considered its noblest part. Burial of the heart was also common practice at the time of the Crusades. In part, this was because the heart was easy to carry back to Europe and also because of its ancient symbolism and biblical references (Daniell 1997:122).

Other parts of the body could also be treated separately. For example, when the Bishop of Hereford died in Italy in 1228, only his heart and head were carried back to England. When Robert de Ros, who was one of the signers of the Magna Carta, died, his heart was buried at Croxton Abbey, and his bowels were buried before the high alter of Kirkham Abbey. The dismemberment of the body during this period appears to be bound up with the triumphant years of the Catholic Church. In 1299, Pope Boniface VIII issued a ban on what he called a cruel and profane practice. His successor, Pope Benedict XI, modified this stance due to the practicalities of burial, and Pope John XXII, being even more practical, found a way to make money by issuing licenses for the division of a body (Daniell 1997:123).

By the fourteenth century, charnel houses or ossuaries became common in England as churches and churchyards became overcrowded with human remains. Galleries piled high with bones were used to hold catechism classes and charity gatherings. Like the many catacombs throughout Europe, these ossuaries stacked and stylized the bones according to various sizes and shapes. Besides fulfilling a practical need, ossuaries were used in some cases to raise the level of spirituality. For example, in western Brittany beginning around AD 1450, the Catholic Church began promoting an existing death cult in order to increase formal participation in the rites of the church.

Sometimes parts of dead bodies are sought and retained. The bodies of saints have been parceled out to cathedrals all over the world. Bits and pieces have been smuggled and bestowed, hoarded and exhibited, and stolen and retrieved for hundreds of years (Quigley 1996:250). Even the smallest holy relics have a physical presence which may provoke or reaffirm one's religious faith. The skull, legs, arms, and breast of St. Agatha, martyred in 251 AD, are encased in an effigy and are exposed three times per year in Catania, Sicily (Quigley 1996:259). Although some relics were conferred upon churches and individuals, some were bought and sold. King Canute of England was said to have paid a significant sum in the eleventh century for the arm of St. Augustine. Impassioned collectors amassed great hoards of relics, and the size of one's collection was a status symbol. Public displays of human anatomy, especially relics, were once fairly common. A ruler of Saxony possessed 17,000 holy relics, and Frederic the Great acquired more than 19,000 holy bones in the fifteenth and early sixteenth centuries (Quigley 1996:261). The zeal for collecting was so intense that when St. Hugh of Lincoln made the pilgrimage to Normandy to pay his respects to St. Mary Magdalene, he took the opportunity to bite off two pieces of her arm to take back to England (Quigley 1996:263).

Just as holy relics have a physical presence which may provoke or reaffirm one's religious faith, the remains of criminals in some cases hold symbolic relevance which can lead to their destruction and the continued punishment of the corpse. The corpses of criminals in Europe have been denied burial, dissected, or dismembered as part of their sentence. During its long history, the Catholic Church has tried the dead for heresy, and those found guilty were disinterred from sacred ground and burned or reburied elsewhere. The Church of England convicted Thomas Becket, archbishop of Canterbury, of high treason 400 years after his death, and his remains were exhumed and publically burned.

The dismemberment of the corpse was often explicitly directed in the legal death sentence. In English law under Edward III, those convicted of high treason were half-hanged, after which their entrails were removed and burnt in front of them, their heads were severed, and their bodies quartered. It was not uncommon for a piece of the felon to be displayed as a warning to others. For example, the skin of Richard de Pudlicote, who was hanged in 1306 for stealing the Crown Jewels, was stretched across the chapel door to discourage would-be thieves (Quigley 1996:281).

In North America, many different indigenous groups routinely disarticulated human remains as part of their mortuary practices. They also engaged in the ritual destruction or mutilation of enemy warriors. The site of La Quemada, a large fortified complex with notable public architecture in Zacatecas, Mexico, demonstrates a wide variety of bone deposits, many of which have extensive modification in the form of perimortem breakage, cutmarks, reduction in size and shape, and burning. Occupied from AD 500 to 900, this Epiclassic site represents part of the northern Mesoamerican frontier during a period when regions in central Mexico were being abandoned (Nelson 1997; Nelson et al. 1992; Trombold 1985).

The variations in the bone assemblages at La Quemada suggest that a multiple approach to the curation of the dead was practiced. There are numerous examples of these types of complex mortuary practices throughout Mesoamerica. This is due in part to the fact that the human body was intricately woven into the ideology of many of these cultures. For example, the Nahuatl believed that specific regions of the body, and in particular bones, held the power to heal or hurt. The belief that a portion of an individual's vital force was housed in their bones is revealed in the following account: The femur of a sacrificed individual was kept in the house of the warrior who had captured him in combat. When the captor returned to battle, his wife hung the relic from the roof, covered it with paper, and offered it copal incense, at the same time asking for her husband's safe return (Durán 1951:167).

This theme of the femur as a bone of importance is seen even today as illustrated in the mural "La Gran Tenochtitlan" by Diego Rivera in Mexico City. In this painting of a busy street scene, there stands an individual looking over a healer holding a human femur. The Nahuatl also believed that in many cases, the transcendental forces that the gods bestowed to humans were stored on the left side of the body (López Austin 1980:165). It is of interest to note that the disarticulated human remains found in, the Temple follows this pattern with a preference for left elements. The ancient Nahuatl also placed a great deal of importance on the joints. It was believed that these regions, known as minor animistic centers, were weak spots through which supernatural forces could enter the bone and cause damage. This could explain why many long bones retrieved from both the banquette and midden have had the epiphyses (ends of long bones) removed.

For the Huichol Indians, whose language was influenced by Central Mexican Nahuatl and who currently inhabit the area near La Quemada, dismemberment and isolated human skeletal elements play a central role in many of their mythologies (Furst 1996; Grimes 1964; Lumholtz 1900; Negrín 1975; Zingg 1938/1977). This is seen in the yarn tablas of Sánchez (1975). In one yarn tablas is the story of Great Grandmother Growth, who, upon her death, had her human body fall to pieces. From these various body parts, new plants and animals were born. Sánchez tells the story of a man who survived the great flood with the help of Great Grandmother Growth. Upon his death, the parts of his body dispersed, and new plants were created from them (Negrín 1975).

In his classic ethnography of the Huichol Indians, Lumholtz (1900) describes their "God of Death," named Tokákami. According to Zingg (1938/1977:365), Tokákami is a "horrid ghoul and a figure of death," not a god as Lumholtz had suggested, but nevertheless, it seems to be another example of disarticulated human skeletal material figuring prominently in Huichol mythology. In an illustration of a statue of the Huichol ghoul from Lumholtz's "Symbolism of the Huichol Indians" (1900:61), there are white lines representing human long bones attached to strings around his waist and over his back (Fig. 9.2).

The variations in these human bone assemblages provide ethnographic support that a multiple approach to the curation of the dead was practiced at La Quemada. Indeed, there is a strong likelihood that ancestor veneration along with the ritualized destruction of enemy remains accounts for the multiple mortuary behaviors present (Nelson et al. 1992; Pérez 2002; Pérez et al. 2000). Many of the remains, particularly skulls and long bones, appear to have been placed on or suspended from racks located in several residential and ceremonial centers throughout the site (Nelson et al. 1992). Thus, although the assemblages as a whole reflect an abundance of evidence that individuals were dismembered and defleshed, analysis of the patterning of the types of bone with cuts revealed differences in both the frequency and

Fig. 9.2 A reproduction of
the statue of the Huichol
Indians from Lumholtz
(1900, 61) who called it the
God of Death–Note the
representation of long bones
attached to a belt around the
waist and over the back

morphology of the cutmarks at each of the deposits, along with the importance
placed on specific elements.

When considering if a particular mortuary practice is part of a veneration process,
it is fairly common to try and differentiate elites from nonelites as defined by or based
on socioeconomics; it is less common to look for individuals who may have held a role
such as shaman. And yet if it is possible to infer ancestor veneration through the materiality of the body, it seems reasonable that research should be able in some cases to
identify the burials of spiritual leaders. It is difficult to precisely define shaman in
anthropology since it is used to capture so many different cultural expressions, but it
generally refers to religious or spiritual individuals who can induce an altered state and
cross into other dimensions, and may mediate between the living and the nonliving.

In the bioarchaeological record, there are some cases where it is argued that based on
the material culture and other defining characteristics, it is possible to see shamans.
Grosman et al. (2008) describe in great detail a Natufian burial from Israel that was

unique in many ways. Limestone slabs lined the rooms, the walls were plastered, large stones were placed on the body, over 50 tortoise shells with consumption breakage were included, and 30 individuals with missing long bones. An adult female aged at 45 from the site had congenital asymmetry in some of her bones that would have given her an odd gait. She had an articulated foot placed above her left leg. Taken as whole, these lines of evidence could suggest an individual who was considered to be held in high esteem from her community. Maher et al. (2011) discuss an unusual burial that include fox remains. The authors discuss possibilities of shifting relationships with the animal world since this burial departs from others in prior time periods. Porr and Alt (2006) analyzed the skeletal remains from the isolated burial of Bad Dürrenberg, in Central Germany, one of the richest Mesolithic graves in Europe. The authors focused on an individual with a pathological variation to the structure in the atlas vertebra and the foramen magnum, which may have created neurological conditions leading to altered states of consciousness. This osteological information combined with grave goods and red ochre found in this adult woman's grave suggests possible shamanistic elements in the materiality of this burial.

When considering the lived experiences of differences, it is important to remember that it is more than just gender or ethnicity, class, age, etc. These cultural phenomena overlap and intersect in one's bodily experience. Bioarchaeology has an important role to play in understanding the intersections of different embodied experiences to construct identity. These different identities in bodily praxis must be seen as and constituted with material culture. It is important not to fixate on a single factor, such as gender, when constructing identity because bodily experience (as seen with ancestor veneration) is as assorted as the material culture used to establish these identities. The concept of the body as material culture provides bioarchaeologists with ways to interpret social construction of identity, in this case spirituality, because it locates material culture as an extension of the body.

9.3.2   Trophy Taking

The practice of taking and displaying the human remains of one's enemy dates to as early as the Pleistocene (Conroy et al. 2000) through modern warfare (Harrison 2006). Culturally modified human remains, often referred to as "trophy taking," can include any part of the human body and has been well documented in both the eastern and western hemispheres (see chapters in Chacon and Dye 2007b). As discussed in Sect. 9.1 of this chapter, the corpse is seen as a transitional object between life and death for both the perpetrators and victims. As such, the cultural significance of trophy taking sends a powerful message by ritually destroying or mutilating the remains of an opponent (Pérez 2012a). A single overarching hypothesis that would seek to explain the practice of trophy taking through time and space would be extremely problematic given that "theoretical discussions of indigenous warfare and ritual behaviors often associated with fighting (such as human trophy taking), necessarily require consideration of the region's tremendous ecological and cultural diversity" (Chacon and Dye 2007a:5). However, for the victim, it seems reasonable to hypothesize that the physical and psychological impact of this type of cultural performance is staggering. This type of violence creates a "spectacle" in which the display of the remains demonstrates the power and strength of the victors and the vanquished group's mortality and creates psychological trauma that impacts the survivors. This is accomplished through the transformation of young healthy men, women, and children into an unrecognizable mass of body parts.

9.4 The Body as a Symbol and the Power of the Dead: Ethical Considerations

Bioarchaeology has a responsibility to create a research context in which problems of bias are exposed. Although the rules for scientific inquiry are a function of its values, namely, an accurate understanding of the natural world, the choice of which areas or aspects of analysis will be carried out is very much a function of the social and cultural values of the researchers.

Social scientists, including forensic anthropologists and bioarchaeologists, attempt to analyze human behavior by (re)constructing events, in part, by using cultural taphonomic signatures left on the bodies. Almost all will tell you that it is much easier to do the work in a dispassionate and clinical way when the events are spatially or temporally removed from one's own culture. The closer one is to the event, the more difficult it is to remain dispassionate and analytical. This is particularly true for bioarchaeologists who work with precontact skeletal remains.

So, how does the body as material culture figure into this? Human remains are a lens through which cultural processes can be examined. How dead bodies are discussed, hidden, and displayed can be used as a point of departure for examining both the living and the dead. Anthropologists need to consider their personal motives and investment in their research in order to limit and expose potential biases. Researchers must continue to struggle to identify the cultural assumptions of their research and ask how answers might be formulated under different assertions. It is in this way that researchers can avoid perspectives that alienate living people and prohibit them from understanding the meaning associated with death and the appropriateness of discussing, viewing, or displaying bodies or parts of bodies.

9.5 Summary

The focus of this chapter has been to illustrate the ways in which the body can be used as an object in rituals that cement social interactions and reaffirm cultural ideology. In Chap. 6, the body was explored in terms of the various ways in which it reflects the individual's lived experience and the ways that the body is shaped by social and political forces. This chapter extends these approaches to understand how the body continues to have agency well after death. From ancestor veneration to trophy taking and ceremonial warfare to large-scale massacres or cannibalism, the body and body parts can take on a multitude of roles and identities. Understanding this reality allows future researchers to conceptualize alternative ways to view and interpret skeletal material. While this approach is obviously a little more abstract and harder to grasp, it provides insight into realms of human existence that are rarely explored from the perspective of the individual or the group. Theorizing the different ways that human remains were used to communicate specific ritual or ceremonial aspects of daily life is important to attempt for the insights that it can reveal.

References

Alt, K. W., & Pichler, S. (1998). Artificial modifications of human teeth. In K. W. Alt, F. Rosing, & M. Teschler-Nicola (Eds.), *Dental anthropology fundamentals, limits and prospects* (pp. 387–415). New York: Springer.

Ames, K. M., & Maschner, H. D. G. (1999). *Peoples of the northwest coast: Their archaeology and prehistory*. London: Thames and Hudson, Ltd.

Arens, W. (1979). *The man eating myth*. Oxford: Oxford University Press.

Ashmore, W., & Geller, P. L. (2005). Social dimensions of mortuary space. In G. F. M. Rakita, J. E. Buikstra, L. A. Beck, & S. R. Williams (Eds.), *Interacting with the dead: Perspectives on mortuary archaeology for the new millennium* (pp. 81–92). Gainesville: University Press of Florida.

Baker, S. A. (1990). *Rattlesnake Ruin (42Sa 18434): A case of violent death and perimortem mutilation in the Anasazi culture of San Juan County, Utah*. Unpublished MA thesis, Brigham Young University, Provo.

Billman, B. R., Lambert, P. M., & Leonard, B. L. (2000). Cannibalism, warfare, and drought in the Mesa Verde region in the twelfth century AD. *American Antiquity, 65*, 1–34.

Blom, D. E. (2005). Embodying borders: Human body modification and diversity in Tiwanaku society. *Journal of Anthropological Archaeology, 24*(1), 1–24.

Blondiaux, J., Fontaine, C., Demondion, X., Flipo, R.-M., Colard, T., Mitchell, P. D., et al. (2012). Bilateral fractures of the scapula: Possible archeological examples of beatings from Europe, Africa and America. *International Journal of Paleopathology*, edited by Helen Codere. *2*(4), 223–230.

Boas, F. (1966). *Kwakiutl Ethnography, edited by Helen Codere*. Chicago: University of Chicago Press.

Brewis, A. A. (1992). Anthropological perspectives on infanticide. *Arizona Anthropologist, 8*, 103–119.

Carbonell, E., Cáceres, I., Lozano, M., Saladié, P., Rosell, J., Lorenzo, C., et al. (2010). Cultural cannibalism as a paleoeconomic system in the European Lower Pleistocene: The case of level TD6 of Gran Dolina (Sierra de Atapuerca, Burgos, Spain). *Current Anthropology, 51*(4), 539–549.

Carmen, R. A., Guitar, A. E., & Dillon, H. M. (2012). Ultimate answers to proximate questions: The evolutionary motivations behind tattoos and body piercings in popular culture. *Review of General Psychology, 16*(2), 134–143.

Chacon, R. J., Chacon, Y., & Guandinango, A. (2007). The Inti Raymi festival among the Cotacachi and Otavalo of Highland Ecuador. In R. J. Chacon & R. G. Mendoza (Eds.), *Latin American indigenous warfare and ritual violence* (pp. 116–141). Tucson: University of Arizona Press.

Chacon, R. J., & Dye, D. H. (2007a). Introduction to human trophy taking: An ancient and wide-spread practice. In R. J. Chacon & D. H. Dye (Eds.), *The taking and displaying of human body parts as trophies by Amerindians* (pp. 5–31). New York: Springer.

Chacon, R. J., & Dye, D. H. (2007b). *The taking and displaying of human body parts as trophies by Amerindians*. New York: Springer.

Codere, H. (1990). Kwakiutl: Traditional culture. In W. Suttles (Ed.), *Handbook of North American Indians, Vol. 7: Northwest coast*. Washington, DC: Smithsonian Institution Press.

Cohen, E. (2008). Southeast Asian ethnic tourism in a changing world. *Asian Anthropology, 7*(1), 25–56.

Conroy, G., Weber, G., Seidler, H., Recheis, W., Nedden, D. Z., & Haile, J. (2000). Endocranial capacity of the bodo cranium determined from three-dimensional computed topography. *American Journal of Physical Anthropology, 113*, 111–118.

Cummings, S. R., Ling, X., & Stone, K. (1997). Consequences of foot binding among older women in Beijing, China. *American Journal of Public Health, 87*(10), 1677–1679.

Daems, A., & Croucher, K. (2007). Artificial cranial modification in prehistoric Iran: Evidence from crania and figurines. *Iranica Antiqua, 42*, 1–21.

Daniell, C. (1997). *Death and burial in medieval England, 1066–1550*. London: Routledge.

Darling, A. J. (1998). Mass inhumation and the execution of witches in the American southwest. *American Anthropologist, New Series, 100*(3), 732–752.

Douglas, M. (1992). *Purity and danger: An analysis of the concepts of pollution and taboo*. New York: Routledge. (Original work published 1966)

Durán, D. (1951). *Historia de las Indias de Nueva España y Islas de la tierra firme* (D. Heyden & F. Horcasitas, Trans.). Mexico City: Editora Nacional.

Flower, W. H. (1881). *Fashion in deformity: As illustrated in the customs of barbarous and civilised races*. London: MacMillan.

Furst, P. T. (1996). Myth as history, history as myth: A new look at some old problems in Huichol origins. In S. B. Schaefer & P. T. Furst (Eds.), *People of the peyote: Huichol Indian history, religion, and survival* (pp. 26–60). Albuquerque: University of New Mexico Press.

Gaither, C. (2012). Cultural conflict and the impact on non-adults at Puruchuco-Huaquerones in Peru: The case for refinement of the methods used to analyze violence against children in the archeological record. *International Journal of Paleopathology, 2*(2–3), 69–77.

Gardner, R. (1963). *Dead birds*. Watertown, MA: Documentary Educational Resources.

Grimes, J. E. (1964). *Huichol syntax*. The Hague: Mouton and Company.

Grosman, L., Munro, N. D., & Belfer-Cohen, A. (2008). A 12,000-year-old Shaman burial from the southern Levant (Israel). *Proceedings of the National Academy of Sciences of the United States of America, 105*(46), 17665–17669.

Harrison, S. (2006). Skull trophies of the Pacific War: Transgressive objects of remembrance. *The Journal of the Royal Anthropological Institute, 12*(4), 817–836.

Heinz, D. (1999). *The last passage: Recovering a death of our own*. New York: Oxford University Press.

Hubert, J. (2000). *Madness, disability and social exclusion: The archaeology and anthropology of "difference"*. New York: Routledge.

Hulme, P. (1986). *Colonial encounters: Europe and the native Caribbean, 1492–1797*. London: Methuen.

Ichord, L. F. (2000). *Toothworms and spider juice: An illustrated history of dentistry*. Brookfield: Millbrook.

Jiménez-Brobeil, S. A., du Souich, P., & Al Oumaoui, I. (2009). Possible relationship of cranial traumatic injuries with violence in the south-east Iberian Peninsula from the Neolithic to the Bronze Age. *American Journal of Physical Anthropology, 140*(3), 465–475.

Judd, M. A. (2002). Ancient injury recidivism: An example from the Kerma period of ancient Nubia. *International Journal of Osteoarchaeology, 12*, 89–106.

Kuckelman, K. A. (2007). *The archaeology of Sand Canyon Pueblo: Intensive excavations at a late-thirteenth-century village in southwestern Colorado* [HTML title]. Cortez: Crow Canyon Archaeological Center. Retrieved July 27, 2010, from www.crowcanyon.org/sandcanyon

Kuckelman, K. A., Lightfoot, R. R., & Martin, D. L. (2002). The bioarchaeology and taphonomy of violence at Castle Rock and Sand Canyon Pueblos, Southwestern Colorado. *American Antiquity, 67,* 486–513.

Lambert, P. M. (1999). Human skeletal remains. In B. R. Billman (Ed.), *The Puebloan occupation of the Ute Mountain Piedmont, Vol. 5: Environmental and bioarchaeological studies.* Phoenix: Publications in Archaeology, No. 22, Soil Systems.

Lambert, P. M. (2000). Violent injury and death in a Pueblo II-III sample from the southern Piedmont of sleeping Ute Mountain, Colorado. *American Journal of Physical Anthropology Supplement, 30,* 205.

LeBlanc, S. A. (1999). *Prehistoric warfare in the American southwest.* Salt Lake City: The University of Utah Press.

Liénard, P., & Boyer, P. (2006). Whence collective rituals? A cultural selection model of ritualized behavior. *American Anthropologist, 108*(4), 814–827.

Logan, M. H., & Qirko, H. N. (1996). An evolutionary perspective on maladaptive traits and cultural conformity. *American Journal of Human Biology, 8,* 615–629.

López Austin, A. (1980). *The human body and ideology: Concepts of the ancient Nahuas.* Salt Lake City: University of Utah Press.

Lucero, L. J., & Gibbs, S. A. (2007). The creation and sacrifice of witches in classic Maya society. In V. Tiesler & A. Cucina (Eds.), *New perspectives on human sacrifice and ritual body treatments in ancient Maya society* (pp. 45–73). New York: Springer.

Lumholtz, C. (1900). Symbolism of the Huichol. *Memoirs of the American Museum of Natural History, III.* New York.

Maher, L. A., Stock, J. T., Finney, S., Heywood, J. J. N., Miracle, P. T., & Banning, E. B. (2011). A unique human-fox burial from a pre-Natufian cemetery in the Levant (Jordan). *PLoS One, 6*(1), e15815.

Martin, D. L. (1997). Violence against women in the La Plata River Valley (A.D. 1000–1300). In D. L. Martin & D. W. Frayer (Eds.), *Troubled times: Violence and warfare in the past* (pp. 45–75). Amsterdam: Gordon and Breach.

Martin, D. L., Harrod, R. P., & Fields, M. (2010). Beaten down and worked to the bone: Bioarchaeological investigations of women and violence in the ancient Southwest. *Landscapes of Violence 1*(1), Article 3.

Martin, D. L., Harrod, R. P., & Pérez, V. R. (2012). Introduction: Bioarchaeology and the study of violence. In D. L. Martin, R. P. Harrod, & V. R. Pérez (Eds.), *The bioarchaeology of violence* (pp. 1–10). Gainesville: University of Florida Press.

Martin, D. L., & Osterholtz, A. J. (2012). A bioarchaeology of captivity, slavery, bondage, and torture. New directions in bioarchaeology, special forum. *The SAA Archaeological Record, 12*(3), 32–34.

Maschner, H. D. G. (1997). The evolution of northwest coast warfare. In D. L. Martin & D. W. Frayer (Eds.), *Troubled times: Violence and warfare in the past* (pp. 267–302). Amsterdam: Gordon and Breach.

Meiklejohn, C., Agelarakis, A., Akkermans, P. A., Smith, P. E. L., & Solecki, R. (1992). Artificial cranial deformation in the proto-Neolithic and Neolithic near east and its possible origin: Evidence from four sites. *Paléorient, 18*(2), 83–97.

Miller, J. (2000). *Tsimshian culture: A light through the ages.* Lincoln: University of Nebraska Press.

Milner, G. R., & Larsen, C. S. (1991). Teeth as artifacts of human behavior: Intentional mutilation and accidental modification. In M. A. Kelley & C. S. Larsen (Eds.), *Advances in dental anthropology* (pp. 357–378). New York: Wiley.

Negrín, J. (1975). *The Huichol creation of the world.* Sacramento: E. B. Crocker Art Gallery.

Nelson, B. A. (1997). Chronology and stratigraphy at La Quemada, Zacatecas, Mexico. *Journal of Field Archaeology, 24*, 85–109.

Nelson, B. A., Darling, J. A., & Kice, D. A. (1992). Mortuary practices and the social order at La Quemada, Zacatecas, Mexico. *Latin American Antiquity, 3*(4), 298–315.

Neumann, G. K. (1942). Types of artificial cranial deformation in the Eastern United States. *American Antiquity, 7*(3), 306–310.

Nickens, P. R. (1975). Prehistoric cannibalism in the Mancos Canyon, Southwestern Colorado. *Kiva, 40*(4), 283–293.

Ogilvie, M. D., & Hilton, C. E. (2000). Ritualized violence in the prehistoric American southwest. *International Journal of Osteoarchaeology, 10*, 27–48.

Osterholtz, A. J. (2012). The social role of hobbling and torture: Violence in the prehistoric Southwest. *International Journal of Paleopathology, 2*(2–3), 148–155.

Pearson, M. P. (1993). The powerful dead: Relationships between the living and the dead. *Cambridge Archaeological Journal, 3*, 203–229.

Pérez, V. R. (2002). La Quemada tool induced bone alterations: Cutmark differences between human and animal bone. *Archaeology Southwest, 16*(1), 10.

Pérez, V. R. (2012a). The politicization of the dead: Violence as performance, politics as usual. In D. L. Martin, R. P. Harrod, & V. R. Pérez (Eds.), *The bioarchaeology of violence*. Gainesville: University of Florida Press.

Pérez, V. R. (2012b). The taphonomy of violence: Recognizing variation in disarticulated skeletal assemblages. *International Journal of Paleopathology, 2*(2–3), 156–165.

Pérez, V. R., Martin, D. L., & Nelson, B. A. (2000). Variations in patterns of bone modification at La Quemada. *American Journal of Physical Anthropology Supplemental, 30*, 248–249.

Pickering, T. R., White, T. D., & Toth, N. P. (2000). Brief communications: Cutmarks on a plio-pleistocene hominid from Sterkfontein, South Africa. *American Journal of Physical Anthropology, 111*, 579–584.

Porr, M., & Alt, K. W. (2006). The burial of Bad Dürrenberg, Central Germany: Osteopathology and osteoarchaeology of a Late Mesolithic shaman's grave. *International Journal of Osteoarchaeology, 15*(5), 395–406.

Quigley, C. (1996). *The corpse: A history*. Jefferson: McFarland and Company, Inc.

Rives, J. (1995). Human sacrifice among pagans and Christians. *Journal of Roman Studies, 85*, 65–85.

Sánchez, J. B. (1975). *Tablas*. Los Angeles: Ankrum Gallery.

Scott, G. R., & Turner, C. G., II. (1988). Dental anthropology. *Annual Review of Anthropology, 17*, 99–126.

Sofaer, J. R. (2006). *The body as material culture: A theoretical osteoarchaeology*. Cambridge: Cambridge University Press.

Stein, J. R., & Fowler, A. P. (1996). Looking beyond Chaco in the San Juan basin and its peripheries. In M. A. Adler (Ed.), *The prehistoric Pueblo world, A. D. 1150–1350* (pp. 114–130). Tucson: University of Arizona Press.

Stewart, T. D. (1937). Different types of cranial deformity in the Pueblo area. *American Anthropologist, 39*(1), 169–171.

Tiesler, V. (2007). Funerary and nonfunerary? New references in identifying ancient Maya sacrificial and postsacrificial behaviors from human assemblages. In V. Tiesler & A. Cucina (Eds.), *New perspectives on human sacrifice and ritual body treatments in ancient Maya society* (pp. 14–45). New York: Springer.

Tiesler, V. (2012). Studying cranial vault modifications in ancient Mesoamerica. *Journal of Anthropological Sciences, 90*, 1–26.

Tiesler, V., & Cucina, A. (2007). *New perspectives on human sacrifice and ritual body treatments in ancient Maya society*. New York: Springer.

Torres-Rouff, C. (2002). Cranial vault modification and ethnicity in middle horizon San Pedro de Atacama, North Chile. *Current Anthropology, 43*, 163–171.

Trinkaus, E. (1982). Artificial cranial deformation in the Shanidar 1 and 5 Neanderthals. *Current Anthropology, 23*, 198–199.

Trombold, C. D. (1985). A summary of the archaeology of the La Quemada region. In M. S. Foster & P. C. Weigand (Eds.), *The archaeology of West and Northwest Mesoamerica* (pp. 327–352). Boulder: Westview.

Tung, T. A. (2007). Trauma and violence in the Wari empire of the Peruvian Andes: Warfare, raids, and ritual fights. *American Journal of Physical Anthropology, 133*, 941–956.

Turner, C. G., II. (1993). Cannibalism in Chaco Canyon: The charnel pit excavated in 1926 at small house ruin by Frank H.H. Roberts, Jr. *American Journal of Physical Anthropology, 91*, 421–439.

Turner, C. G., II, & Turner, J. A. (1999). *Man corn: Cannibalism and violence in the prehistoric American southwest*. Salt Lake City: The University of Utah Press.

Weyl, N. (1968). Some possible genetic implications of Carthaginian child sacrifice. *Perspectives in Biology and Medicine, 12*(1), 69–78.

White, T. D. (1992). *Prehistoric cannibalism at Mancos 5MTUMR-2346*. Princeton: Princeton University Press.

Wilcox, D. R., & Haas, J. (1994). The scream of the butterfly: Competition and conflict in the prehistoric southwest. In G. J. Gumerman (Ed.), *Themes in southwest prehistory* (pp. 211–238). Santa Fe: School of American Research Press.

Wilson, A. S., Taylor, T., Ceruti, M. C., Chavez, J. A., Reinhard, J., Grimes, V., et al. (2007). Stable isotope and DNA evidence for ritual sequences in Inca child sacrifice. *Proceedings of the National Academy of Sciences of the United States of America, 104*(42), 16456–16461.

Zingg, R. M. (1977). *The Huichols: Primitive artists*. Report of the Mr. and Mrs. Henry Pfeiffer expedition for Huichol ethnography. Millwood: Kraus. (Original work published 1938)

Chapter 10
Relevance, Education, and the Future

In an article that took up the challenges of interdisciplinarity, Brewer (1999:330) wrote that "… the world has problems, but universities have departments…." This was an early call for researchers to break out of their disciplinary silos and venture forth to form collaborative projects with individuals outside their academic departments. For the last 15 years or more there has been a recognition that solving the complex problems in the world today will require social and natural scientists to work with others outside of their own field. Bioarchaeology is *de facto* interdisciplinary because of its historical trajectory. Starting out in anatomy and the biological sciences, it evolved into biological anthropology, then broadening out to include archaeological and cultural theory, it has blossomed into a spectacular vision of interdisciplinary possibilities for collaboration. It is at its core an anthropological enterprise which means it takes a multiple-field approach to situating humans in both time and space as a given but includes environmental, geological, historical, biomedical, and other perspectives in innovative ways. Bioarchaeology as a discipline implicitly recognizes that the world (and its peoples) have *always* had problems and it could provide important perspectives on today's problems to understand how past people survived, succumbed to, or transformed challenges to survival in the past. Putting a longer time frame on problems can reveal the kinds of complex adaptations that humans are capable of making. It also permits an examination of what went wrong when civilizations collapsed. Bioarchaeological data explicitly can reveal what the limits of human adaptation are to extremes in the cultural or physical environment. In solving problems, in teaching in innovative and engaging ways, in providing guidelines for an ethical science responsive to historical and contemporary issues, bioarchaeology encapsulates it all.

D.L. Martin et al., *Bioarchaeology: An Integrated Approach to Working with Human Remains*, Manuals in Archaeological Method, Theory and Technique, DOI 10.1007/978-1-4614-6378-8_10, © Springer Science+Business Media New York 2013

10.1   Applied Bioarchaeology: Moving Past the Analysis of the Bones

Research involving human skeletal remains like any other field of anthropology must be applicable to helping address problems in the world today. As discussed in Chap. 4, forensic anthropology is one critical way in which research is applied. However, forensic anthropology is not the only way that researchers looking at the bones can address problems that people face today. Many bioarchaeologists use human remains as an entry point (or foot in the door) to being part of addressing human behavior writ large. The bones and teeth are only an access point; the heart of the research is in the contextualization and integration with other data sets. When students say to us that they are interested in working with old bones, we often ask them: What is your bigger question? What kinds of things about the human condition are you interested in? Teaching an integrated, engaged, and ethical bioarchaeology is one way to provide students with a way to get beyond the bones and into research that is relevant to today.

10.1.1   From Global Health and Nutrition to Violence and Conflict: Understanding the Living via the Dead

Research involving human skeletal remains has proven to be valuable for the study of diet and disease in the past (as was discussed in previous chapters). However, what we learn about past health and nutrition can be useful for understanding the spread of epidemics (Barrett et al. 1998) or the consequence of the reemergence of diseases such as tuberculosis (Roberts and Buikstra 2003). This research is also useful in understanding how environmental conditions affect the long-term health of a population over a short period of time, which is advantageous because utilization of modern clinical literature requires relying on medical records that were often discontinuous and incomplete and long-term studies that lasted decades but still did not reveal the disease pattern (Roberts 2010).

One way that the study of human skeletal remains can improve our understanding of health is that it compensates for an increasing trend toward not performing autopsies in modern society. Referred to as clinical or conventional autopsies, the rates throughout the United States, Canada, European Union, and Australia, have dropped significantly due to a number of factors, such as ideological views that prohibit dissection, misconceptions about the method and usefulness of autopsies, and the time constraint associated with performing them (Burton and Underwood 2007). In the United States and United Kingdom, the rate is estimated to be between 10 and 15% (Ayoub and Chow 2008; Burton and Underwood 2007). The loss of autopsy information is problematic because many diseases and afflictions can only be observed fully with postmortem dissection and often an autopsy will reveal undiagnosed conditions or invalidate a clinical diagnosis (Roulson et al. 2005; Burton and Underwood 2007). Bioarchaeological research, especially on historic populations with well-documented medical histories, might offer a means of retaining at least some of the information that clinical autopsies can provide.

Like health and disease, violence is still present in the modern world and its causes are still likely to be found in social inequality as well as an unequal access to and limited availability of resources. Through the analysis of patterns of violence in the past, it is possible for researchers to aid in better interpreting future patterns of violence as well as provide more accurate understanding of the causes for violence and conflict.

In terms of interpreting violence more efficiently, data on the patterning, severity, and location of trauma can provide two distinct lines of evidence for suggesting how violence affects individuals by being able to answer these kinds of questions: (1) What were the long-term behavioral consequences based on the relationship between point of impact and potential for traumatic brain injury for the victim? (2) Does the patterning infer information on the perpetrators? In recent years, health professionals have noted that there is a portion of the population that is at risk for suffering from repeated injuries resulting in re-admittance to hospitals and clinics a number of times within a relatively short period of time. This reoccurring pattern of trauma is referred to as injury recidivism. In clinical literature this phenomenon is often talked about as an outcome of proximate causes (e.g., alcohol or low social inequality). The value of an anthropological approach to looking at injury recidivism is that through an evolutionary and biocultural lens it becomes apparent that injury recidivism can affect anyone in a population. Given that these factors are linked with a significantly high percentage of the prison, homeless, and unemployment populations that cannot afford treatment, the morbidity and mortality rate of injury recidivism is higher than necessary and has unnecessarily high costs on the community.

An example of how understanding violence in the past can aid in better identifying causes of violence is accented in a forthcoming book by Harrod and Martin entitled *The Bioarchaeology of Climate Change and Violence*. There has been research published that suggests that resource scarcity due to climate change leads to violence (Raleigh and Urdal 2007). This is based on the fact that there are often a number of droughts at or around the time of an increase in violence. The problem is that the correlation of violence and fluctuations in the climate does not necessarily imply a causal relationship. By exploring violence in the past and looking at factors beyond climate, Harrod and Martin demonstrate that other sociopolitical reasons such as migration are more important to understand. This is important critical because before policies are made to prevent violence as a result of climate change, it would be more productive to consider the role that both climatic *and* cultural factors play in the development and maintenance of violence.

The fields of forensic anthropology and bioarchaeology will only continue to grow in the future as they offer insights into the physical consequences of a particular person's lived experience, such as their social position within the society, lifestyle, nutrition, health, and even activity level, which are characteristics that are beyond the scope of many other disciplines and research approaches.

10.1.2   Ethnobioarchaeology: Understanding the Dead via the Living

Ethnobioarchaeology first suggested by Walker et al. (1998:389) is a term used to describe projects where bioarchaeologists collaborate with cultural anthropologists or human behavioral ecologists working in the field to collect data that can be used to better understand changes to the skeleton. The idea is that through the study of living populations, bioarchaeologists can developed better models with which to interpret behavior in the past. Walker was the pioneer in this field. Prior to even putting a name to this approach, he was conducting collaborative research with Hewlett who was working with hunter-gatherer groups in Central Africa.

Recently, Harrod (2012) has argued for more bioarchaeologists to incorporate ethnobioarchaeology into their research. The point of the commentary was that archaeologists have been studying the living to better interpret the lives of the dead for over half a century (Kleindienst and Watson 1956), and yet only a small amount of this type of work has been or is being conducted by researchers who analyze human remains. Harrod highlights projects that incorporate an ethnobioarchaeological approach. The first is focused on dental health and is a collaboration between Walker and a cognitive anthropologist Lawrence Sugiyama and a cultural anthropologist Richard Chacon (Walker et al. 1998) (Fig. 10.1).

This project explored dental health among the Yanomamö, Yora, and Shiwiar horticultural groups in the Amazonia Basin. The importance of this research was that it demonstrated that the relationship between dental change and nutrition is more complicated than previous research had implied. Additionally, the researchers found that behavior or cultural practice had a significant impact on dentition showing the important influence of culture on dental pathologies.

The second project was a collaboration between Harrod and Martin and a cognitive anthropologist Pierre Liénard (Harrod et al. 2012) (Fig. 10.2). This project looked at patterns of violence among the Turkana pastoralist of Eastern Africa. The importance of this project is that it found that violence was deeply intertwined with ideology and was far more complicated than many past bioarchaeological studies of violence have suggested.

The importance of enthobioarchaeology is not simply that it provides ethnographic analogies that can help researchers infuse behavior and ideology into their studies of human skeletal remains. The partnership between bioarchaeologists and ethnography-focused researchers is a two-way street. Bioarchaeologists can also offer insight into ethnographic research as they often are analyzing hundreds of individuals over long periods of time (Chap. 7), which is not typically possible when working with extant populations. In this way it is similar to what the study of ancient human remains can offer research on modern disease.

Fig. 10.1 Dr. Richard Chacon standing next to an adult Yanomamö man hunting with bow and arrow near the village of Poremababopateri (Photo by Lawrence Sugiyama)

10.2 Teaching Bioarchaeology: The Importance of a Four-Field Approach and the Need for More Theory

As discussed in Chap. 2, bioarchaeology research needs to be guided by an ethos of responsibility and ethics, and that can only be sustained if social theory and ethics are embedded in the ways that it is taught to future generations of bioarchaeologists. There is no doubt that students need to be proficient in bioarchaeological methods, but an additional focus needs to be on understanding the theoretical underpinning of the discipline. The question "Why is this important and whom does it benefit?" should be the first thing asked when students begin their journey into the subfield of bioarchaeology research. What is required in bioarchaeology is that students understand that it is biocultural in approach, and that the real heart of the research goes beyond the bones. A biocultural approach helps to structure and contextualize the research strategy so that data are collected in ways that best explain human behavior.

The approach is simple: No empirical analysis is without theoretical assumptions, or without a framework of analysis. There is no interpretation that is simply a neutral selection of "data speaking to us." Related to this is the dual character of theoretical knowl-

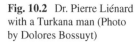

Fig. 10.2 Dr. Pierre Liénard
with a Turkana man (Photo
by Dolores Bossuyt)

edge: It is both explanatory and constitutive (Smith 1995:27–28). Bioarchaeological
theories derived from a biocultural paradigm are the result of knowledge giving a common, more general and coherent explanation for a variety of specified cases. Without an
emphasis on theoretical concepts, students often become lost in or mesmerized by the
empirical data (i.e., human remains) and fail to see the benefits and the validity of applying the data to larger contexts. Theories are not just the result but also the precondition
for the possibility of bioarchaeological empirical knowledge. In rethinking the educational goals of bioarchaeology, it is not a course in osteology. Training future practitioners and researchers in the constitutive function of theories fulfills the crucial role of a
more time-independent intellectual education. In other words, teaching bioarchaeology
with a combined focus on theory, method, and data produces critical thinking, which is
a useful skillset that never becomes obsolete.

The teaching of bioarchaeology has always been about more than the transmission
of facts, and many curricula have focused on drawing students in via forensics and an
appeal to the exotic. A better approach to bioarchaeological pedagogy takes the stu-
dent away from read-and-forget activities and replaces it with substantive curricula
that encourage students to become critical thinkers. Bioarchaeological research is
responsible for representing aspects of others' lives to students. In essence, this
research speaks for others as it offers explanatory models of cultural behaviors. Often
the researcher does not engage in the behaviors they are studying, but they are attempting to communicate their cultural and emotional significance.

The teaching of bioarchaeology needs to incorporate an ethical representation of the individuals and cultures that become the subject of the courses being taught. The aim should be to present the information in a broad cultural context to allow students to develop a nuanced understanding of how individuals exist as members of a broader cultural population. This is made easier when the students are grounded in a four-field anthropological education and understand the theoretical framework and empirical knowledge of the field of bioarchaeology. Because the data cannot speak for itself, all observation is theory-dependent. When coupled with the fact that the observation can itself have an effect on the very reality it is supposed to describe, it is fundamental that researchers and students of bioarchaeology be trained to become aware of their own and others' assumptions.

Bioarchaeology offers the unique opportunity to explore the underpinning of all societies and is fundamental to understanding cultural identity and nation building. The discipline offers a unique opportunity to create cultural heritage or the combination of tangible objects (sites, landscapes, structures, artifacts, and archives) and intangible values (the ideas, customs, and knowledge that gave rise to them) through the study of the body as material culture and exploration of past lifeways. Bioarchaeology is an essential element of cultural heritage determination and curation (e.g., NAGPRA) and is increasingly incorporated into the forensic sciences which are important to many global peace-keeping missions. Moreover, bioarchaeology is an employment growth area, and this is likely to continue. The rapid expansion of the cultural heritage management industry accompanying a booming economy over the last decade has created unprecedented demand for graduates with bioarchaeological skills.

10.2.1 A Model for Interdisciplinary Learning

Bioarchaeology is a discipline which crosses the traditional Arts/Science divide and has a greater breadth and more diversity than most other professional disciplines. Students need to be able to apply a wide range of techniques to bioarchaeological problems and be armed with the appropriate detailed knowledge of where to go for collaborative support for technical applications in the field and the laboratory. One way to provide students with these needed skill sets could be for universities and colleges to explore collaborative practices, such as joint teaching programs, particularly across specialist subfields, as well as the sharing of facilities or equipment where practical. One of the potential strengths of bioarchaeology is its diversity, and any efforts which develop collaboration while respecting that diversity add to the range and depth of education for students.

It has been hypothesized that students who learn by inquiry-based teaching strategies will show a greater understanding of content and concept acquisition than students learning through expository learning. Bioarchaeological researchers are increasingly turning their attention to the fertile regions that lie between the traditional boundaries of disciplines. Students are increasingly realizing the need to prepare themselves for a world where they may have to move between traditional disciplines over the course of their working lives.

The interdisciplinary component of a bioarchaeology education puts emphasis on the process and design of solutions, instead of the solutions themselves. This approach allows students to explore concepts like mathematics and science in a more personalized context, while helping them to develop the critical thinking skills that can be applied to all facets of their work and academic lives. This method of teaching bioarchaeology allows students to utilize the skill sets they are learning for discovery, exploration, and problem-solving. Bioarchaeology by design, by its very nature, is a pedagogical strategy that promotes learning across disciplines.

10.2.2   Hands-On Learning

One of the most common approaches to teaching bioarchaeology is combining traditional lectures and readings with laboratory practicals. This pedagogical philosophy, though useful, does not allow for students to achieve their full potential. Bioarchaeology lends itself to a problem-based learning model. Problem-based learning is a student-centered instructional strategy in which students collaboratively answer questions and solve problems and then reflect on their experiences (inquiry). Learning is driven by challenging, open-ended problems. Students work in small collaborative groups and teachers take on the role as "facilitators" of learning.

One exciting approach to teaching bioarchaeology is the use of cooperative learning groups. Cooperative groups operate as self-sustaining and self-regulating "learning teams" where the members provide support, encouragement, and advice to each other so that everyone on the team can succeed. Bauer-Clapp and colleagues, in their article *Low Stakes, High Impact Learning: A Pedagogical Model for a Bioarchaeological and Forensic Anthropology Field School* (2012), speak to the usefulness of this approach. This course provides a low-stakes opportunity to learn excavation and laboratory analysis methods, as the field school utilizes "burials" and "crime scenes" built with plastic skeletons and props. The course features a peer education component, with students working as a team to collectively determine the best course of action to excavate, analyze field data, and write a research report. The "learning teams" approach involves the students taking collective responsibility for identifying their own research goals and planning how these might be addressed. This is a vital learning-how-to-learn skill that also provides students with leadership roles and conflict-management skills that enhance their learning outcomes.

Although it may seem counterintuitive, the best groups (those that produce the most original work) are those that have been deliberately formed on the basis of heterogeneity. When creating long-term bioarchaeological projects, instructors should try to create groups with students who have different backgrounds, skills, abilities, and attitudes. Homogeneous groups may work together more smoothly, but they fail to learn important lessons about group dynamics that they will need to work with people of greater diversity productively in the future. Homogeneous groups are also predisposed to "groupthink." Bioarchaeological courses lend themselves to this unique pedagogy. In order to maximize the potential of the group's tasks and to insure that the diversity of knowledge, skills, and perspectives is distributed equitably, the instructor needs to take the time to assess the students' abilities.

10.2.3   A Tool for Teaching Ethics

As discussed above and in Chap. 2, ethics must be at the forefront of a bioarchaeological teaching pedagogy. Bioarchaeologists have a responsibility to uses their academic training and research experience to teach the general public and, in particular, their students about the ethics of cultural heritage. This is crucial because most college students are honestly enthusiastic about what they think bioarchaeology is, but few of them have been exposed to it as an academic discipline. Their knowledge of bioarchaeology, and by extension cultural heritage in general, comes from the media (e.g., Bones, CSI, Indiana Jones, Laura Croft, The History Channel, or the Discovery Channel) and perhaps a visit to a natural history museum. Students who enter courses that deal with cultural heritage, like bioarchaeology, bring with them preconceived notions and powerful misconceptions.

Dismantling the belief systems that students bring with them is often one of the most difficult challenges faced by instructors. This requires bioarchaeology to adopt a decolonizing pedagogy that allows students to actively reflect on and critique the discipline while working against existing forms of discrimination and exploitation. This approach prepares them for the concrete demands of the educational and/or professional spaces that bioarchaeologists occupy.

Bioarchaeology is now recognizing that its practice necessarily involves engagements with non-bioarchaeologists or local stakeholders. Bioarchaeology courses that incorporate the use of biometric data derived from the analysis of skeletal remains need to focus on how the collection is viewed by the various stakeholders in order to further the students' understanding of how the "data" that are being used in their scientific inquiry impacts others. Even if the collection comes from outside the United States and is not subject to NAGPRA or other international laws, it is important for the students to understand the collections' relationship to the university, the department, and community. Even if there is no identifiable descendant community, it is important to consider how the collection came to be made accessible to them. Given that many collections are non-European, bioarchaeologists need to understand the challenges of teaching using these collections within the context of post-racial ideology which informs students' expectations and experiences in the classrooms. This is particularly important to undergraduates who are often not used to considering their own privilege and power within their academic education.

10.2.4   The Role of Engagement in Bioarchaeology

As universities and colleges have become committed to renewing and coordinating their outreach and engagement missions, it has changed what it means to be a researcher. It is now clear that to be a scholar one must integrate teaching, research, and service, engage with those outside the academy, and synthesize and use what can be learned as a catalyst for change. Bioarchaeology is uniquely suited for training students because it asks them to step back from their research and look for connections to build bridges between theory and practice. Bioarchaeology and anthropology departments have often struggled with the tensions that can come from trying to balance research, teaching, and community outreach. This is particularly true when the scholar moves from the relative safety of the academy into complex and often convoluted realm of activist research.

The criticism that follows many of those who dare to engage in these often politically charged research agendas is that activist research lacks objectivity, is often simplistic, underproblematized, and undertheorized. Yet, if researchers are asking their students to follow the guiding question "why is this important and whom does it benefit," the discipline must come to realize that in the relationship between science and society there can be no such thing as research beyond politics. The celebrated historian Howard Zinn's book title offers the following reminder: "You Can't be Neutral on a Moving Train." The problem facing scholars and specifically biological anthropologists is not how to be out of the world but how to be in it. By embracing a scholarship of engagement, bioarchaeology can improve research, teaching, and integration thus incorporating reciprocal practices of civic engagement into the production of knowledge. This provides for more inclusiveness and truly collaborative projects that benefit all parties.

All of the above should also be regularly reinforced by, and intertwined with, an active bioarchaeological research environment. Students should be exposed to the excitement of research of the highest standards and given every opportunity to take part in a wide range of field and laboratory work, as well as internships in government, museum, private organizations, and with community partners where practical.

10.3   The Future of Bioarchaeology: The Bioarchaeology of "Me"

Bioarchaeology is an exciting, innovative, and relevant subdiscipline of anthropology, and it is experiencing a fluorescence that has never been seen before. Newly minted PhDs in bioarchaeology are producing a body of scholarship (books and journal articles) that is growing rapidly. University Press of Florida produces a book series entitled "Bioarchaeological Interpretations of the Human Past: Local, Regional, and Global Perspectives," edited by Clark

Spencer Larsen of Ohio State University. The series is described as a focus on bioarchaeology that highlights "... biocultural responses to stress, health, lifestyle and behavioral adaptation, biomechanical function and adaptive shifts in human history, dietary reconstruction and foodways, biodistance and population history, warfare and conflict, demography, social inequality, and environmental impacts on population...." [http://www.upf.com/seriesresult.asp?ser=bioarc]. As of 2013, there are 12 books in this series with titles such as *Bioarchaeology of Identity in the Americas* (Knudson and Stojanowski 2009), *Bioarchaeology of the Human Head* (Bonogofsky 2011), *Bioarchaeology of Individuals* (Stodder and Palkovich 2012), *Bioarchaeology of Climate Change* (Robbins Schug 2011), *Violence, Ritual and the Wari Empire: A Social Bioarchaeology of Imperialism in the Ancient Andes* (Tung 2012), and *Bioarchaeology of Violence* (Martin et al. 2012). These and other volumes in the series emphasize the integrative, interdisciplinary, and biocultural aspects of using bioarchaeology method, theory, and data to look at a variety of social and political-economic conditions. A new book series is being initiated by Springer Publications called Bioarchaeology and Social Theory (editor, Debra L. Martin from the University of Nevada, Las Vegas). This series will emphasize the use of theory in developing explanations for human behavior in a wide variety of ancient and historic settings.

Engaging blogs and websites are on the rise that underscore the connections of the past to the present. Kristina Killgrove's "Powered by Osteons" website connects bioarchaeology to popular culture, contemporary issues, and critiques of the good and bad in bioarchaeology research [http://www.poweredbyosteons.org/]. Christina Cartaciano's blog post entitled "What's in These Bones? The Bioarchaeology of Me" [http://www.theposthole.org/read/article/99] invites readers to imagine their bones being found in the not too distant future by bioarchaeologists. What would their toolkit of available observational and special application methods be able to reveal? She speculates on the use of isotopes to reveal that although she may have died in England, she was born and raised on the island of Guam. But unlike the ancestors, there will be complications for future bioarchaeologists doing isotopic dietary analyses. Drinking bottled water and eating imported meat could confound the isotopic signature being examined. Cartaciano further muses about DNA analyses—would they be able to reveal that her mother was born in the Philippines and her father in the United States?

All of these and more can now be found by Bing-ing or Google-ing "bioarchaeology" and this was emphatically not the case prior to 2005 or so. In "Bones Don't Lie" a PhD candidate at Michigan State University [http://www.bonesdontlie.com/] keeps up with new scholarship in bioarchaeology and mortuary studies and puts her own unique twist on what it all means. "Past Thinking" pulled together a long list of websites that relate to archaeology, and many of them include bioarchaeology [http://www.pastthinking.com/links/]. For example, Rosemary Joyce maintains an engaging blog on "Ancient Bodies, Ancient Lives" that often includes insights on bioarchaeology and mortuary studies [http://ancientbodies.wordpress.com/].

10.3.1 There are Jobs in Bioarchaeology

In 2013, there were 15 academic jobs announced for tenure-track assistant professors who can teach bioarchaeology and/or forensic anthropology. The interesting trend to note though is that most of the job descriptions pay homage to the fact that bioarchaeology is interdisciplinary. The job descriptions link bioarchaeologists with being able to offer courses in forensic anthropology, archaeology, cultural anthropology, paleoanthropology, and others. This tacit assumption being made is that bioarchaeology is necessarily interdisciplinary and relevant within departments of anthropology, and this is a recent phenomenon.

Beyond academic employment, bioarchaeologists are increasingly hired by Cultural Resource Management (CRM) firms that conduct small and large survey and excavation projects. Bioarchaeologists can easily work for county or state medical examiner's offices working on forensic cases. There is also a need for bioarchaeologists to work on international human rights teams that exhume recent and not so recent massacre victims of civil and secular wars. Museum work is perfectly suited for bioarchaeologists, and increasingly they are hired to curate and organize archaeological collections.

10.3.2 Bioarchaeology is Practiced in the United States

In the post-NAGPRA era, many claims were made that bioarchaeology is no longer a viable activity in the United States because Native American consultants and representatives would likely not permit excavation of burials. While it is true that field schools and large-scale excavations conducted by non-tribal archaeologists are largely not engaged in burial retrieval or analysis, there is a growing trend within some tribes to have their own archaeological team, and in some cases, burials are encountered and they are analyzed. One good example of this is Northern Arizona University, Department of Anthropology, which has a very large cohort of Native American anthropology majors at the undergraduate level as well as graduate students. The Society of American Archaeology (SAA) Record (published quarterly) has focused regularly on collaborations among archaeologists, bioarchaeologists, and indigenous groups, both in the United States and in other countries.

Large national and state repositories such as the American Museum of Natural History in New York City and the Smithsonian Institute in Washington, DC, continue to accept proposals to conduct projects on the skeletal collections under their stewardship (Fig. 10.3).

There are an increasing number of projects where historic cemeteries need to be relocated, and these efforts open up new avenues for understanding conditions of life in early historic North America. Anne Grauer, professor and chair of the Department of Anthropologyist at Loyola University in Chicago served as the former chief bioarchaeologist for the excavation and analysis of the Peoria City Cemetery in Illinois. Burials from the 1800s needed to be removed and analyzed due to a public works projects involving the expansion of a library (Fig. 10.3).

Fig. 10.3 Dr. Anne Grauer in her teaching laboratory

Fig. 10.4 Dr. Jerry Rose at the site of Umm Qais (Ancient Gadera) in Jordan

Jerry Rose, professor of Anthropology at the University of Arkansas, Fayetteville, was the director of a large project that helped excavate and analyze a large historic African-American cemetery that was going to be flooded due to a Corps of Engineers project. He has conducted skeletal and dental research in the Lower Mississippi Valley and Trans-Mississippi South working with both precontact and historic human remains. He directed a yearly bioarchaeology field school in Jordan from 1995 to 2007 (Fig. 10.4). Dr. Rose his been directing a bioarchaeological field school in Egypt since 2008 that is still ongoing.

Carlina de la Cova, assistant professor of Anthropology and African American Studies, University of Indiana, Bloomfield, has focused her research in bioarchaeology

Fig. 10.5 Dr. Carlina de la Cova in her teaching laboratory (Photo by Chris English)

on examining health disparities among African-American and European-American indigents from the 1800s onward. She incorporates theory about race, identity, ideology, culture, and socioeconomic status. In addition to the information from human skeletal remains, she integrates data from primary historical sources such as census records, family papers, government documents, and medical records. She also has an ongoing study of the large skeletonized cadaver populations that have been used in many studies. These large skeletal collections are used as reference populations because their identity was thought to be known. Her research on these remains and into their history and background has revealed that many of these collections (housed in medical schools and research institutions) were largely comprised of individuals that were poor, from prisons or mental institutions or otherwise destitute (Fig. 10.5).

Michael Blakey is the National Endowment for the Humanities Professor of Anthropology at the College of William and Mary in Williamsburg, Virginia. He was the lead director of the African Burial Ground project. A construction project revealed the existence of a cemetery populated with individuals who had been formerly enslaved Africans and African-Americans during the seventeenth and eighteenth century in New York City. He led an analysis of these human remains, as well as contributing to the reinterment ceremony called the Rites of Ancestral Return orchestrated by the Schomburg Center in Black Culture in New York City. He continues to work on the bioarchaeology of the African Diaspora and the ethics of publicly engaged research (Fig. 10.6).

These bioarchaeologists and many more are working with their students with human remains, continuing to push the boundaries of engaged and ethical research and leading the next generation of bioarchaeologists into new areas of research and scholarship. To those who think that NAGPRA (and similar legislation around the world) has brought an end to the analysis of human remains, the bioarchaeological scholarship discussed above and throughout this text sends a clear message that this area of study is more relevant today to a much broader group of people than it was 20 years ago.

Fig. 10.6 Dr. Michael
Blakey in his teaching
laboratory (Photo by Cecelia
Moore)

10.4 Summary

This text was never envisioned as covering every aspect of what bioarchaeology is. And, there is recognition that the author's presentation of bioarchaeology is just one of several bioarchaeological approaches practiced in the United States. The authors' vision for bioarchaeology is broad and rooted in the practice of theory, engagement, and ethical consideration. This presentation was envisioned as an overview of *possibilities* within the field for answering important questions about the human condition, for engaging with people outside of academia, for developing an ethos (and set of ethical protocols) that are not shaped solely by laws and public perceptions, and for inviting students and others to take bioarchaeological approaches into new areas with innovation and creativity.

References

Ayoub, T., & Chow, J. (2008). The conventional autopsy in modern medicine. *Journal of the Royal Society of Medicine, 101*(4), 177–181.

Barrett, R., Kuzawa, C. W., McDade, T. W., & Armelagos, G. J. (1998). Emerging and re-emerging infectious diseases: The third epidemiologic transition. *Annual Review of Anthropology, 27*, 247–271.

Bauer-Clapp, H. J., Pérez, V. R., Parisi, T. L., & Wineinger, R. (2012). Low stakes, high impact learning: A pedagogical model for a bioarchaeology and forensic anthropology field school. *The SAA Archaeological Record, 12*(3), 24–28.

Bonogofsky, M. (2011). *The bioarchaeology of the human head: Decapitation, decoration, and deformation*. Gainesville: University Press of Florida.

Brewer, G. D. (1999). The challenges of interdisciplinarity. *Policy Sciences, 32*(4), 327–337.

Burton, J. L., & Underwood, J. (2007). Clinical, educational, and epidemiological value of autopsy. *Lancet, 369*(9571), 1472–1480.

Harrod, R. P. (2012). Ethnobioarchaeology. New directions in bioarchaeology, special forum. *The SAA Archaeological Record., 12*(2), 32–34.

Harrod, R. P., Liénard, P., & Martin, D. L. (2012). Deciphering violence: The potential of modern ethnography to aid in the interpretation of archaeological populations. In D. L. Martin, R. P. Harrod, & V. R. Pérez (Eds.), *The bioarchaeology of violence* (pp. 63–80). Gainesville: University of Florida Press.

Kleindienst, M. R., & Watson, P. J. (1956). "Action Archaeology" the archaeological inventory of a living community. *Anthropology Tomorrow, 5*, 75–78.

Knudson, K. J., & Stojanowski, C. M. (2009). *Bioarchaeology and identity in the Americas*. Gainesville: University Press of Florida.

Martin, D. L., Harrod, R. P., & Pérez, V. R. (2012). The bioarchaeology of violence. In C. S. Larsen (Ed.), *Bioarchaeological interpretations of the human past: Local, regional, and global perspectives*. Gainesville: University Press of Florida.

Raleigh, C., & Urdal, H. (2007). Climate change, environmental degradation and armed conflict. *Political Geography, 26*, 674–694.

Robbins Schug, G. (2011). Bioarchaeology and climate change: A view from south Asian prehistory. In C. S. Larsen (Ed.), *Bioarchaeological interpretations of the human past: Local, regional, and global perspectives*. Gainesville: University Press of Florida.

Roberts, C. A. (2010). Adaptation of populations to changing environments: Bioarchaeological perspectives on health for the past, present and future. *Bulletins et Mémoires de la Société d'anthropologie de Paris., 22*(1–2), 38–46.

Roberts, C. A., & Buikstra, J. E. (2003). *The bioarchaeology of tuberculosis: A global perspective on a re-emerging disease*. Gainesville: University Press of Florida.

Roulson, J., Benbow, E. W., & Hasleton, P. S. (2005). Discrepancies between clinical and autopsy diagnosis and the value of post mortem histology; a meta-analysis and review. *Histopathology, 47*, 551–559.

Smith, S. (1995). The self-images of a discipline. In K. Booth & S. Smith (Eds.), *International relations theory today* (pp. 1–37). Oxford: Polity Press.

Stodder, A. L. W., & Palkovich, A. M. (2012). The bioarchaeology of individuals. In C. S. Larsen (Ed.), *Bioarchaeological interpretations of the human past: Local, regional, and global perspectives*. Gainesville: University Press of Florida.

Tung, T. A. (2012). *Violence, Ritual, and the Wari Empire: A social bioarchaeology of imperialism in the ancient Andes*. Gainesville: University Press of Florida.

Walker, P. L., Sugiyama, L. S., & Chacon, R. J. (1998). Diet, dental health, and cultural change among recently contacted South American Indian Hunter-Horticulturists. In J. R. Lukacs (Ed.), *Human dental development, morphology, and pathology: A tribute to Albert A. Dahlberg* (pp. 355–386). Eugene: University of Oregon Anthropological Papers, No. 54.

Index

CPSIA information can be obtained at www.ICGtesting.com
Printed in the USA
LVOW01s0423150115

422902LV00013B/785/P

9 781493 921195